Coastal Disaster Surveys and Assessment for Risk Mitigation

This collection covers essential concepts in the management of coastal disasters, outlining several field surveys of such events that have taken place in the 21st century, including the Indian Ocean Tsunami, the Tohoku Earthquake and Tsunami, and the storm surges generated by Hurricane Katrina, Cyclone Nargis, and Typhoon Haiyan. Measurements of flood heights, distributions of structural destruction, and the testimonies of residents are reported, with the results being analysed and compared with past events and numerical simulations to clarify and reconstruct the reality of these disasters. The book covers the state-of-the-art understanding of disaster mechanisms and the most advanced tools for the simulation of future events:

- Uniquely explains how to use disaster surveys along with simulations to mitigate risk
- Combines pure scientific studies with practical research and proposes procedures for effective coastal disaster mitigation

Coastal Disaster Surveys and Assessment for Risk Mitigation is ideal for students in the field of disaster risk management, as well as engineers who deal with issues related to tsunamis, storm surges, high wave attack and coastal erosion.

Coastal Disaster Surveys and Assessment for Risk Mitigation

Edited by
Tomoya Shibayama
Miguel Esteban

CRC Press
Taylor & Francis Group
Boca Raton London New York

CRC Press is an imprint of the
Taylor & Francis Group, an **informa** business

First edition published 2023
by CRC Press
6000 Broken Sound Parkway NW, Suite 300, Boca Raton, FL 33487-2742

and by CRC Press
4 Park Square, Milton Park, Abingdon, Oxon, OX14 4RN

CRC Press is an imprint of Taylor & Francis Group, LLC

Library of Congress Cataloging-in-Publication Data
Names: Shibayama, Tomoya, editor. | Esteban, Miguel, editor.
Title: Coastal disaster surveys and assessment for risk mitigation / edited by Tomoya Shibayama and Miguel Esteban.
Description: First edition. | Boca Raton : CRC Press, 2023. | Includes bibliographical references and index.
Identifiers: LCCN 2022020981 | ISBN 9780367741280 (hardback) | ISBN 9780367741297 (paperback) | ISBN 9781003156161 (ebook)
Subjects: LCSH: Coastal engineering. | Coastal zone management. | Natural disasters--Simulation methods. | Hazard mitigation--Simulation methods. | Environmental risk assessment. | Hydrological surveys. | Social surveys.
Classification: LCC TC205 .C549 2023 | DDC 627/.58--dc23/eng/20220819
LC record available at https://lccn.loc.gov/2022020981

ISBN: 978-0-367-74128-0 (hbk)
ISBN: 978-0-367-74129-7 (pbk)
ISBN: 978-1-003-15616-1 (ebk)

DOI: 10.1201/9781003156161

Typeset in Sabon
by SPi Technologies India Pvt Ltd (Straive)

Contents

Preface

Coastal disasters such as tsunamis, storm surges and high wind waves frequently attack the shorelines of many countries around the world. These events can affect large sections of the coastline, sometimes hundreds of kilometres in length, such as in the case of the *2011 Tohoku Tsunami* and the *2004 Indian Ocean Tsunami*, or be smaller events, like the 2014 storm surge caused by an extratropical cyclone in Nemuro, Japan, which flooded a relatively small area of the coastline. However, when speaking to local residents who have been affected, many say they never thought that their area was at risk of being inundated by seawater. This clearly highlights the role of researchers in this field to raise awareness amongst the local society on how they should implement protection countermeasures to build more resilient societies against such events.

In this book, I draw on the scientific knowledge that I have accumulated over 45 years of experience in observing and studying coastal disasters to explain why they happen and how we can help protect residents. I started studying coastal disaster prevention in the late 1970s since I vividly remembered the *1961 Ise Bay Typhoon* and the *1960 Chilean Tsunami* during my childhood. At that time, while knowledge of coastal disaster prevention was shared among professionals, it was not widespread among the general public. Later, after the *2004 Indian Ocean Tsunami* and the *2011 Tohoku Tsunami*, knowledge and awareness of such events started to spread worldwide to all sectors of society.

It is often said that Japan possesses advanced technology and science to help with disaster preparedness. I think the main reason for this is that we live in a country prone to natural disasters. This has led many young scholars worldwide to come to Japan to study disaster risk management. 33 PhDs have been trained so far from my laboratory, including 25 international students from 17 different countries, and many of them have returned to their homelands to work as university academics in the field of coastal disaster management (at a total of 20 universities, 13 overseas and 7 in Japan). This robust international network for disaster research has frequently collaborated in post-disaster surveys and regularly interact with each other even during normal times. Taking advantage of this solid network that I have created and fostered over the years, I asked the various team members to contribute to publish our research results and proposals to mitigate disasters in the present book.

This book introduces the mechanisms of coastal disasters and the current state of disaster prevention and mitigation, using actual examples. We have tried to make the book as easy to read as possible. However, more details regarding other literature that we have published can be found in the references, which can be referred to for the full results of the field surveys, simulations, and formulas derived.

I sincerely hope that this book will help the reader to understand the nature of disasters, internalizing such knowledge and helping to foster an interest in helping society work together to mitigate them. As such, it is my wish that it will trigger an enhanced willingness to expand knowledge on how to protect against natural hazards to enhance the resilience of coastal settlements, leading to long-term sustainable management of the coastline.

Tomoya Shibayama
Professor of Civil and Environmental Engineering
Waseda University
Professor Emeritus, Yokohama National University
March 2022

Acknowledgements

The editors would like to appreciate the efforts made by the various authors and editorial staff who assisted in publishing this book. In particular, the contributions from our colleagues in the WASEDA-YNU Advanced Coastal Environment and Management Group (WAYCEM) were key to making this manuscript a reality.

The contents of this book are based on field surveys, hydraulic laboratory experiments and numerical simulations that have spanned several decades of academic work. It has been over 45 years since the lead editor started his research activities in the field of coastal disasters. Since then, over 300 undergraduate and graduate students have contributed to the creation of a wide body of knowledge, much of which is summarized here. Thus, the editors would like to sincerely appreciate the contribution of past students to academia in general and coastal disaster research in particular. A true academic continuously learns from their own students.

We would also like to take the opportunity to appreciate the financial support that has been received over the years from diverse funding organizations and private companies. A non-exhaustive list includes the Japanese Ministry of Education, Culture, Sports and Science and Technology (MEXT), the Japan Society for the Promotion of Science (JSPS), the Japan Science and Technology Agency (JST), Penta Ocean Co. Ltd., and New CC Construction Consultants Co., Ltd. During the writing process of this manuscript, JSPS KAKENHI Grant Number JP20KK0107 and the Japan Science and Technology Agency (JST), through the Belmont Forum Grant Number JPMJBF2005, also supported us. Finally, the editors would like to mention that the editing and writing process of this book was performed as part of the activities of the Research Institute of Sustainable Future Society, Waseda Research Institute for Science and Engineering, Waseda University.

About the editors

Dr. Tomoya Shibayama is a Professor of Coastal Engineering at the Department of Civil and Environmental Engineering, Waseda University in Tokyo, Japan. He is also a Professor Emeritus at Yokohama National University. He received his B.E., M.E. and PhD in Civil Engineering from the University of Tokyo. Formerly, he was an Associate Professor at the University of Tokyo and at the Asian Institute of Technology, and a Professor at Yokohama National University. He is one of the top leaders in disaster risk mitigation in Japan, particularly regarding tsunamis and storm surges, through the use of a wide range of methodologies that include hydraulic laboratory experiments, field surveys and numerical simulations. He has served as the leader of survey teams for all major tsunami and storm surge events since the *2004 Indian Ocean Tsunami*. In 2019 he won the Hamaguchi International Award for Enhancement of Tsunami/Coastal Disaster Resilience.

Dr. Miguel Esteban is a former PhD student of Prof. Shibayama, and is now also a Professor of Coastal Engineering at the Department of Civil and Environmental Engineering, Waseda University in Tokyo, Japan. He received his PhD in Civil Engineering from Yokohama National University, and his MEng. from the University of Bristol. Formerly, he was an Associate Professor at the University of Tokyo and Waseda University, and a post-doc fellow at Kyoto University and the United Nations University. His research deals with all aspects of coastal disaster risk management and adaptation to climate change and sea-level rise, through the use of sustainability science multidisciplinary approaches to complement coastal engineering. He has followed Prof. Shibayama in all field surveys of major disasters since 2010.

Contributors

Hendra Achiari
Bandung Institute of Technology
Bandung, Indonesia

Rafael Aránguiz
Universidad Catolica de la
    Santisima Concepcion,
    Concepcion, and Research
    Center for Integrated Disaster
    Risk Management (CIGIDEN)
Santiago, Chile

Le Van Cong
Vietnam Administration of Seas
    and Islands (VASI)
Hanoi, Vietnam

Nguyen Danh Thao
Ho Chi Minh City University of
    Technology (HCMUT)
Ho Chi Minh City, Vietnam

Cheki Dorji
Royal University of Bhutan
Phuentsholing, Bhutan

Leki Dorji
Royal University of Bhutan
Phuentsholing, Bhutan

Miguel Esteban
Waseda University
Tokyo, Japan

Naoki Hoshiyama
Waseda University
Tokyo, Japan

Naoto Inagaki
Waseda University
Tokyo, Japan

Ravindra Jayaratne
University of East London
London, UK

Jae-Hun Jeong
Sekwang Engineering Consultants
    Co., Ltd.
Seoul, Korea

Michael B. Kabiling
Taylor Engineering, Inc.
Jacksonville, FL, USA

Yil-Seob Kim
Sekwang Engineering Consultants
    Co., Ltd.
Seoul, Korea

Wudhipong Kittitanasuan
Wishakorn Co., Ltd.
Bangkok, Thailand

Kristina Knüpfer
Waseda University
Tokyo, Japan

Hyuck-Min Kweon
Sungkyunkwan University
Seoul, Korea

Sun-Yong Lee
Sekwang Engineering Consultants
   Co., Ltd.
Seoul, Korea

Shaowu Li
Tianjin University
Tianjin, China

Martin Mäll
Yokohama National University
Yokohama, Japan

Tasnim Khandker Masuma
Weathernews
Chiba, Japan

Ryo Matsumaru
Toyo University
Tokyo, Japan

Takahito Mikami
Tokyo City University
Tokyo, Japan

Philip Mzava
University of Dar es Salaam
Dar es Salaam, Tanzania

Ryota Nakamura
Niigata University
Niigata, Japan

Sittichai Naksuksakul
PTT Research and Technology
   Institute
Bangkok, Thailand

Nguyen Ngoc An
New CC Construction
   Consultants Co., Ltd.
Ho Chi Minh City,
   Vietnam

Shinsaku Nishizaki
Waseda University
Tokyo, Japan

Ioan Nistor
University of Ottawa
Ottawa, Canada

Joel Nobert
University of Dar es Salaam
Dar es Salaam, Tanzania

Koichiro Ohira
Chubu Electric Power Company,
   Incorporated
Nagoya, Japan

Zahra Ranji
K.N. Toosi University of
   Technology
Tehran, Iran

Thamnoon Rasmeemasmuang
Burapha University
Saen Suk, Thailand

Winyu Rattanapitikon
Thammasat University
Pathum Thani, Thailand

Tomoya Shibayama
Waseda University
Tokyo, Japan

Mohsen Soltanpour
K.N. Toosi University of
  Technology
Tehran, Iran

Jacob Stolle
Institut National de la Resercher
  Scientifique
Québec City, Canada

Takayuki Suzuki
Yokohama National University
Yokohama, Japan

Tomoyuki Takabatake
Kindai University
Osaka, Japan

Hiroshi Takagi
Tokyo Institute of Technology
Tokyo, Japan

Karma Tempa
Royal University of Bhutan
Phuentsholing, Bhutan

Nguyen The Duy
Ho Chi Minh City University of
  Technology (HCMUT)
Ho Chi Minh City, Vietnam

Thit Oo Kyaw
Taisei Rotec's Research Institute
Saitama, Japan

Justin Valdez
Waseda University
Tokyo, Japan

Nimal Wijayaratna
University of Moratuwa
Moratuwa, Sri Lanka

Chapter 1

Introduction

What are coastal disasters?

Tomoya Shibayama and Miguel Esteban
Waseda University, Tokyo, Japan

Ioan Nistor
University of Ottawa, Ottawa, Canada

CONTENTS

1.1 INTRODUCTION

Tsunamis, storm surges, and extreme waves are some of the hazards that can affect the inhabitants of coastal communities throughout the world, potentially turning into disasters if various unfavourable circumstances occur simultaneously. Such extreme events have a wide range of return periods, from tsunamis which can occur once every hundred or thousand years, to storm surges and extreme waves that can affect coastal zones of the world almost every year. Chapters 2 and 3 of this book provide a more detailed account of some of the major events that occurred in recent times, with a general explanation of these three different types of hazards being provided below. It is important to note that the study of such events is not only important to understand how to protect coastal areas, but can also inform the study of other sudden extreme hydrodynamic events, such as glacial lake outburst floods in countries such as Bhutan (Section 9.2). Also, it should not be forgotten that communities located at river mouths can also suffer from fluvial flooding (and the problems that cities in Tanzania suffer due to such events are detailed in Section 9.7). Thus, while the main focus of the present book is on coastal hazards, it is crucial to remember that these extreme

DOI: 10.1201/9781003156161-1

1

events do not happen in isolation, and that there is a wide interdependency between different types of phenomena, forcing a disaster risk manager to think holistically and consider all aspects of such occurrences. By doing so, it is possible to increase the overall resilience of a given community, which can ensure it performs better against future hazards and ensure the long-term sustainability of the socio-economic environment it supports.

1.2 TSUNAMIS

Geomorphic changes under the sea constitute one of the most common generating mechanism leading to the occurrence of tsunamis (though not the only one). Many tsunamis, including the *2004 Indian Ocean Tsunami* (Section 2.1), are caused by the sudden movement of the Earth's tectonic plate boundaries along the faults separating them. Prior to the 2004 event, the majority of the world's population did not know the meaning of the word *tsunami*. However, images of powerful waves ravaging the coastlines of many countries in the region have quickly raised awareness about the threat posed by these phenomena. Many other large tsunamis that followed, including the 2006 Java (Section 2.2), 2009 Samoa (Section 2.3), 2010 Chile (Section 2.4), and 2010 Mentawai (Section 2.5) tsunamis, were "local" in nature (in that they mostly affected the countries in which they took place, and did not propagate across the ocean). The other major event that made headlines around the world was the *2011 Tohoku Tsunami*, which was generated by an extensive movement of the seafloor at a faultline between adjacent plate boundaries (Section 2.6).

However, it is not only the movement along faultlines that can generate tsunamis. For example, the *2018 Sulawesi Tsunami* was triggered by an earthquake that caused a submarine landslide which generated waves that went on to devastate the city of Palu and the gulf around it (Section 2.7). Similarly, the *2018 Sunda Strait Tsunami* in Indonesia was generated by the eruption of the Krakatoa volcano, which caused large amounts of volcanic debris to fall into the sea, generating a tsunami (Section 2.8). One common denominator of tsunami waves is that, irrespective of how they are generated, they can propagate over significant distances through the sea and eventually reach a coastline. Depending on the magnitude and local nearshore coastal bathymetry and topography, they can generate significant overland inundation, causing severe damage to infrastructure and killing many.

1.2.1 Tsunami measurements and damage mechanisms

The authors have been conducting tsunami post-disaster surveys for many years. When these events take place, it is necessary to measure the spatial extent of the inundation as well as the run-up heights in order to understand the magnitude of tsunami waves. Specifically, it is necessary to find the water-marks left behind on structures, vegetation, soil, etc., in the affected area and determine the maximum depth of the flood, i.e. the maximum height above the ground that the water reached at a given location. Then, using surveying techniques, one can compare the flood water level with the level of the nearest free water body (adjusted for astronomical tide differences) and determine the final inundation height of the tsunami. Inland flood runs up topographic features in the proximity of the shoreline and eventually stops moving when it reaches a point where all its kinetic energy is converted into potential energy (see Section 5.1). The height of inundation at this point is referred to as the maximum run-up height. The spatial distribution of the run-up height along a coastline can provide a rough estimation of the total energy of the tsunami. However, it is not only important to record the tsunami inundation heights and run-ups, but to also clarify the mechanisms through which it damages coastal infrastructure. This includes how tsunami waves can damage coastal dikes, flood over land (see Section 6.1), scour around structures (see Section 6.2) or transport debris that can accumulate around structures (debris damming) and increase the forces acting on them (see Section 6.3).

As already explained, most of the tsunamis that have caused significant disasters were generated by the sudden displacement of tectonic plate boundaries. This is especially the case of the Japanese islands, where four tectonic plates (the Pacific, Philippine Sea, Eurasian and North American plates) are interlocked, with the different faultlines at their boundaries frequently generating tsunamis. In particular, the Pacific coast of Honshu Island is prone to such events, given that it intersects these four plate boundaries, though other locations in the world (notoriously the Indonesian and Chilean coastlines, see Sections 9.1 and 9.5) also sit at plate boundaries.

1.2.2 Tsunami propagation

When a tsunami is generated, waves can propagate over a wide ocean area until they reach a coastline. Since the energy of a tsunami is not dissipated during its propagation, waves generated, for instance, off the coast of Chile can travel across the Pacific Ocean to the north-eastern coast of Japan. Similarly, a tsunami generated off the north-eastern Japanese coast will travel to Chile (see Section 7.1). The *1960 Chilean tsunami* (see Section 9.5) caused significant damage to the coastline of Japan, as the event was not expected to result in high waves along the northeast coast of the country. As a result of such consequences, tsunamis are currently monitored at the Pacific Tsunami Warning Center in Hawaii and Alaska, and a network of buoys placed in the ocean (at either its surface or bottom) can record the passing waves and provide forecasts. This provides information on their potential heights along the shore, which can help inform warning systems and evacuation (see Section 8.1).

Due to their long period and wavelength (relative to water depth; compared to wind waves), tsunamis are classified as "long waves". Under the assumption of long wavelength, their wave speed can be calculated using only the water depth along the path which they travel (see also Section 7.1). Thus, when the location of an earthquake can be determined with some degree of accuracy, the prediction of the arrival time of a potential tsunami can be calculated in advance. However, determining *a priori* the height of the tsunami is more challenging, as it requires assumptions regarding the size as well as the motion (vertical, horizontal, or a combination of both) of the fault. This is why information regarding the arrival time of tsunamis can be disseminated by scientists and relevant authorities relatively quickly, often within a few minutes of an earthquake. However, information on the potential height of the wave is often revised as more information gradually becomes available.

1.2.3 Recent advances in understanding tsunamis

The image that many researchers used to have of a tsunami was that of a single large wave, often referred to as a solitary wave. This was especially the case since there was little video information on such waves, forcing academics to rely on the description of eyewitnesses and their own interpretations of those. Since the 1984 *Central Japan Sea Earthquake Tsunami*, researchers have been able to witness the propagation of tsunamis from the time they reached the coastline due to the popularisation of portable video recorders and cameras attached to cell phones (and, recently, of smartphones). Therefore, researchers have now a more realistic image of tsunamis, which can manifest in slightly different forms: 1) *solitary waves* (soliton splitting waves), 2) *hydraulic bores*, similar to waves generated by a dam break, or 3) surges, when the water level rises on the seaside and floods the land side continuously without the presence of a highly turbulent bore front. Nowadays, tsunami researchers can collect video images of tsunamis at the time of their arrival, posted on the web by local people in the affected areas, which can be compared to research of the destruction captured during field surveys in order to understand disaster mechanisms.

Recent research has highlighted that many areas that were not previously thought of as being at risk of tsunamis are, in fact, hazardous. For example, the southwest part of the Sea of Japan has never been mentioned by Japanese tsunami researchers as a potential area at risk. However, this does not mean that there is no risk of one occurring at a certain moment in time; rather, it means that the risk has not been properly investigated. Other such previously unknown risks include the hazards due to sloshing inside closed water systems (see Section 5.3).

In order to examine whether tsunamis have occurred and/or impacted a particular area, one of the first things to do is to study ancient historical documents. In the case of Japan, there are official government documents that narrate ancient history and the private records of aristocrats, monks, and rich farmers, together with diaries or letters (see Section 9.1). All of these can provide accounts of earthquakes and even tsunamis that have taken place in the past, and many other countries also have a history of such events (see for example Section 9.4. on Canada). Also, it is important to conduct geotechnical surveys employing soil cores obtained through boring to assess the inland extent of past tsunamis (a field of study known as pale-otsunami research). In the case of the North Pacific side of the Japanese islands, it was reported that the *869 Jogan Tsunami* brought sediment from the sea bottom inland (as was the case of the *2011 Tohoku Earthquake and Tsunami*, which also brought sand and mud that covered the inundated areas). The layer of sediment left behind by historical tsunamis is rather distinctive, and can be carbon-dated to determine when such events took place. Therefore, even for an area where there is little knowledge of tsunami in recent times, the study of old documents and bore surveys can provide some evidence of the return periods of such events.

1.2.4 Tsunami reconstruction and *build back better* philosophy

Finally, it is important to consider how to build back better after a tsunami strikes an area, in order to improve the resilience of coastal communities against such events. Different countries have followed a variety of recon-struction paths (see Section 8.2. for the case of Indonesia, or 9.2 about Sri Lanka), resulting in different increases in the levels of resilience of settle-ment. The study of such lessons, identifying what worked and what did not during reconstruction, is crucial to learn from the past and formulate better strategies to protect and reconstruct areas after future events. In order to help to create a society where the risks due to natural hazards are mini-mized, it is necessary to holistically consider all the different types of disas-ters that may affect it, so that local residents can make informed choices about where they should live (Section 8.3).

1.3 STORM SURGES AND HIGH WAVES

Storm surges and high (extreme) waves are two different physical phenom-ena, but they occur simultaneously at different scales, as they are caused by tropical cyclones and other types of low-pressure systems. They can have important socio-economic effects on settlements at risk, leading to major economic damage if societies have a low level of resilience against them. While some countries such as China have reported decreasing levels of dam-age due to increased preparedness (see Section 9.1), in general the risk posed by them throughout the planet is increasing, due to the challenges posed by climate change and the increase in population growth in coastal areas.

1.3.1 Storm surges

Storm surges are mainly caused by a combination of the upward suction that results from a drop in atmospheric pressure and wind shear stress, which further pushes the mass of seawater towards the coastline (see Section 5.2). When storm surges are high compared to the coastal topography they can flood the land, with the risk of flood levels increasing if they coincide with high tides. The behaviour of storm surges is somewhat similar to that of tsunamis, in that a large volume of seawater with high momentum is driven over land, potentially damaging structures (see Section 9.1 for an example of the types of forces that can be exerted by storm surges on school buildings in the Philippines).

As stated earlier, under a low-pressure system or tropical cyclones, the atmospheric pressure drops, which in turn leads to the generation of strong winds. Each drop in pressure of 1 hPa causes an average water level change of about 0.99 cm – moreover, the strong wind shear on the surface leads to mass water transport in the wind direction. At the shoreline, the water level rises as the edge of the land will reflect back the seawater pushed by the pressure and wind. The suction effect is typically not as significant as the effect of the wind, as even if the atmospheric pressure drops to 950 hPa, the difference between this and normal pressure (1013 hPa) is 63 hPa, so the static rise in water level would only be 63 cm. However, the wind-driven components of the surge are highly dependent on the local topography and can reach several metres in some cases, such as in bays that narrow towards their inner side or very flat coastal plains. Therefore, predicting a storm surge requires an accurate prediction of the wind field of a given storm.

In the past, a storm surge was considered by engineers only in terms of its inundation depth. This was due to the need to design the height of the storm surge barriers constructed along the shoreline. Tokyo Bay, Osaka Bay, Ise Bay and all other inland bays in Japan that are vulnerable to storm surges are protected by concrete walls (storm surge barriers), which have been constructed to the level of the estimated height of a potential storm surge. Recently, residents have provided eye-witness accounts of the storm surge due to cyclone Sidr in Bangladesh (see Section 3.2), and eye-witness accounts and video recordings of typhoon Haiyan in the Philippines (see Section 3.5), showing that these can travel as fast-moving bore waves, with their top breaking when they propagate over shallow water or land. Therefore, the general image of the danger represented by a storm surge has been changed from that of water level change to a bore-type wave attack.

Many countries around the world have been affected by storm surges in recent times. The dramatic images of flooding of New Orleans in the USA after Hurricane Katrina in 2005 (Section 3.1) are still reminisced by many. Such scenes were repeated in 2012 in New York during the passage of hurricane Sandy, which flooded many parts of the city, including its iconic subway system (Section 3.4). Other notable events not mentioned already include cyclone Nargis in 2008, which flooded low-lying areas of Myanmar (Sections 3.3 and 9.1). This last event is particularly interesting given that, generally, Bangladesh is the country bordering the Indian Ocean that is considered to be most at risk from storm surge flooding (see Section 9.2). However, it is not only tropical cyclones that cause storm surges: extra-tropical cyclones have also been known to flood countries in Europe such as the UK, Germany or Estonia (see Section 9.3). Even Arctic regions can be affected by low-pressure systems (Section 9.6).

Japan is generally considered to be well-protected against storm surges, as the country is often affected by such events, such as for example typhoon Jongdari in 2018 (Section 4.1) and typhoon Faxai in 2019 (Section 4.2). Nevertheless, some recent events have challenged that notion and caused localised damage to some urban areas, including the storm surge that flooded Nemuro in 2014 (Section 3.6) and typhoon Jebi in 2018 (Section 3.7).

1.3.2 High waves

High (wind) waves are generated when strong winds blow over the sea surface due to the presence of low-pressure systems (as described earlier), effectively transferring the energy of the wind to the sea surface, causing ripples. In water surfaces subjected to strong winds (where waves are being generated), the surface of the sea appears to rise and fall irregularly (also known as "choppy" seas). Each of the waves generated moves in various directions and have different periods and heights, overlapping each other. This irregularity can be broken down into regular component waves using spectral analysis. In the open sea, short-period waves quickly decay. In contrast, longer-period waves move ahead at a faster speed so that their period gradually concentrates in the range of around 3–15 seconds and propagate over long distances as a swell. Swell waves can be observed on beaches at any time of the year, although waves with heights of over 10 metres can hit the coast during the passage of a tropical cyclone.

In recent years, changes in the behaviour of typhoons in the vicinity of the Japanese islands have been increasingly reported due to rising sea surface temperatures. Typhoons gather their energy from warm seas and it is thus logical that as the surface temperature increases, the intensity and thus wind speeds at the heart of the storm will increase, in turn generating higher waves and storm surges (see Section 7.2). There is some evidence that this may have started to happen already in some areas such as the Gulf of Oman (see Section 9.3). There are also issues caused due to the slowing down of the translation speeds of the typhoon as it moves over the sea. Since the wind field is a combination of the translation speed of the storm and the wind generated by it, the slower the typhoon, the lower the wind speed due to its movement, and the smaller the storm surge. On the other hand, as the wind blows for a longer time, the energy transmission from wind to ocean surface increases, and thus the storm surge and wind wave heights are also expected to increase (see Section 7.3). When the latter effect is greater, storm surges and waves may be higher due to the reduced speed of the typhoon (Inagaki et al. 2021).

It is important to note that waves and storm surges can lead to coastal erosion, which can have a long-term effect on land usage in the affected areas (with all the socio-economic problems associated with it) and increase the vulnerability to future disasters. A number of different types of counter-measures can be employed, and this book provides detailed examples of some such schemes in Korea, Thailand, Vietnam (see Section 9.1) and the USA (Section 9.4).

REFERENCE

Inagaki, N., Shibayama, T., Esteban, M. & Takabatake, T. (2021): Effect of translate speed of typhoon on wind waves, *Natural Hazards*, 105, 841–858. doi:10.1007/s11069-020-04339-4

Chapter 2

Field surveys of tsunami disasters around the world

CONTENTS

DOI: 10.1201/9781003156161-2

2.1 THE 2004 INDIAN OCEAN TSUNAMI

Takayuki Suzuki

Yokohama National University, Yokohama, Japan

Tomoya Shibayama

Waseda University, Tokyo, Japan

2.1.1 Introduction

December 26, 2004, at 7:58 am local time (0:58 UTC), a major earthquake of magnitude 9.1 occurred off the eastern coast of Sumatra, Indonesia, and generated a tsunami that killed at least 225,000 people across a dozen countries, with Indonesia, Sri Lanka, India, Maldives, and Thailand particularly badly hit (see also Section 9.1). The tsunami was recorded by tidal gauges not only in countries facing the Indian Ocean but also in Antarctica, Japan, and others. Many countries facing the Indian Ocean lost many of their citizens due to the tsunami flooding, which took the inhabitants of coastal areas by surprise due to both a lack of awareness and warning (given that there was no ground shaking to alert them).

This *2004 Indian Ocean Tsunami* (also known sometimes as the *2004 Boxing Day Tsunami* in the west), introduced many people around the world to the term "tsunami" for the first time, disaster risk managers started to consider such events more seriously, and efforts were made to investigate the causes and mechanism of damage. In addition, the construction of an alarm network started to be promoted by researchers and governments in each country, e.g., the Pacific Tsunami Warning Center (PTWC), which already existed before the 2004 event and expanded towards the Indian Ocean after this event, and the new development of the Indonesia Tsunami Early Warning System (Ina TEWS).

2.1.2 Survey at Banda Aceh, Indonesia

A field survey was conducted after the tsunami attack between February 12 and 14, 2005, in Banda Aceh, at the northern end of Sumatra Island, which was severely damaged by the earthquake and tsunami, and Aceh Besar Province, on the west coast of Aceh. The ground and tsunami trace heights were measured using an auto-level (Sokkia, B21) and survey staff, with the results being shown in Figure 2.1.1. Latitude and longitude were also

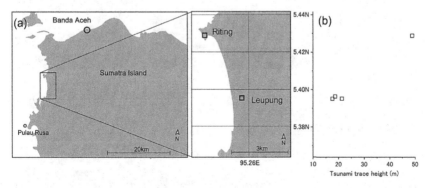

Figure 2.1.1 Summary of post-tsunami field surveys: (a) surveyed locations at the northern part of the Sumatra Island, Indonesia, and (b) surveyed tsunami trace height above water level.

recorded at each measurement point using GPS (GARMIN, eTrex Legend). Tsunami trace elevations were corrected to the tidal level at 8:30 am on December 26, 2004, in local time at Pulau Rusa (Tsuji et al., 2005), the nearest point where estimated tidal data exists (see Figure 2.1.1). Interviews with local residents were also conducted, and full details of all data gathered can be found in Shibayama et al. (2007).

2.1.2.1 Damage patterns along the northern coast of Banda Aceh

On the northern coasts of Banda Aceh (see circle in Figure 2.1.2(a)), erosion countermeasures had been constructed along the coast, though the tsunami caused a retreat of the shoreline position of between 50 m to 100 m. The stone wall that had been erected to a height of approximately 3 m above sea level collapsed due to the tsunami (Figure 2.1.2(a)), and a stone tomb in Syiah Kuala was also destroyed.

Around this low-lying area, shrimp farms were located on the land side of the shoreline (Figure 2.1.2(b)). Thus, the tsunami probably retained a high momentum as it propagated over the shrimp farms, then swiftly moved to attack the city centre, completely destroying all houses and flooding up to about 4 km inland from the coastline.

2.1.2.2 Riting

Riting is a village located on the west coast of Banda Aceh, where the peninsula is protruding to the west. As a consequence of the unique combination of the shape of the peninsula and local topography, the energy density of the

Figure 2.1.2 Survey locations in the Sumatra Island: (a, b) Banda Aceh, (c, d) Riting, and (e, f) Leupung.

tsunami was concentrated in the narrow saddle between two small mountains, with run-up reaching a maximum height of 48.9 m (Figure 2.1.2(c), (d)). This was the highest record of tsunami run-up anywhere as a consequence of the *2004 Indian Ocean Tsunami*.

2.1.2.3 Leupung

Leupung is a sub-district located slightly south of Riting, and here it was confirmed that the tsunami reached a mountain cliff located about 900 m from the coastline, with a run-up of 18.6 m being measured (Figure 2.1.2(e)).

In addition, at a location about 340 m from the coastline, trace heights 21.4 m and 17.5 m high were left on the upper part of the tree. The tsunami likely hit the mountain cliff and reflected, leading to a superposition of the incoming and outgoing waves. Based on interviews with residents of the area, the tsunami destroyed the village, and only 400 of its 8,000 inhabitants survived. Generally, the area was characterized by a scattering of low-altitude hills, with the flat low-lying land closer to the water being completely erased by the tsunami (Figure 2.1.2(f)).

2.1.3 Survey at southern coasts of Sri Lanka

Field surveys were conducted from January 7 to 9, 2005, along the south coastline of Sri Lanka, from Galle to Kirinda, covering the main cities and characteristic terrains in the area (including capes and estuaries). See also Section 9.2. for additional details regarding risk management in Sri Lanka before and after this event. In the survey, the tsunami trace heights were measured, and observations were made about the human and structural damages at each location. The overall methodology was similar to that outlined for the case of Banda Aceh, with interviews also being carried out with local residents.

Survey locations and the results of the observed tsunami trace height are shown in Figure 2.1.3. The highest tsunami trace was observed at the harbour area of Hambantota, representing a 10.9 m run-up height. The following subsections will provide additional details of the observations made at several points along the coastline.

2.1.3.1 Kirinda fishery port

Kirinda fishery port was constructed with assistance from Japan's Official Development Assistance (ODA) in 1984. However, after its construction the port gradually filled with sand due to the littoral drift. Although an extension of the breakwaters had been constructed and sand dredging had been made, siltation had progressed to a level where only small fishing boats could enter the port before the tsunami.

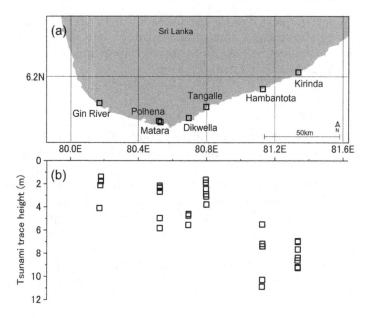

Figure 2.1.3 Summary of the survey in Sri Lanka: (a) survey locations along the southern coastline of Sri Lanka and (b) surveyed tsunami trace height.

Kirinda was the point of Sri Lanka that was located closest to the epicentre of the earthquake (among those surveyed), and thus the maximum run-up height reached 9.2 m (see Figure 2.1.4(a)). Figure 2.1.5(a) shows a house located 166 m from the port, where the edge of its roof was broken due to the run-up due to the tsunami. Sand had also accumulated around the loading/unloading area of the port (Figure 2.1.5(b)), though after the tsunami it was completely gone (according to local people). Since the water depth at the side of the loading/unloading area was about 2 m after the tsunami, most of the sand that had accumulated in the port must have been either carried to land by the forward flow or removed back into the ocean by the backflow.

2.1.3.2 Polhena

There was a wide reef flat in front of Polhena beach, between the edge of the coral reef and the shoreline. Part of the energy of the tsunami was likely reflected at the edge of the coral reef, with the tsunami then losing energy as it propagated over the reef flat, resulting in measured tsunami inundation heights between 2.2 and 2.7 m (Figure 2.1.4(b)). Although the inundation height was low compared to other locations, many houses near the beach were flooded, and even in those that did not collapse many of their residents drowned due to the tsunami (Figure 2.1.5(c), (d)).

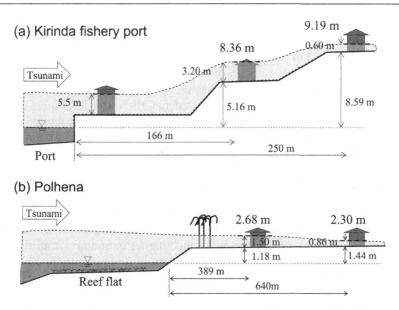

Figure 2.1.4 Cross-shore sketch of the survey locations in Sri Lanka: (a) Kirinda fishery port, and (b) Polhena.

Figure 2.1.5 Survey locations in the southern coast of Sri Lanka: (a, b) Kirinda fishery port, (c, d) Polhena, and (e, f) Gin River.

2.1.3.3 Gin River

At Gintota Bridge, which is located at the mouth of the Gin River, the water level rose up to a height of 4.1 m due to the tsunami (Figure 2.1.5(e)), which advanced along the river and flooded a part of the floodplain in this region (thigh floodplain originally formed near the river mouth during the rainy season and experienced regular small-scale flooding). As a consequence of this, residents were accustomed to experiencing flooding, and the damage was not significant in this area.

About 1.7 km upstream from the river mouth, the water level increased up to 2.1 m (Figure 2.1.5(f)). Moreover, at about 4.6 km from the river mouth, according to interviews with residents, the water level increased to around 1.4 m due to the first tsunami run-up. At this location, about 3 hours after the first tsunami attack time, the second wave ran up the river, and the water level increased again up to a level of around 1.8 m.

2.1.4 Conclusions

The *2004 Indian Ocean Tsunami* caused significant damage to both the northern coast of Sumatra Island and the southern part of Sri Lanka. However, the energy of tsunami was much higher in Sumatra, and the size and intensity of the damage varied depending on the geographical characteristics of the land and local bathymetry, amongst other factors. Thus, it is clear that disaster risk management needs to carefully consider both the external force of a given hazard and the vulnerability of those exposed to it. Even in the case when major events take place it is possible to minimize the loss of life if adequate evacuation countermeasures are in place (see Section 8.1). Therefore, to reduce the potential damage caused by tsunamis, it is not only important to construct structures that can reduce to some extent the energy of the incoming waves, but also for local residents to take appropriate evacuation actions. Effective tsunami evacuation measures include the creation and dissemination of a hazard map, the improvement of awareness about earthquakes and tsunamis, and the implementation of an early warning system that is fit for local conditions.

In the future, it is important to create a disaster prevention system that takes into account both the dimension and population of a given settlement and the availability of funding to improve disaster countermeasures. Failing to take into account such conditions can lead to reconstruction patterns that do not lower the vulnerability of a given settlement (see Section 8.2), thus highlighting the need for residents and governments to cooperate to reduce overall risks.

2.2 THE 2006 JAVA TSUNAMI

Hiroshi Takagi
Tokyo Institute of Technology, Tokyo, Japan

2.2.1 Introduction

On July 17 at 8:19 UTC (3:19 PM local time), 2006, a magnitude Mw 7.7 earthquake occurred off the south coast of Western Java, Indonesia. While the earthquake itself caused no significant damage, it triggered a regional tsunami that flooded about 300 km of the coastline and claimed more than

800 lives (OCHA, 2020). The earthquake was centred on the Java trench, the subduction zone situated between the Australian and Sunda plates. While there was extensive damage and loss of life due to the tsunami, the earthquake caused only moderate shocks. Hence, there was very little warning before the tsunami struck, and people at the shore were not informed of the approaching waves by any warning system of alert by the authorities (OCHA, 2020). Based on these characteristics, this earthquake is considered to be a tsunami earthquake[1] (Ammon et al., 2006). Lavigne et al. (2007) reported that the maximum flow depth reached 5 m, and maximum run-up heights were 15.7 m. The overall tsunami height data measured by multiple survey groups indicated that the average wave height of the event was 4–8 m along at least 200 km of the coastline (Hebert et al., 2012).

2.2.2 Field survey

This chapter presents the results of the fieldwork conducted by the joint survey team of Yokohama National University and Bandung Institute of Technology (Shibayama et al., 2006). The team surveyed an area of about 200 km of the coastline from West to Central Java (Figure 2.2.1) from July 21 to 24 (within one week after the earthquake) in 2006. The main purpose of the survey was to investigate the tsunami inundation height, which was established through physical evidence at each site and interviews with local residents (Table 2.2.1). Laser range finders, prisms and GPS receivers were used to measure the inundation height at each location with reference to the sea level at the time of the survey. The highest tsunami inundation measured by the survey team was 4.7 m, recorded on a beach of Widarapayung Wetan (Figure 2.2.2(a)). According to local residents, the splash due to the tsunami reached at least 7.6 m above sea level (Figure 2.2.2(b)).

Adipala Buton is a small coastal village located between Pangandaran and Widarapayung Wetan. Here the team did not find any visible evidence of

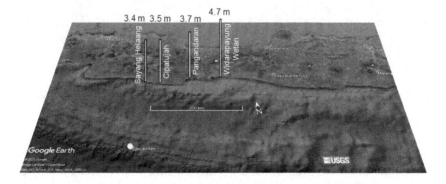

Figure 2.2.1 The distribution of tsunami heights in southwest Java Island, Indonesia.

---

[1] A tsunami earthquake is defined as an event that generates an unusually large tsunami relative to the seismic intensity (Kanamori, 1972).

Table 2.2.1 Tsunami heights measured from the sea level

Place	S	E	Height (m)	Marker
Pangandaran East coast	7:41:36.4	108:39:42.2	3.7	Muddy trace on external wall of house
Pangandaran West coast	7:41:24.0	108:38:47.8	3.7	Muddy trace on external wall of house
Widarapayung Wetan	7:41:49.8	109:15:55.5	7.6	Height of a trace on a tree
Widarapayung Wetan	7:41:49.8	109:15:55.5	4.7	Muddy trace on external wall of house
Cipatujah	7:44:52.3	108:00:41.1	3.5	Muddy trace on external wall of house
Sayang Helaag	7:40:08.3	107:41:47.1	3.4	Height of a sand dune
Sayang Helaag	7:40:06.3	107:41:27.6	3.3	Height of a sand dune

Figure 2.2.2 Photos taken during the field surveys: (a) Widarapayung Wetan, (b) Widarapayung Wetan, (c) Adipala Buton, and (d) Cipatujah.

inundation, as there were no houses or structures that had been located within the area that was flooded by the tsunami (Figure 2.2.2(c)). Even though visibility of the sea was good in this area, several people died at the beach, probably due to the fact that they did not feel a strong tremor that triggered their evacuation (such natural clues that help people to make a decision to evacuate, see Section 8.1). and the speed of the approaching tsunami, which would have been too fast for them to escape.

The coast of Cipatujah is covered with sand dunes, which may have dampened the momentum of the tsunami. However, in sections where the dunes were lower significant scouring was observed (Figure 2.2.2(d)), indicating that the tsunami flowed through narrow gaps and spread to the villages behind. Since the memory of the *2004 Indian Ocean Tsunami* (see Section 2.1) was still fresh in the minds of many, villagers fled while shouting "tsunami, tsunami".

2.2.3 Situation in Pangandaran

The largest loss of life took place in the resort area of Pangandaran, where more than 200 people were killed and over 3,000 houses were severely damaged. The estimated run-up heights from eyewitness accounts were 4–6 m (Mori et al., 2007). The tsunami arrived Pangandaran approximately one hour after the main earthquake. While it is estimated that the *2004 Indian Ocean Tsunami* reached the Banda Aceh coastline at 600 km/h, the 2006 event was slower, and reached Central Java at 256 km/h due to the shallow depth of the ocean in front of the coastline affected (Lavigne et al., 2007). The lack of awareness of people in the area about the possibility of a tsunami arriving could be explained due to the fact that the shaking due to the earthquake was not strong and the tsunami took some time to arrive.

This area forms a typical tombolo, which consists of an island connected to the mainland by a sandbar (see Figure 2.2.3). As the island belongs to a national park few lived there, in contrast to the sandbar, which had developed into a congested tourist area. Waves around this stretch of the coastline are usually calm, given the protection offered by the island. However, diffraction of the approaching tsunami waves took place, resulting in the two separate waves that arrived from both the east and west directions (Figure 2.2.3). Even if people were to see one of the tsunami waves and quickly run away from the shore, there is a high chance that they would have been swallowed by the one coming from the other side.

It is noteworthy that no special hard countermeasures such as coastal dikes or elevated roads have been implemented along this coast since the event (see Figure 2.2.4). Enhanced evacuation measures such as sign boards that indicate evacuation routes have also not been installed. Instead, many new buildings have been built after the earthquake, with their number on Pangandaran Beach increasing from 921 in 2006 to 1052 in 2017 (Mardiatno et al., 2020). All of this indicates that the recovery and expansion of the tourist industry appears to have been prioritized in this area, suggesting that balancing development and disaster risk prevention is not an easy task. This is similar to reconstruction patterns observed in Banda Aceh, Indonesia, following the *2004 Indian Ocean Tsunami* (see Section 8.2).

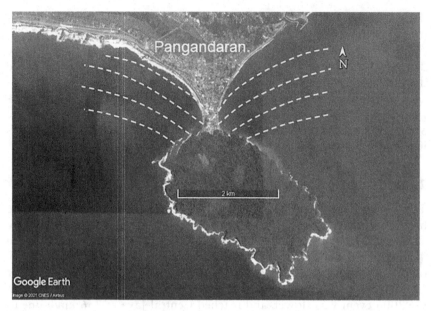

Figure 2.2.3 A famous beach tourist destination, Pangandaran in southwest Java Island, developed on a tombolo. Tsunami diffraction took place in the shadow of the island, flooding the town from both directions.

(a)   (b)

Figure 2.2.4 Eastern beach of Pangandaran: (a) Immediately after the earthquake, July 22, 2006 and (b) present situation, 2019 (© Google Earth).

2.2.4 Conclusions

Given the number of catastrophic earthquakes that Indonesia has historically suffered, the *2006 Java Tsunami* was not necessarily an extreme event in terms of both tsunami magnitude and recorded casualties. Nevertheless, the tsunami directly hit the popular tourist town of Pangandaran, which developed on a narrow area of tombolo. The beach is normally free from high waves thanks to the presence of the offshore headland. However, due to the diffraction of the tsunami, people were caught by two waves approaching from two opposite directions. To make matters worse, the people did not feel the tremor as strongly as during other events, which may have contributed to a false sense of security and reduced their fear that a tsunami might have been approaching.

More than a decade after the disaster, this tourist destination has been rebuilt and many more hotels are present in the area, though there seems to have been little progress in enhancing tsunami preparedness. Perhaps the tourism industry is worried that emphasizing the risk of tsunami will result in fewer visitors, which illustrates the difficulty of attempting to implement disaster prevention countermeasures in tourist areas.

2.3 THE 2009 SAMOAN TSUNAMI

Ryo Matsumaru

Toyo University, Tokyo, Japan

2.3.1 Introduction

At 6:48 on September 29, 2009 (local time, UTC–11), a powerful earthquake of Mw 8.1 took place near the Samoan islands, which generated a large tsunami. Figure 2.3.1 shows the bathymetry around the Samoan archipelago and the epicentre of the main shock. The tsunami struck the coastal

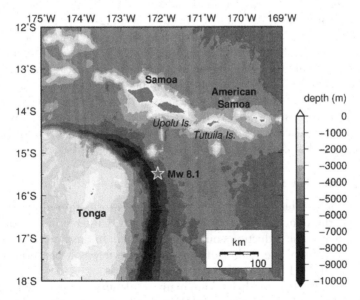

Figure 2.3.1 Bathymetry around the Samoan islands (data from the General Bathymetric Chart of the Ocean, GEBCO) and the location of the earthquake (United States Geological Survey, USGS).

areas of Samoa (officially called the Independent State of Samoa), American Samoa, and Tonga. As a result, at least 149 people in Samoa, 34 people in American Samoa and 9 people in Tonga died (USGS, 2009).

A team consisting of members of Waseda University and Samoa National University conducted two rounds of field survey in Samoa and American Samoa. The first one was conducted in October 2009, one month after the tsunami event. The purpose of the survey was to measure the height of the tsunami, confirm the damage that took place, and understand the potential to formulate future tsunami countermeasures in the island. The second survey was a follow-up to the first survey and was conducted in October 2015, with the purpose to document the status of recovery from the tsunami along the coastal areas and the response of the community to disaster mitigation.

2.3.2 First field survey

2.3.2.1 Tsunami inundation and run-up

The field survey was conducted from October 28 to 31, 2009, in both the Independent State of Samoa and American Samoa. The main areas targeted in this field survey were the south coast of Upolu Island in the Independent State of Samoa and the west coast of Tutuila Island in American Samoa, where the tsunami damage was most severe. The tsunami inundation and run-up heights were measured using methodologies similar to those described in Section 2.1, with the measurements taken being shown in Figure 2.3.2.

After the event, several international survey teams also conducted tsunami field surveys on the islands (Arikawa et al., 2010; Namegaya et al., 2010; Okal et al., 2010). All teams reported inundation heights of over 5 m in a wide area of the coast of Upolu and Tutuila islands, with the tsunami attacking not only the southern but also the northern coasts of the islands. In Upolu Island the measured run-up heights were over 10 m along the southeastern coast, and gradually decreased towards the west. In Tutuila Island, higher run-up heights were observed on the western coast, with the tsunami height decreasing towards the east. Localized high run-up heights were measured at the tips of capes and the inner part of bays.

2.3.2.2 Interview survey to residents

The team also interviewed residents to collect information about the tsunami and the damage that took place, trying to gather opinions on how future tsunami risk reduction could be formulated. In this book only a brief outline of some of the findings from these interviewswill be outlined (see Mikami et al. (2011) for further details).

Regarding the tsunami behaviour, many residents saw the wave breaking on top of a coral reef platform which was about 2 km wide at its most extensive point. This provided cues to villagers about the approaching tsunami and acted as a trigger for residents to evacuate early in some places,

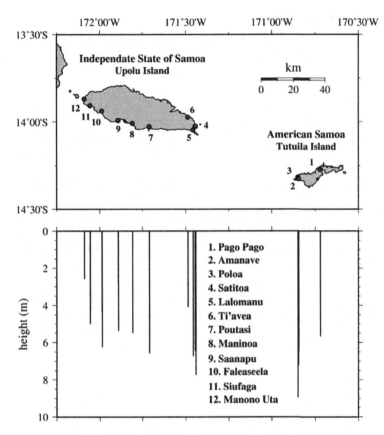

Figure 2.3.2 Surveyed points and distribution of measured tsunami inundation heights in Upole and Tutuila Islands (Mikami et al., 2011).

likely contributing to saving lives (as early evacuation is widely recognized as an important factor for the decrease in mortality rate, see Section 8.1). Breaking on the coral reef also decreased the kinetic energy of the tsunami in one of the surveyed villages, similarly to the reports of residents in the Maldives or the Solomon Islands during other events (Fujima et al., 2006; Tomita et al., 2011).

During the interviews in Ulutogia village, residents explained that the community had already decided to relocate to a higher place. For this, some meetings were held within one month after the tsunami, and then the Matai (local word for the chief of the village) decided to relocate the village to higher ground. The relocation site was about 1.5 km inland from the coast. In Samoa, village territories generally extend inland from the shore to the hills (Crichton et al., 2020), and this means it is not difficult to relocate residents without having to negotiate with other villages if the Matai shows strong leadership.

2.3.2.3 Numerical simulation

Numerical simulations of the tsunami propagation were also carried out to understand the general behaviour of the waves around the Samoan Islands. The governing equations are the non-linear shallow water equations and finite difference method, and a leap-frog scheme was employed to solve the equations (see also Section 5.1).

It should be noted that there were two limitations when conducting these simulations. The bathymetry data, with a grid size of 30 seconds obtained from GEBCO, was too coarse to accurately reproduce the coral reef around the islands, though this was employed as it was the only available data at that time. The fault parameters estimated by Japan Meteorological Agency (2009) were used in the simulation, though it is not clear how accurately they represent the actual earthquake that took place. Nevertheless, despite these limitations the simulations appeared to be able to simulate well the general behaviours of the tsunami, including the wave patterns, heights, and arrival times.

The result of the simulation is shown in Figure 2.3.3. The first wave reached the coast of the Samoan Islands between 15 and 20 minutes after the earthquake, with there being about 10 minutes between the arrivals of each consecutive wave. Figure 2.3.3 clearly shows that the wave diffracted around the eastern and western edges of the islands.

2.3.3 Second field survey

The second field survey was conducted from October 5 to 10, 2015. The team revisited all the sites, including American Samoa, surveyed in 2009 to check the reconstruction status and the level of preparedness against future tsunamis.

As six years had passed since the tsunami disaster, the team confirmed that the infrastructure damaged along the coastal areas had been restored

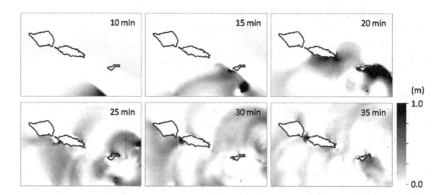

Figure 2.3.3 Tsunami height distributions between 10 and 35 minutes after the earthquake (Mikami et al., 2011).

Figure 2.3.4 Left. Evacuation Route to a higher area. Right. Interviews at relocated community (Ulutogia Village).

and that evacuation routes to higher ground had been constructed at some locations (Figure 2.3.4).

Ulutogia Village, which had already decided to relocate to a higher place in October 2009, had completed its relocation and resettlement (Figure 2.3.4). During interviews with local residents carried out in 2015, nobody expressed any particular difficulties with their livelihoods after the relocation. This may be due to the relocation site being close to its original location, and that the entire community relocated (which contrasts markedly to reconstruction after tsunamis in other areas such as Indonesia, see Section 8.2). It could be said that this is a good example of how the strong will of community leaders led to the relocation of the entire community, which will likely greatly increase their resilience to a future tsunami, but whether or not this situation will be maintained will require further investigation in the future.

2.3.4 Conclusion

A number of conclusions can be drawn from the two field surveys that were conducted. The tsunami height exceeded 5 m in a wide area of Upole and Tutuila Islands and the maximum run-up height was over 20 m in Poloa, American Samoa, with the simulation results indicating that the first wave attacked the coast of the Samoan Islands between 15 to 20 minutes after the earthquake. An extensive coral reef platform in front of the coast affected the tsunami behaviour, with the wave breaking on top of the reef being seen by many and acting as a trigger for evacuation.

The social structure of Samoa, characterized by the strong leadership of Matai (the chief of the village), strongly influenced the decision-making process of residents. Thus, some villages decided to relocate within one month of the tsunami, and nobody returned to the initial location of the village (at least up to 2015). Due to widespread education about tsunami disasters, and regardless of whether the tsunami warning arrived (Ulutogia) or not (Satitoa, Poloa), many people indicated that they felt danger when the tsunami passed over the coral reef and decided to evacuate. Thus it is clear that global awareness about tsunami disasters, following the 2004 *Indian Ocean Tsunami in December* (see Section 2.1.), is increasing. This is a welcome development, and should hopefully help to improve disaster risk management in countries at risk, and enhance the long-term resilience of coastal communities.

2.4 THE 2010 CHILE TSUNAMI

Rafael Aránguiz

Universidad Catolica de la Santisima Concepcion, Concepcion, and Research Center for Integrated Disaster Risk Management (CIGIDEN), Santiago, Chile

2.4.1 Introduction

Chile is known to be one of the most earthquake-prone countries in the world, due to the fact that it is located along the subduction zone of the Nazca plate beneath the South American Plate. An analysis of earthquakes of magnitude above 6.0 in the period 1562 to 1997 identified a total of 102 tsunamis along the Chilean coast (Lagos, 2000). Moreover, 4 additional events have taken place in the last 20 years, namely the Maule 2010, Iquique 2014, Illapel 2015 and Chiloe 2016 earthquakes (Aránguiz, 2018). The largest earthquake ever recorded instrumentally around the world took place in 1960 in southern Chile and had a magnitude of 9.5 (see also Section 9.5), generating a large tsunami that affected not only the Chilean coast, but was also recorded in California, Alaska, New Zealand and Japan (Takahashi, 1961). After this event, mitigation measures such as tsunami breakwaters started to be constructed in Kamaishi and Ofunato in Japan (Murata et al., 2009), and the Pacific Tsunami Warning Center was implemented in Hawaii.

On the 27th of February 2010 at 3:34 local time (6:34 UTC) an earthquake of magnitude Mw 8.8 took place in central Chile (Delouis et al., 2010; Lorito et al., 2011). The earthquake generated a destructive tsunami along the coast (33–39°S) and caused loss of life and significant damage to housing and public infrastructure (Contreras and Winckler, 2013; Fritz et al., 2011). In addition, since the tsunami took place in the summer season, it was found that most of the casualties were not residents of coastal zones, but tourists (Fritz et al., 2011). Moreover, the wave appeared to have a diverse impact along the coast; while some areas recorded a run-up of 29 m (Fritz et al., 2011), other places such as the eastern shore of the Gulf of Arauco experienced inundation heights of less than 3.5 m (Aránguiz et al., 2019).

The 2010 Maule earthquake had an estimated rupture area of 450 km by 150 km, extending from Arauco Peninsula in the south (37°S) to Pichilemu in the north (34°S) with an average slip of 10 m. The hypocenter was found to be at 36.208°S, 72.963°W at a depth of 32 km (Delouis et al., 2010). It is important to note that this earthquake could have filled the slip deficit accumulated during 175 years since the 1835 earthquake (Delouis et al., 2010). Therefore, the probability of a large earthquake in the same zone in the near future is rather low, due to the fact that the plates move at a velocity of ~8 cm/year (Saillard et al., 2017). The earthquake was felt as far as Antofagasta in the north (23.5°S) and Puerto Montt in the south (41.5°S), a total extension of 2,000 km. The total affected area comprised 75% of the Chilean population and the economic loss was estimated to be 30 billion US$, which is equivalent to 18% of the country's GDP (Contreras and Winckler, 2013). The history of large earthquakes along Chilean territory has motivated the development and update of seismic design codes during the course 20th century. Therefore, despite the large magnitude of the earthquake and extension of the affected area, the damage was limited (Contreras and Winckler, 2013) and only 521 victims of the earthquake and tsunami were reported (Fritz et al., 2011). Of these, only 124 died due to the tsunami, concentrated in the coastal regions of Maule (69) and Biobío (33), though victims were also found at Robinson Crusoe Island (18) and Mocha Island (4) (Fritz et al., 2011). Robinson Crusoe Island is located 600 km northwest of the 2010 earthquake rupture area, meaning that the quake was not felt there and those who perished were not aware that a tsunami could be approaching. Moreover, the tsunami warning system did not work properly, and the warning was cancelled at 4:49 local time, when two tsunami waves have already arrived at Robinson Crusoe Island.

2.4.2 Field surveys

Several field survey teams visited the tsunami-affected area and documented tsunami run-up, flow depth, inundation height and conducted eyewitnesses interviews (Fritz et al., 2011; Mikami et al., 2011; Sobarzo et al., 2012, Vargas et al., 2011). Figure 2.4.1 shows examples of tsunami impact on roads and houses. Figure 2.4.2 shows the spatial distribution of tsunami run-up measurements along the coast of Central Chile. At most places, the maximum tsunami wave arrived during low tide, somewhat attenuating the damage, with field survey data being corrected by the tide at the time of maximum inundation. The maximum tsunami run-up took place at Constitución and reached a level of up to 29 m at a steep coastal cliff (Fritz et al., 2011). The tsunami surged at least 15 km into Maule River, though the city of Constitución, located on the southern river bank, experienced relatively lower run-up and a horizontal inundation of up to 500 m. Orrego Island, which is located in the middle of the river near the river mouth, was completely submerged by the tsunami, with the flow depth reaching 10 m. The island was often used as a campground, and many people were washed away with no possibility of evacuation (Fritz et al., 2011).

Figure 2.4.1 Tsunami impact along the coast of Biobio region. (a) Tsunami scour
on the coastal road at Dichato. (b) Remains of the foundations of a
house and a displaced boat at Dichato. (c) Damaged two-story house
at Coliumo. (d) Displaced house at Tumbes, Talcahuano. (e) Remains
of the foundations of a house at Mocha Island.

North of Constitución (from 35°S to 33°S), the run-up distribution exhib-
ited typical heights of between 5 and 10 m, with a few peaks of around
14 m near the San Antonio area. For instance, small fishing towns, such as
Illoca and Duao, were flooded with tsunami run-ups of 5.8 m and 7.0 m,
respectively. The small low-lying village of Llolleo, at the northern bank of
Maipo river, was completely flooded and all lightweight material houses
were completely destroyed (maximum inundation height was 4.6 m). South
of Constitución (35–39°S), a significant variation in the impact due to the
tsunami was observed along the coast, and run-ups ranged from 5 to 15 m.
Some low-lying campgrounds were also flooded, such as the ones located
at Pelluhue and Curanipe, where the tsunami reached run-ups of 12.4 and
11.2 m, respectively.

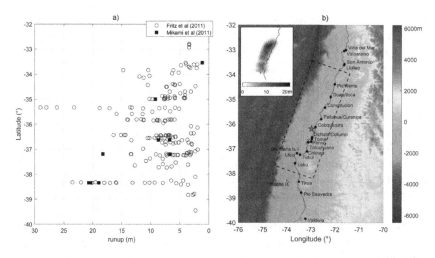

Figure 2.4.2 (a) Tsunami run-up measurements along the Chilean coast. (b) Locations affected by the 2010 Chile tsunami. The black dashed line shows the rupture area, the star indicates the epicentre. The inset shows the 2010 finite fault slip model from Moreno et al. (2012).

In a few cases the run-up decreased below 5 m, such as around the area near Cobquecura (the epicentral area), where the run-up was only 2.8 m (Fritz et al., 2011). In a similar manner, low inundation heights were measured at the east coast of both the Bay of Concepción and the Gulf of Arauco (Fritz et al., 2011; Mikami et al., 2011). For example, a run-up of 3.7 m was measured at Tomé, located at the northeastern side of the Bay of Concepción, while inundation heights of 6.5 and 6.7 m were measured at the southern shore of the bay, at Penco and Talcahuano, respectively. In a similar manner, no inundation was recorded at the 2 km wide Biobio River, and the inundation height was in the order of 2 m along the coast of San Pedro de la Paz and 3.0 m at Colcura, all of them located at the eastern shore of the Gulf of Arauco. Nevertheless, large run-ups were observed at the southern shore of the gulf, with 8.4 and 13.4 m measured at Tubul and Llico, respectively. It is interesting to note the large run-up measured in a coastal cliff at the southern shore of Tirua river, which reached up to 20 m. South of Tirua, the run-up rapidly decreased and did not exceed more than 5 m.

2.4.3 Analysis of 2010 tsunami behaviour

It is interesting to notice the low run-up at Cobquecura (~36.2°S) and the eastern shore of the Gulf of Arauco (~37°), in which maximum measured run-ups were less than 5 m. This appears to have been due to the characteristics of the seabed morphology and the presence of submarine canyons (Aránguiz and Shibayama, 2013). For the case of submarine canyons, the most important variables that influence tsunami propagation are its width and length, such that the canyons induce a diffraction–refraction phenomenon, and thus the wave fronts propagate in a direction more normal to the edges of the canyon. This effect can cause a spatial variation of the wave amplitude along the coast, such that the run-up in front of the canyon (Biobio river mouth or Cobquecura) is always lower than in the case where there is a regular bathymetry (i.e., with no underwater canyon), though the wave amplitude increases at both sides of the canyon (Aránguiz and Shibayama, 2013). This phenomenon was enhanced due to the presence of Santa María Island inside the Gulf of Arauco, and thus large tsunami run-ups were observed at the southern shore of the Gulf, which reached up to 18 m at Llico (Mikami et al., 2011).

Another interesting phenomenon is the generation of edge waves (see also Section 5.1). These waves are typically generated by continental subduction zone earthquakes and are affected by refraction, propagating parallel to the shoreline (Geist, 2013). Since the 2010 earthquake rupture area occurred beneath a wide and shallow continental shelf, significant edge wave activity was observed. Moreover, edge waves may be reflected by steep coasts as well as by the shelf-slope boundary, and thus become trapped on the continental shelf. Therefore, different tsunami wave directions can be explained by oppositely propagating edge waves (Morton et al., 2011).

Resonance also plays an important role in tsunami wave amplification, such that the behaviour of the waves is greatly influenced by the local bathymetry rather than the characteristics of the source (Power and Tolkova, 2013; Rabinovich, 1997). In fact, Yamazaki & Cheung (2011) found that coupled oscillations of dynamic systems at different geographic scales such as the continental shelf, the Gulf of Arauco, and the Bay of Concepción were responsible for multiple long-period tsunami waves in Talcahuano during the *2010 Chile tsunami*. The modes at the continental shelf scale may have also been responsible for the large tsunami amplification and late arrival of the wave to Concepción. Moreover, Aránguiz et al. (2019) found that the Bay of Concepción shows complex oscillation structures and that most of these structures contain Talcahuano, implying that the superposition of several modes results in large amplitudes in that location. Subsequently, the generation of edge waves and resonance explains the multiple waves and late arrival reported in both historical and recent tsunami events at Talcahuano (Aránguiz et al., 2019). Moreover, it has been demonstrated that larger resonators can modify the period of the smaller resonator, suggesting resonance coupling between multiple resonators and not only with shelf oscillations. For instance, there is an influence of the Gulf of Arauco on the Bay of Concepción, as well as an influence of the Bay of Concepción on Coliumo Bay. The spatial pattern of natural oscillation modes is also affected by morphological features such as islands, peninsulas, and submarine canyons. Therefore, some areas located close to nodal lines of regional modes are less affected by those modes and consequently would experience less tsunami amplification (Aránguiz et al., 2019). This is the case with San Vicente Bay (near Biobio Canyon) and Cobquecura (near Itata Canyon). On the contrary, Coliumo Bay is strongly influenced by shelf modes, which explains the significant inundation at Dichato during the 2010 tsunami. In a similar manner, long-period oscillations of the Gulf of Arauco and shelf modes are coupled with the pumping mode of the Bay of Concepción, resulting in large tsunami wave amplification at Talcahuano and Penco (Aránguiz et al., 2019).

2.4.4 Conclusions

The 2010 magnitude 8.8 earthquake generated a tsunami that affected a large portion of the central coast of Chile, though a variety of inundation heights, run-ups and wave characteristics could be observed along the shoreline. The generation of edge waves over the wide and shallow continental shelf helps to explain the different tsunami wave directions. The steep continental slope at the boundary of the shelf also facilitated the reflection of edge waves, trapping and helping them propagate along the coastline. In addition, geographical features, such as submarine canyons, caused a spatial variation in the tsunami wave amplitude along the coast, such that the run-up in front of the canyon is lower than to its sides. Finally, constructive interference of edge waves and resonance can also explain the late arrival of the largest tsunami waves in some areas, whereas other nearby places experienced much smaller waves that caused only minor damage.

2.5 THE 2010 MENTAWAI TSUNAMI

Takayuki Suzuki

Yokohama National University, Yokohama, Japan

2.5.1 Introduction

On October 25, 2010, at 21:42 local time (14:42 UTC), a significant earthquake of magnitude 7.7 occurred off the Mentawai Islands in Indonesia, an archipelago located on the west side of the Sumatra Island. The earthquake generated a tsunami that caused damage to the coastal area of islands nearby and went on to propagate through the Indian Ocean. According to the Indonesian National Board for Disaster Management (Badannasionar Penang Grangan Benkana [BNPB], in Bahasa Indonesia), the earthquake and tsunami caused 509 casualties (with a further 21 missing) and heavily damaged 550 houses in the Mentawai Islands.

To measure the tsunami trace heights and gather information from the residents' field surveys and interviews were conducted after the event by several international survey teams (Tomita et al., 2011; Hill et al., 2012; Satake et al., 2013). In this chapter survey results of the south coast of Sipora island are presented, together with some insights on how to improve preparedness against future tsunami threats on these remote islands. A more detailed report of all survey data is given in Mikami et al. (2014).

2.5.2 Survey methods and results

2.5.2.1 Methodology

The field survey was conducted around the south coast of Sipora island on November 19–20, 2010, about a month after the event. Four of the tsunami-affected villages, Beriulou, Masokut, Bosua, and Gobik, were surveyed with the aim of observing the distribution of tsunami trace heights and analysing damage patterns. The tsunami trace height was measured at each location using a laser ranging instrument (IMPULSE, Laser Technology Inc.) and survey staff, with the precise location of each point being recorded using GPS. Traces of the tsunami were identified by broken branches, tree debris, watermarks left on structures, and the testimonies of eyewitnesses. The tide level correction was conducted based on the estimated tide level (WXtide32) at Siberut island, located to the north of Sipora island.

Also, questionnaire surveys were conducted in several villages to understand the behaviour of residents during the event and their awareness about tsunami risk. This survey was translated into Indonesian and conducted by native Indonesian speakers.

2.5.2.2 Tsunami height distribution

The tsunami trace heights measured at four villages in Sipora island are shown in Figure 2.5.1. Inundation heights exceeded 3 m in all villages surveyed and were over 6 m in two locations. The inundation heights

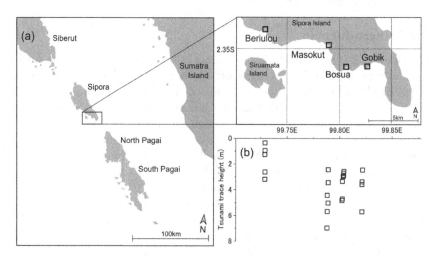

Figure 2.5.1 Tsunami trace height distribution along the south coast of Sipora island.

measured in Beriulou were smaller than those measured in the other villages. According to the dataset of Hill et al. (2012), inundation heights of 9.05 m and 5.55 m were measured at the small Siruamata island, located 3 km offshore from Beriulou, which would have shielded Beriulou from some of the energy of the tsunami.

The tsunami trace heights recorded along the south coast of Sipora and the North and South Pagai islands were less than 10 m (Tomita et al., 2011; Hill et al., 2012; Satake et al., 2013). The highest tsunami run-up height (16.9 m) was recorded on Sibigau island, off the west coast of South Pagai island, and the tsunami height along the north coast of Sipora and Siberut islands was less than 2 m (Hill et al., 2012). Therefore, the most severely damaged areas observed were along the shorelines of the Indian Ocean side of Pagai island and the southern part of Sipora island.

2.5.2.3 Damage to villages

2.5.2.3.1 Damage at Masokut

A river runs through the northern part of the village, with a sand dune shielding it from the sea (the height of the dune was about 1.5 m above sea level). In this village, inundation heights were measured at six locations: 4.42, 5.67, 6.96 m on the north side, 2.43 m in the centre of the village, 3.43, 5.01 m on the south side.

Many houses near the coast were severely damaged (Figure 2.5.2(a)). The houses that had been built on higher ground showed less damage, though their edges suffered about 1 m depth of scouring (Figure 2.5.2(b)). Due to

Figure 2.5.2 Damage patterns at surveyed locations in Sipola Island: (a, b) Masokut, and (c, d) Beriulou.

the absence of any road to the hills, it was difficult for residents to evacuate (in general, many of these villages were characterized by having poor road networks which, when they existed, consisted of narrow paths with a concrete surface just wide enough for a motorbike or two to traverse. In many cases transport around the islands is by boat).

2.5.2.3.2 Damage at Beriulou

In this village, a road connected the coastline to the hinterland behind it (Figure 2.5.2(c)), with the tsunami propagating along it and reaching more than 300 m from the coastline. Houses were severely damaged in an area up to 100 m from the coastline, with their seaside suffering particularly badly (Figure 2.5.2(d)). Inundation heights were 2.61 and 3.18 m along the most heavily damaged part of the village, gradually decreasing as the tsunami propagated inland, and a height of 0.35 m was measured 350 m away from the coastline.

The road allowed people to evacuate inland, though there were no hills or high buildings to eventually reach safety if a larger tsunami affects the area in the future.

2.5.3 Risk management strategy

2.5.3.1 Geographic conditions

In order to consider a tsunami mitigation strategy, it is important to understand the geographical characteristics of villages in this archipelago. The most effective way to mitigate future tsunami damage would be to move homes to higher ground (as what happened in Samoa following the 2009 tsunami, see Section 2.3). However, this is not always a realistic proposition, given that in many cases there are no suitable hills nearby, or residents may want to remain close to the coastline given that their livelihoods are often in that area. In such cases, residents need to consider the possibility of building tsunami shelters and evacuation routes to higher ground, which can allow them to quickly evacuate if the need arises (see Sections 8.1 and 8.2).

However, it can be difficult from an engineering and financial point of view for small villages to build and maintain a tsunami shelter. Therefore, securing and expanding evacuation routes should be considered as a disaster risk management countermeasure, as these can represent no-regret options (see Section 9.2. in Bangladesh), allowing the economic development of the island during normal conditions. Nevertheless, it would be important to prepare hazard maps and tsunami evacuation signs, which can raise awareness and inform people as to what to do in the event of an earthquake. This type of maps and signs also helps maintain long-term awareness about tsunamis (Figure 2.5.3).

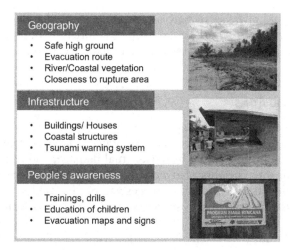

Figure 2.5.3 Tsunami vulnerability education (Modified based on Mikami et al. (2014)).

2.5.3.2 Resilience of buildings and infrastructures

Generally speaking, the quality of the buildings and infrastructures in the Mentawai Islands was low. Many of them did not use reinforced beams or columns, and when they did, the concrete was of poor quality. Therefore, they possess little lateral stability against a tsunami or other flow and could easily be swept away, particularly given that tsunamis generally carry with them abundant debris (see Section 6.2). Some of the buildings in the area fared well, particularly those built by richer residents or used for religious purposes, which had reinforced beams made of better quality concrete. Moreover, some of these buildings benefited from not having sidewalls, allowing water to flow through the building and limiting structural damage.

2.5.3.3 Preparedness against future tsunamis

To prepare for a future tsunami attack it is important to raise awareness amongst residents of the risks these hazards pose. Disaster preparedness plans should include three major components: frequent tsunami training, education for children, and relevant and accurate information about the risks and countermeasures being disseminated. Based on the questionnaire survey that was conducted, tsunami training had not been implemented in all villages, though children were indeed educated about such risks in some villages.

Tsunami evacuation training has been recognized as an effective way to help residents understand what to do during such an event (e.g., Dengler, 2005), and should continue regularly in areas which already have such exercises, and establishd in others that do not hold them at present.

Residents should always evacuate after feeling an earthquake or receiving an evacuation order by authorities (which requires the implementation of early warning centres, such as that in the Indian Ocean following the 2004 Indian Ocean Tsunami, see Section 2.1).

2.5.4 Conclusions

The present chapter summarised the field surveys conducted around the southern part of Sipora after the *2010 Mentawai Islands Tsunami* to clarify the distribution of inundation heights, actual damage, and resident awareness. Tsunami trace heights of over 6 m were recorded at two locations. While the residents of some of the villages had participated in tsunami evacuation exercises, not all villages shared the same level of tsunami awareness. Furthermore, in some villages it was difficult to evacuate immediately after the earthquake due to the lack of evacuation routes to hills or high buildings.

Due to the relative poverty of the islanders and the lack of financial resources it is likely that it will take time to improve the resilience of the villages to future events. In this regard the islands possess a number of geographical and socio-economic disadvantages that could hamper the development of tsunami mitigation countermeasures, including their remoteness, lack of resilience of buildings and other infrastructure, and low level of tsunami awareness and training.

2.6 THE 2011 TOHOKU TSUNAMI

Takahito Mikami
Tokyo City University, Tokyo, Japan

Tomoya Shibayama
Waseda University, Tokyo, Japan

2.6.1 Introduction

At 14:46 on March 11, 2011 (local time), a large earthquake of moment magnitude Mw 9.0 took place off the Pacific coast of Tohoku, Japan. The earthquake generated a major tsunami that caused severe damage to coastal areas in Iwate, Miyagi, and Fukushima Prefectures. The official figures as of March 2021 report 15,899 deaths and 2,526 missing (National Police Agency of Japan, 2021), with more than 99 % of them being in the three prefectures mentioned above. The results of the investigation showed that around 90 % of casualties were caused by drowning (National Police Agency of Japan, 2012), indicating that the tsunami was the main cause of death.

The Pacific coast of Tohoku faces the Japan Trench, where the oceanic plate subducts beneath the continental plate, with the region having generated many earthquakes and their associated tsunamis in the past. The oldest tsunami event documented in this region was the Jogan Tsunami in 869 with Nihon Sandai Jitsuroku (an ancient Japanese historical text) mentioning that a tsunami inundated coastal areas and around 1,000 people drowned. The Jogan Tsunami has been investigated by geologists since around 1990 by analysing tsunami deposits in sediment layers (e.g., Abe et al., 1990; Minoura et al., 2001). These paleotsunami studies, together with numerical simulations, have revealed that this tsunami reached several kilometres inland from the shoreline of the Sendai Plain (with the extent of inundation being comparable to that of the 2011 event). After the Jogan Tsunami, the region experienced a number of tsunami events of various magnitudes, among which the Keicho Tsunami in 1611 is considered to be one of the largest (Hatori, 1975).

Since the Meiji Era (from the year 1868), there were three major tsunami events in this region, namely the *1896 Meiji Sanriku Tsunami*, the *1933 Showa Sanriku Tsunami*, and the *1960 Chile Tsunami*. The *Meiji Sanriku* and *Showa Sanriku Tsunamis* were generated by earthquakes off the Sanriku Coast, causing the loss of 22,000 and 3,000 lives, respectively. The tsunami in 1960 was triggered by a large earthquake in Chile (see Section 9.5), which took more than 22 hours to reach the Pacific coast of Japan. At that time the warning system against a far-filed tsunami did not exist, and thus coastal areas suffered heavy damage. Over 100 casualties were reported throughout Japan, most of which were in the Sanriku Coast (comprehensive surveys for this event were conducted and the results were compiled in The Committee for Field Investigation of the Chilean Tsunami of 1960 (1961)). In 2010, another large earthquake occurred in Chile (see Section 2.4) and a tsunami again reached the Pacific coast of Japan. The tsunami heights measured along the Sanriku Coast were around 2 m (Tsuji et al., 2010), which were smaller than those in the 1960 event and, although fishery and aquaculture suffered severe damage, no casualties was recorded.

Figure 2.6.1 shows a comparison of the tsunami heights along the Pacific coast of Tohoku for the four major events described above. The Sanriku Coast experienced tsunamis of around 5 m as a consequence of the *1960 Chile Tsunami*, and of more than 20 m in the other three events. Regarding the coast of the Sendai Plain and Fukushima, there is no data available for the *1896 Meiji Sanriku* and *1933 Showa Sanriku Tsunamis*. In this region, 2–3 m waves were observed during the *1960 Chile Tsunami*, and 5–10 and 10–20 m in the Sendai Plain and Fukushima, respectively, during the *2011 Tohoku Tsunami*. Although the data for the older tsunami events is limited, it can be nevertheless said that among these four events only the *2011 Tohoku Tsunami* caused severe damage to the entire stretch of the Pacific coast of Tohoku.

It should also be noted that there is a regional difference of past tsunami experiences along the Pacific coast of Tohoku. The north region (Sanriku Coast) experienced two very large tsunamis over 30 m (*1896 Meiji Sanriku* and *2011 Tohoku Tsunamis*) in 115 years, and thus generally residents in this region maintained alive the memory of these large events. On the other

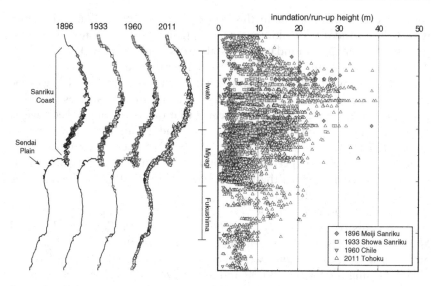

Figure 2.6.1 Comparison of the tsunami heights along the Pacific coast of Tohoku
for four major events: *1896 Meiji Sanriku, 1933 Showa Sanriku, 1960
Chile,* and *2011 Tohoku Tsunamis* (Data source: Japan Tsunami Trace
Database, https://tsunami-db.irides.tohoku.ac.jp/tsunami/).

hand, in the south region (Sendai Plain and Fukushima coast), the last large
tsunami event was the Jogan Tsunami in 869, indicating that there was
almost no memory of a large tsunami event taking place amongst society
there. It can be considered that this regional difference in memory greatly
affects disaster preparedness, evacuation behaviour, and the rehabilitation
process.

2.6.2 Inundation and run-up heights in the 2011 tsunami

In order to measure tsunami heights along the coast affected by the 2011
tsunami, the Tohoku Earthquake Tsunami Joint Survey Group was organized. In total, 299 researchers belonging to 64 universities/institutes joined
this group, and inundation and run-up heights were measured at 5,247
points (The Tohoku Earthquake Tsunami Joint Survey Group, 2011). The
dataset of these measured heights is available on the website of the group
and also included in the Japan Tsunami Trace Database.

Figure 2.6.2 shows the distributions of the tsunami inundation and run-
up heights measured along the Pacific coast of Tohoku according to the
results of the survey group. Generally speaking, the patterns of inundation
and run-up heights can be divided into three distinct groups, the northern,
middle, and southern parts. In the northern part (the Sanriku Coast), the
run-up heights were larger than the inundation heights. In the middle part

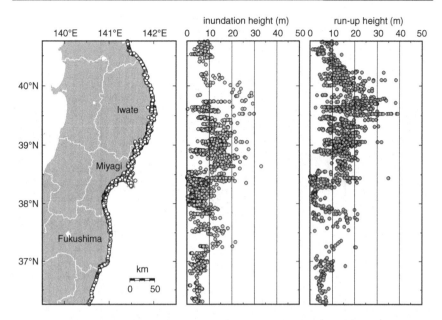

Figure 2.6.2 Distribution of the tsunami inundation and run-up heights along the Pacific coast of Tohoku as a consequence of the 2011 tsunami (Data source: Japan Tsunami Trace Database, https://tsunami-db.irides. tohoku.ac.jp/tsunami/).

(the Sendai Plain), the inundation heights were larger than the run-up heights. In the southern part (Fukushima), the run-up heights were slightly larger than, or almost the same as, the inundation heights.

This difference mainly resulted from the geographical features of each section of the coastline. The northern part is made up of rias, which are characterized by an indented coastline with many bays and the land rising rapidly from the sea, meaning that a tsunami can run up a hill that is situated just behind the shore without much energy being dissipated. The middle part has the Sendai Plain situated next to the coastline, with the shoreline being relatively straight and there being a wide low-lying ground behind it. This allows a tsunami to penetrate far inland, though this process causes the wave to gradually lose its energy and results in lower run-ups. The southern part is similar to the middle part, but the low-lying ground is narrower. While the behaviour of the tsunami is similar to that in the Sendai Plain, as the low-lying coastal plain is less wide the wave loses less energy, which results in a higher run-up.

2.6.3 Field survey of the 2011 tsunami

The authors conducted several field surveys as part of a larger survey group (Shibayama et al., 2011; Mikami et al., 2012). One of the surveys was conducted in the coast of Miyagi and the northern part of Fukushima

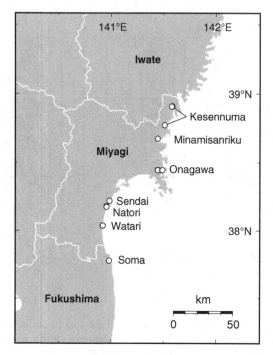

Figure 2.6.3 Locations visited during the field surveys from the 25th to 28th of March 2011.

from the 25th to 28th of March 2011, which covers the rias (Kesennuma, Minamisanriku, and Onagawa) and the coastal plain areas (Sendai, Natori, Watari, and Soma), as shown in Figure 2.6.3. The results and findings in each area of this survey are described below.

2.6.3.1 Rias coastal area

Kesennuma City is located inside Kessennuma Bay, which is narrow and long, and opens to the Pacific Ocean at its southern part. The main part of the city and Kesennuma Port are located at the innermost part of the bay. The inundation heights measured around the port were 7.79 m and 11.84 m. Many fishing vessels were carried by the tsunami and left stranded on land (Figure 2.6.4(a)), together with sludge with a thickness of around 10 cm. These findings clearly show that the water carried with it a variety of sediments and objects from the sea, which might have caused an additional force to act on houses and other structures, and hindered relief operations.

Minamisanriku Town is situated in Shizugawa Bay, which has a wider mouth than that of Kesennuma Bay and opens to the Pacific Ocean at its eastern part. This town is one of the areas which had suffered the most severe damage during the *1960 Chile Tsunami*, and thus some monuments

Figure 2.6.4 Photos taken during the authors' survey: (a) fishing vessels left stranded in Kesennuma City, (b) tsunami evacuation building in Minamisanriku Town, (c) overturned building in Onagawa Town, (d) elementary school building in Arahama, Sendai City, (e) coastal dyke failure in Watari Town, and (f) coastal dyke failure in Soma City.

had been built to remind residents of that event, though these were all destroyed during the 2011 event. The main part of the town is located at the northwestern part of the bay, though it is slightly far from the surrounding hills and thus some tall buildings were designated as tsunami evacuation buildings. The authors surveyed one of the tsunami evacuation buildings (which was four stories high and survived the tsunami with almost no structural damage) and interviewed one resident of that building (Figure 2.6.4(b)).

The resident explained that he went to the rooftop of the building after the long ground shaking, with the tsunami reaching it just after he made it to the rooftop. The water level eventually reached the rooftop, but the building could function as an evacuation site, as the water only reached up the waist level of those at the very top (inundation heights were 15.41 m and 14.30 m, measured at a point on the rooftop and on the fourth floor of the landward side, respectively).

Onagawa Town, situated in the innermost side of Onagawa Bay, has a complex coastline that opens to the Pacific Ocean at its eastern part, with a number of fishing villages located around its circumference. In Onagawa Town, as in other locations, there was widespread destruction of wooden buildings and weaker constructions. Amongst the sturdier buildings, the town hall and a hospital were inundated, and four reinforced concrete buildings were overturned. One of the overturned buildings had pile foundations, but these failed in shear (Figure 2.6.4(c)), indicating that the combination of lateral loading by the tsunami, buoyancy during the inundation, and liquefaction due to the ground shaking caused serious structural damage. The overturned buildings were designed according to a pre-1981 design code, which was modified in 1981 to strengthen the seismic design, and thus after that buildings were designed to be stronger against horizontal earthquake loading (and accordingly also against the strong horizontal forces due to a tsunami). The inundation heights measured in the main part of the town were 13.52 m, 13.27 m, and 17.39 m. The port was protected against wind waves by a breakwater, but this was completely destroyed by the tsunami, with its caissons being carried away by the force of the waves.

2.6.3.2 Coastal plain area

Sendai City, the capital of Miyagi Prefecture, has paddy fields and residential areas situated along its coastline. One of these residential areas, Arahama, was one of the worst affected parts of the city, suffering widespread damage and inundation. This stretch of the coastline had a coastal forest, the Teizan Canal and a coastal dyke situated along it, and although these features can sometimes act as a buffer against a tsunami (Watanabe et al., 2015), the waves destroyed almost all houses and building and penetrated several kilometres inland. One exception was an elementary school building (Figure 2.6.4(d)) that was located around 700 m far from the shoreline, which survived and functioned as an evacuation site for the residents. The inundation heights measured near the shoreline of this area were 9.32 m and 8.43 m.

Natori City is located on the right side of Natori River. The area around the mouth of the river is known as Yuriage, which has both a residential area and a port. Many houses and buildings were destroyed by the tsunami, though at the centre there was a small hill, the highest elevation in the area, but even this hill was overtopped by the tsunami. This highlights that it was difficult for the residents to find an evacuation site, as the inundation height measured around the port was 8.81 m.

Watari Town is located on the right side of Abukuma River. Around the mouth of the river there is a brackish water lake, which has a port on its north side, with the residential area behind it suffering serious damage due to the tsunami. A coastal dyke built along the coastline was overtopped by the tsunami, with severe scouring taking place behind it that lead to it being severely damaged (Figure 2.6.4(e)). The inundation heights measured in this area were 7.56 m and 4.82 m.

Soma City has a lagoon, which is separated by a sand bar from the Pacific Ocean. Along the coastline of this sand bar there was a coastal dyke, which had additional concrete blocks in front of it for protection, and a coastal forest behind. The tsunami overtopped the dyke and in some sections both the concrete panel and inner core were washed away (Figure 2.6.4(f)). A port was located at the south side of the lagoon, and houses and buildings behind the port were also damaged by the tsunami. The inundation height measured in this area was 6.78 m, with the tsunami then advancing inland over the paddy fields, up to a point 1.8 km far from the shoreline.

2.6.4 Mitigation strategy and reconstruction after the 2011 tsunami

After the 2011 Tohoku Tsunami, there were many discussions among the Japanese coastal engineering community on how to improve preparedness against future tsunami events. One of the results of the discussions can be found in the new concept of separating events into two different levels (Shibayama et al., 2013). In this idea, Level 1 and 2 events are distinguished according to their expected return period. A Level 1 tsunami represents a tsunami with a return period of several decades to a hundred years or so, and coastal protection structures should be designed with this height in mind. For the case of a Level 1 tsunami, the main role of protection structures is defined as preventing seawater from inundating inland areas. A Level 2 tsunami indicates an event with a return period of several hundreds to a thousand years. It is difficult and costly to construct protection structures against such events, and thus the main role of protection structures for this level is to delay or reduce the tsunami water intrusion inland to help the evacuation of residents. In addition to protection structures, communities should formulate an appropriate evacuation strategy based on the characteristics of each coastal area in case a tsunami overtops protection structures. Also, evacuation routes and evacuation sites (tsunami evacuation buildings, temporal shelters, and/or artificial hills) should be prepared and maintained, and other actions, such as land-use planning and evacuation drills, are needed in order to minimize casualties in the event of a Level 2 tsunami.

Based on the concept behind these two tsunami levels many efforts have been made in Tohoku and other coastal areas to improve the resilience of coastal communities. Some coastal areas have reconstructed coastal dykes based on the tsunami height expected for a Level 1 event. Tsunami hazard maps have been updated reflecting the latest findings of historical tsunami events in each coastal area. These hazard maps provide people with an image of the inundation that could be caused by a Level 2 tsunami in each area. In addition to these efforts, some structures in coastal areas affected by the 2011 tsunami have been preserved as memorial ruins, to show what happened during the event and to share lessons learned. For example, the aforementioned elementary school building in Arahama, Sendai City now functions as a memorial ruin, as well as a tsunami evacuation building.

The reconstruction process differs from place to place, according to the conditions of each area, such as topography, population, and major industries. Based on the authors' observations after the 2011 event, there are four different patterns of reconstruction, as shown in Figure 2.6.5. Figure 2.6.5(a) shows that a new coastal dyke, which was higher than the old one, was constructed in Taro, Miyako City. This new protection structure now protects the low-lying area behind it. Figure 2.6.5(b) shows that an artificial hill was

Figure 2.6.5 Photos showing different reconstruction patterns after the 2011 tsunami: (a) new and old coastal dykes in Taro, Miyako City, (b) artificial elevated ground in Rikuzentakata City, (c) new residential and old downtown areas in Onagawa Town, and (d) old residential area in Arahama, Sendai City.

constructed over the former low-lying parts of the downtown of Rikuzentakata City. In this city, residential areas are now situated on top of these new hills, which cover most of the downtown of the city, and adjacent hills. Figure 2.6.5(c) shows that a new residential area was constructed on a natural hill in Onagawa Town. The old downtown in the low-lying area is now used as an open space and residents were moved to new residential areas around nearby hills. Figure 2.6.5(d) shows that there are now no houses around a street in Arahama, Sendai City. Here, all residents were moved to a new area located further inland.

2.6.5 Conclusions

The *2011 Tohoku Tsunami* caused severe damage to large stretches of the Pacific coastline of Tohoku in Japan. The tsunami heights reached more than 20 m in the northern area (rias coastal area), and 5–20 m in the southern area (coastal plain area), which were the largest among those recorded. Field surveys conducted by the authors (led by the lead editor of this book, Tomoya Shibayama) revealed that there were various types of damage mechanisms induced by the tsunami, including damage to buildings, breakwaters, and coastal dykes. After the 2011 tsunami, many improvements and efforts to make coastal areas more resilient were carried out based on the lessons learned from the event. One of the important concepts that was adopted after the event is the idea of differentiating between Level 1 and Level 2 tsunamis. At present in Japan, tsunami countermeasures and evacuation strategies are designed considering the return periods of these two different levels of potential events in each coastal area. At the same time, reconstruction works are underway in the affected settlements, with their patterns varying according to the conditions of each area. In some areas a new artificial hill is being used to improve the resilience of residential areas, and in other areas most of the population is being relocated closer to the mountains.

2.7 THE 2018 SULAWESI TSUNAMI

Tomoyuki Takabatake

Kindai University, Osaka, Japan

2.7.1 Introduction

At around 18:02 local time (10:02 UTC) on September 28, 2018, a significant earthquake with a moment magnitude (M_w) of 7.5 took place on Sulawesi Island, Indonesia. The epicentre of the earthquake was estimated to be at 0.256°S, 119.846°E, 20 km deep (United States Geological Survey 2018). Following the ground shaking, unexpectedly large tsunami waves struck the central-western part of Sulawesi Island. As a result, coastal communities, including Palu City and Donggala Regency, suffered devastating damage. According to the Indonesian National Disaster Management Agency (2018), the total death toll caused by the earthquake and accompanying disasters (tsunamis, landslides, and liquefaction) was over 4,300.

Given that significant tsunamis are seldom generated by strike-slip earthquakes, many researchers visited the affected areas and conducted field surveys to clarify the generation mechanism of the waves, measure the inundation depths and investigate the damage caused to coastal communities. Observations from the field surveys and associated numerical simulations conducted by a number of research groups indicated that the waves were generated by multiple landslides triggered by the earthquake (e.g., Muhari et al., 2018; Heidarzadeh et al., 2019; Omira et al., 2019; Sassa and Takagawa, 2019; Takagi et al., 2019).

An international research team, led by the editor of this book (Prof. Tomoya Shibayama), also visited Sulawesi Island, and conducted post-disaster field surveys (Mikami et al., 2019; Stolle et al., 2020; Krautwald et al., 2021; Harnantyari et al., 2020). Numerical simulations that verified that the submarine landslides identified would indeed produce waves that would reach each of the communities affected within the timeframes reported by local residents were also performed by this team (Aránguiz et al., 2020). This chapter summarises the results obtained from this field survey.

2.7.2 Post-disaster field survey

An international team, composed of researchers from Japan, Indonesia, Germany, Canada, and the USA, visited the areas of Palu Bay affected by the tsunami from the 27th to 31st of October 2018 (around 1 month after the event). A variety of field investigation methods were employed, including measurement of tsunami inundation and run-up heights, observation of structural damage (including the effects of tsunami debris), questionnaire surveys of residents affected, and a bathymetry survey. Figure 2.7.1 shows some of the photographs taken during the field survey to document the damage to structures located along the shore.

The records of the measured tsunami heights are reported in Mikami et al. (2019), indicating that relatively large inundation and run-up heights (i.e., more than 4 m) were recorded in the inner part of Palu Bay, especially around Palu City (see Figure 2.7.2). In fact, severe damage to coastal settlements was reported in this area. In contrast, inundation depth and run-up heights were found to be less than 3 m in the northern part and entrance of Palu Bay, and the observed tsunami damage was more limited there. Mikami et al. (2019) also reported that, at its greatest extent, the tsunami reached a maximum distance inland of 200 m. This relatively short inundation distance suggests that the wavelength of the tsunami waves was shorter than those normally observed during co-seismic tsunami events. As a shorter wavelength is one of the distinctive characteristics of landslide tsunamis (Takabatake et al., 2019,

Figure 2.7.1 Damage to structures located along the shoreline of Palu Bay.

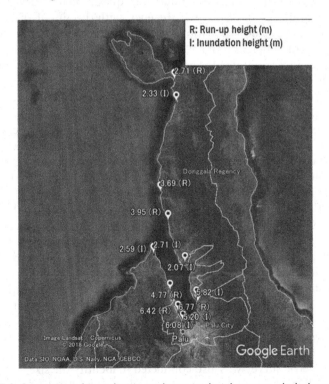

Figure 2.7.2 Summary of inundation and run-up heights recorded along Palu Bay (Mikami et al., 2019).

2020, 2021), this observation also supports the theory that landslides were the main contributor to the tsunamis generated inside Palu Bay.

Stolle et al. (2020) reported their field investigations on tsunami loads on structures, especially those induced by debris entrained within the inundating tsunami. A variety of debris, such as vehicles, marine vessels, and a range of other negatively buoyant objects, were found along Palu Bay, and these clearly caused impact loading on structures. In addition, debris damming, which refers to the accumulation of debris on the front of a structure (contributing to increasing drag forces and elevating the water depth, see Stolle et al., 2017; Stolle et al., 2018), was also observed in several places. Based on their observation, Stolle et al. (2020) highlighted a number of topics that need further research in the future, including analysing debris transport over a complex topography, negatively and neutrally buoyant debris, and debris damming. The necessity of further improving numerical modelling to simulate solid–fluid interactions was also identified. Krautwald et al. (2021) reported on the structural performance of buildings and infrastructure and soil-related issues (e.g., scouring around buildings), which lead to structural failure of the structures suffering from it. The findings were also compared with the ASCE 7 Chapter 6 standards, which deals with tsunami design loads (ASCE, 2016).

The international team also conducted a questionnaire survey of those affected in Palu City and Donggala Regency, with the aim of characterising tsunami awareness and evacuation behaviour of local residents during the event. A total of 200 questionnaire survey sheets were administered and 197 valid answers were obtained. The collected results were statistically analysed, with Harnantyari et al. (2020) noting that the evacuation behaviour observed during the event was comparable to that in other recent coastal disasters (e.g., Takabatake et al., 2018), though not entirely similar. Overall, most respondents had a relatively high level of tsunami awareness, with more than 80% being afraid that a tsunami could arrive after they felt ground shaking. Given that significant tsunamis typically arrive at coastal areas within several minutes of ground shaking (Takagi et al., 2019), such a high level of awareness of local residents likely saved many lives. Harnantyari et al. (2020) also reported that 83% of the respondents initiated their evacuation after seeing others evacuating, and this social trigger likely contributed to reducing casualty levels (given that no official evacuation warning was issued). It is also worth mentioning that many respondents reported that they experienced road congestion while evacuating, highlighting the importance of establishing an effective evacuation plan to ensure that all residents can swiftly evacuate in the event of a future tsunami. To do so, it would be useful to utilize an agent-based evacuation simulation model that could simulate evacuation behaviour in detail (e.g., Takabatake et al., 2017, 2018, 2020, 2020, 2020).

A bathymetric survey was also conducted to identify the size and position of the landslides. As a result, seven small and one large landslides were identified along the western and eastern shorelines of Palu Bay, respectively. It was also found that most of the landslides were located near river mouths. The obtained bathymetric information was used as an input into the tsunami simulations reported by Aránguiz et al. (2020), where GEOWAVE was used to create the initial profile of the water level displaced by the landslides and FUNWAVE (which is based on fully non-linear Boussinesq equations) was used to simulate the propagation process of the generated waves. The simulation results showed that the tsunami arrived at Palu City within 4–10 min after the earthquake, which coincides with the testimonies of local residents (Takagi et al., 2019). Aránguiz et al. (2020) also highlight the importance of conducting further assessments to clarify the risks due to complex landslide tsunamis to many other coastal areas of the world.

2.7.3 Conclusions

The *2018 Sulawesi Earthquake and Tsunami* represented one of the most devastating tsunami events in the recent history of Indonesia. In the present chapter, the results of the field survey conducted by the international team led by Prof. Tomoya Shibayama, the chief editor of this book, consisting of researchers in Japan, Indonesia, Germany, Canada, and the USA are presented. The team used a variety of methods, including the measurement of tsunami inundation and run-up heights, observation of structural damage (including the effects of tsunami debris), a questionnaire survey of the people in the affected areas, and a bathymetry survey and associated numerical landslide tsunami simulations. Essentially, comparatively fewer studies have been made on landslide tsunamis than those generated by an earthquake. Further experimental and numerical research on landslide tsunamis would be crucial to ensure the safety of coastal communities from such tsunami waves.

2.8 THE 2018 SUNDA STRAIT TSUNAMI

Miguel Esteban
Waseda University, Tokyo, Japan

Tomoyuki Takabatake
Kindai University, Osaka, Japan

Ryota Nakamura
Niigata University, Niigata, Japan

Takahito Mikami
Tokyo City University, Tokyo, Japan

Naoto Inagaki and Tomoya Shibayama
Waseda University, Tokyo, Japan

2.8.1 Introduction

At 21:30 local time (UTC+7h) on the 22nd of December 2018, the coast-lines around the Sunda Strait, Indonesia, were hit by tsunami waves. The flooding caused extensive damage that resulted in 437 casualties, 31,943 injuries, 10 still missing and over 16,000 people displaced (National Disaster Management Agency (BNBP) (2019)). The tsunami was generated by the flank collapse of the Anak Krakatau volcano (Robertson et al., 2018; Takabatake et al., 2019), situated approximately at the middle of the Sunda Strait, between the Southern Sumatra and western Java islands.

This was not the only event historically to have been generated in this area, as the explosion of Krakatau island in August 1883 destroyed most of the island and generated a major tsunami 37 m in height, which resulted in over 36,000 casualties (Pararas-Carayannis, 2003). The Krakatau volcano has been active ever since and in 1927 Anak Krakatau (literally "the child of Krakatau" in Bahasa Indonesia) emerged from the sea once again (Hoffmann-Rothe et al., 2006), with Figure 2.8.1 showing what was left of the island after the flank collapse in 2018 (see also Esteban et al., 2021).

The present chapter describes the field surveys performed to survey the damage due to the tsunami, both to the Krakatau archipelago and the surrounding coastline of the Sunda Strait. These surveys documented the inundation heights and run-up heights due to the tsunami, parameters of vital importance for coastal engineers and disaster risk managers to design appropriate coastal structures that can withstand such events, helping to protect the lives and possessions of individuals in at-risk areas.

2.8.2 Methodology

Takabatake et al. (2019) performed a field survey of the affected areas to clarify the hazard mechanism, evacuation patterns and run-up and inundation heights in the aftermath of the *2018 Sunda Strait tsunami*, which was

Figure 2.8.1 Remains of the Anak Krakatau island after the flank collapse.

steered by the lead editor of this book, Prof. Tomoya Shibayama. This was complemented by a second survey eight months after the volcanic eruption that generated the tsunami, between the 14th and 18th August 2019, focusing on the four islands that form the Krakatau archipelago. A variety of complementary methods were employed in these surveys, including measuring tsunami run-up heights on the side of the mountains using a laser ranging instrument (IMPULSE 200LR (minimum reading: 0.01 m), Laser Technology Inc.), a prism and staffs (similar to other chapters of this book, See Sections 2.1 to 2.7). All heights were established using the sea water level as a reference point, and were corrected to the heights above the estimated tidal level at the time of the arrival of the tsunami (estimated as 21:00 local time on the 22nd of December 2018) using a global tidal prediction software (WXtide32). The precise location of each of the survey points was recorded using GPS instruments (GPSMAP 64sc J, Garmin Ltd., see Esteban et al., 2018).

Bathymetric surveys of the islands in the Krakatau archipelago were also conducted using two Garmin GPSMAP 585 echosounders, which were mounted on the two boats used during the survey, with 360° degree panoramic videos of the surface being simultaneously taken using two Nikon KeyMission 360° cameras. An aerial drone survey of Anak Krakatau island (and part of the outer islands) was conducted using a Phantom 4 Pro+ unmanned aerial vehicle (UAV). The photogrammetric processing of the images was conducted using Metashape 1.5.0 Professional Edition (AgiSoft). The DEM of the island prior to the volcanic eruptions that caused the flank collapse were obtained from the three datasets, ALOS World 3D (AW3D: Takaku and Tadono, 2017), Seamless DEM dan Batimetri Nasional (DEMNAS: Badan Informasi Geospasial, 2018).

Finally, a number of informal non-structured key informant interviews with local government officers (tasked with guarding the area) and questionnaire surveys with local residents affected (to ascertain evacuation patterns) were also conducted.

2.8.3 Results

2.8.3.1 Run-up heights

The survey covered a portion of the southern island of Sumatra, between Sinar Agung village (Kiluan Bay) and Kahai beach (South Lampung Regency), and the western coast of Java facing the Sunda Strait, as shown in Figure 2.8.2. Takabatake et al. (2019) showed that inundation heights were more than 4 m high along the coastline of Sumatra island (situated to the north-north-east of Anak Krakatau), while less than 4 m were measured along the north-western direction. In Java island, inundation heights of over 10 m were measured at Cipenyu Beach (south-south-eastern direction from Anak Krakatau).

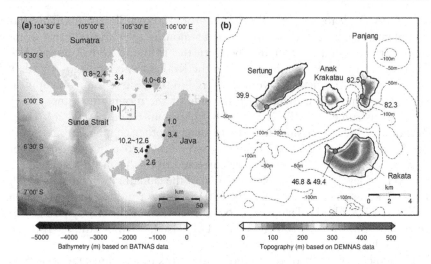

Figure 2.8.2 Maps of the areas surveyed by Takabatake et al. (2019) and Esteban et al. (2021). Bathymetry and topography are based on BATNAS (resolution: 6-arcsecond) and DEMNAS (resolution: 0.27-arcsecond) data of Badan Informasi Geospasial (2018), respectively.

Figure 2.8.3 Left: Survey of Sertung island, which was overtopped by the tsunami. Right: Rakata and Panjang islands were not overtopped by the tsunami.

Esteban et al. (2021) report that the maximum run-up heights in Rakata island were 46.8 and 49.4 m. As this island was much higher than the others surrounding it, with steep slopes facing Anak Krakatau, it was not overtopped (see Figure 2.8.3). The maximum measured run-up in Kecil island (local name: Panjang island) was 82.5 m, and parts of the island were actually overtopped. In Sertung island the maximum run-up was 39.9 m. The geographical distribution of the islands around Anak Krakatau helps to explain the variation in tsunami heights recorded in the shorelines of the Sunda Strait by Takabatake et al. (2019).

2.8.3.2 Volume of the collapsed slope

The aerial photographs obtained using the UAV were stitched together to form a map of the surface of the Anak Krakatau island at the time of the survey (see Figure 2.8.4). From these images a Digital Elevation Model (DEM) was obtained, with Figure 2.8.5 showing two transects of the island before and after the collapse of the volcano that triggered the tsunami. The difference between the former and present-day DEMs allowed Esteban et al. (2021) to calculate the volume of the material that was displaced by the eruption (which formed the landslide that generated the tsunami), which was between 0.286 and 0.574 km^3.

Figure 2.8.4 Stitched map of the Anak Krakatau island from drone photography.

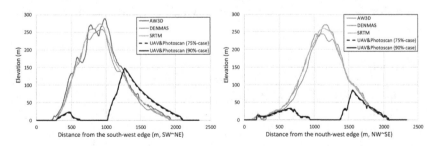

Figure 2.8.5 Transects of the flanks of Anak Krakatau, showing the historical evolution over time and the profile after the 2018 volcanic eruption (from DEM data). AW3D, DENMAS and SRTM represent the data available prior to the tsunami, with UAV data representing the surveys by Esteban et al. (2021).

2.8.4 Discussion

The field results have important implications for the simulation of the tsunami waves that could be generated by the future flank collapses of any of the islands in the Krakatau archipelago, and the importance of source directionality. Within the Krakatoa archipelago, the maximum run-up heights were 165 m and 168 m, comparable to that of 157 m reported in LCDV (2015). These islands partly blocked the tsunami, meaning that the waves were prevented from travelling in certain directions, with shallow bathymetries in some directions also hindering them (though the slightly deeper bathymetry between Panjang and Rakata would not have dampened the waves travelling out through that passage).

TDMRC (2019) and Takabatake et al. (2019) reported how the 2018 tsunami resulted in higher inundation and run-up heights to the north-north-eastern direction from Anak Krakatau, with a maximum surveyed run-up height of 6.8 m. In Java island, significantly higher inundation and run-up heights were measured at Cipenyu Beach, located to the south-south-eastern from Anak Krakatau (11.3 m inundation). Takabatake et al. (2019) indicated how local residents reported two waves (separated by between 2 and 5 min), with the second one being higher than the first. This would indicate a short wavelength, which would make this similar to the 2018 Palu tsunami (Mikami et al., 2019; Takagi et al., 2019, see also Section 2.7). Given that these events were also caused by landslides (for the case of Palu, submarine landslides), it appears that a clearer distinction should be made between earthquake and landslide generation mechanisms, especially in terms of tsunami periods. This ties in with reports of how residents evacuated (see Takabatake et al., 2019), where in the town of Lantera many first performed an initial evacuation to nearby buildings that were perceived to be sturdy (such as mosques), to then later flee to designated evacuation buildings after the passage of the first wave. While such an approach could be valid for earthquake-induced tsunamis (with long intervals between the first and subsequent waves), it might not represent the best evacuation procedure for landslide-induced tsunamis (which can have smaller intervals between them).

2.9 CONCLUSIONS

This chapter describes the field surveys that were conducted in the Krakatau group of islands and the Sunda Strait with the intention of clarifying the inundation and run-up heights that took place after the 2018 Sunda Strait tsunami. Run-up heights of at least 82.5 m, 49.4 m and 39.9 m were measured along Panjang, Rakata and Sertung islands, respectively, which shielded the Sunda Strait from much of the energy of the waves. As a result, there was considerable variation in inundation and run-up heights along the coastal settlements around the strait, highlighting the importance of source directionality in the estimation of the damage that can be caused by tsunamis. It is clear that the study of tsunamis originating from landslides and volcanic eruptions has received less attention than those generated by co-seismic seafloor displacements, and this area of research should be prioritized in the future.

REFERENCES

Badan Informasi Geospasial. (2018). Seamless Digital Elevation Model (DEM) dan Batimetri Nasional. Accessed on 11 November 2019. http://tides.big.go.id/DEMNAS/

Abe, H., Sugeno, Y., & Chigama, A. (1990). Estimation of the Height of Sanriku Jogan 11 Earthquake-Tsunami (AD 869) in the Sendai Plain, *Zisin (Journal of the Seismological Society of Japan)*, 43(4), 513–525 (in Japanese with English abstract).

Ammon, C. J., Kanamori, H., Lay, T., & Velasco, A. A. (2006). The 17 July 2006 Java Tsunami Earthquake, *Geophysical Research Letters*, 33(24), L24308.

Aránguiz, R., & Shibayama, T. (2013). Effect of Submarine Canyons on Tsunami Propagation: A Case Study of the Biobio Canyon, Chile, *Coastal Engineering Journal*, 55(4), 1350016. https://doi.org/10.1142/S0578563413500162

Aránguiz, R. (2018). *Tsunamis en la Región del Biobio desde una mirada multidisciplinaria*. Universidad Católica de la Santísima Concepción, Chile.

Aránguiz, R., Catálan, P. A., Cecioni, C., Bellotti, G., Henriquez, P., & González, J. (2019). Tsunami Resonance and Spatial Pattern of Natural Oscillation Modes with Multiple Resonators. *Journal of Geophysical Research, Oceans*. https://doi.org/10.1029/2019jc015206

Aránguiz, R., Esteban, M., Takagi, H., Mikami, T., Takabatake, T., et al. (2020). The 2018 Sulawesi Tsunami in Palu City as a Result of Several Landslides and Coseismic Tsunamis. *Coastal Engineering Journal*, 62(4), 445–459. https://doi.org/10.1080/21664250.2020.1780719

Arikawa, T., Tatsumi, D., Matsuzaki, Y., Tomita, T. (2010). Field Survey on 2009 Samoa Islands Tsunami. Technical Note of the Port and Airport Research Institute, No. 1211, p. 26 (in Japanese).

ASCE. (2016). *Minimum design loads for buildings and other structures*. American Society of Civil Engineers, Reston, VA. https://doi.org/10.1061/9780784412916

Contreras, M., & Winckler, P. (2013). Pérdidas de vidas, viviendas, infraestructura y embarcaciones por el tsunami del 27 de Febrero de 2010 en la costa central de Chile. *Obras y Proyectos*, 14, 6–19. https://doi.org/10.4067/S0718-28132013000200001

Crichton, R., Esteban, M., & Onuki, M. (2020). Understanding the Preferences of Rural Communities for Adaptation to 21st Century Sea-Level Rise: A Case Study from the Samoan Islands, *Climate Risk Management*, 30, 100254.

Delouis, B., Nocquet, J. M., & Vallée, M. (2010). Slip Distribution of the February 27, 2010 Mw = 8.8 Maule Earthquake, Central Chile, from Static and High-Rate GPS, InSAR, and Broadband Teleseismic Data. *Geophysical Research Letters*. https://doi.org/10.1029/2010GL043899

Dengler, L. (2005). The Role of Education in the National Tsunami Hazard Mitigation Program. *Natural Hazards*, 35(1), 141–153.

Esteban, M., Takabatake, T., Achiari, H., Mikami, T., Nakamura, R., Gelfi, M., Panalaran, S., Nishida, Y., Inagaki, N., Chadwick, C., Oizumi, K., & Shibayama, T. (2021). Field Survey of Flank Collapse and Run-Up Heights due to the 2018 Anak Krakatau Tsunami. *Journal of Coastal and Hydraulic Structures*, 1 (1).

Esteban, M., Takagi, H., Mikami, T., Bahbouh, L., Becker, A., Nurse, L., Shibayama, T., & Nagdee, M. (2018). How to Carry Out Bathymetric and Elevation Surveys on a Tight Budget: Basic Surveying Techniques for Sustainability Scientists. *International Journal of Sustainable Future for Human Security*, 5(2), 86–91. https://doi.org/10.24910/jsustain/5.2/8691

Fritz, H., Petroff, C., Catalán, P., Cienfuegos, R., Winckler, P., Kalligeris, N., Weiss, R., Barrientos, S., Meneses, G., Valderas-Bermedo, C., Ebeling, C., Papadopoulos, A., Contreras, M., Almar, R., Dominguez, J., & Synolakis, C. (2011). Field Survey of the 27 February 2010 Chile Tsunami. *Pure and Applied Geophysics*, 168, 1989–2010. https://doi.org/10.1007/s00024-011-0283-5

Fujima, K., Shigihara, Y., Tomita, T., Honda, K., Nobuoka, H., Hanzawa, M., Fujii, H., Ohtani, H., Orishimo, S., Tatsumi, M., & Koshimura, S. (2006). Survey Results of the Indian Ocean Tsunami in the Maldives. *Coastal Engineering Journal*, 48, 81–97.

Geist, E. L. (2013). Near-Field Tsunami Edge Waves and Complex Earthquake Rupture. *Pure and Applied Geophysics*, 170(9), 1475–1491.

Harnantyari, A., Takabatake, T., Esteban, M., Valenzuela, P., Nishida, Y., et al. (2020). Tsunami Awareness and Evacuation Behaviour during the 2018 Sulawesi Earthquake Tsunami. *International Journal of Disaster Risk Reduction*, 40, 101389. https://doi.org/10.1016/j.ijdrr.2019.101389

Hatori, T. (1975). Tsunami Magnitude and Wave Source Regions of Historical Sanriku Tsunamis in Northeast Japan, *Bulletin of the Earthquake Research Institute-University of Tokyo*, 50(4), 397–414 (in Japanese with English abstract).

Hebert, H. Burg, P. E., Binet, R. Lavigne, F., Allgeyer, S., & Schindele, F. (2012). The 2006 July 17 Java (Indonesia) Tsunami from Satellite Imagery and Numerical Modelling: A Single or Complex Source?, *Geophysical Journal International* 191, 1255–1271.

Heidarzadeh, M., Muhari, A., & Wijanarto, A. B. (2019). Insights on the Source of the 28 September 2018 Sulawesi Tsunami, Indonesia Based on Spectral Analyses and Numerical Simulations. *Pure and Applied Geophysics*, 176(1), 25–43. https://doi.org/10.1007/s00024-018-2065-9

Hill, E. M., Borrero, J. C., Huang, Z., Qiu, Q., Banerjee, P., Natawidjaja, D. H., Elosegui, P., Fritz, H. M., Suwargadi, B. W., Pranantyo, I. R., Li, L., Macpherson, K. A., Skanavis, V., Synolakis, C. E., & Sieh, K. (2012). The 2010 Mw 7.8 Mentawai Earthquake: Very Shallow Source of a Rare Tsunami Earthquake Determined from Tsunami Field Survey and Near-Field GPS Data. *Journal of Geophysical Research - Solid Earth*, 117(B6).

Hoffman-Rothe, A., Ibs-Von Seht, M.,Knieβ, R., Faber, E., Klinge, K., Reichert, C., Purbawinata, M. A., & Patria, C. (2006). Monitoring Anak Krakatau Volcano in Indonesia. *Eos, Transactions American Geophysical Union*, 87(51), 581–586. https://doi.org/10.1029/2006EO510002

Indonesian National Disaster Management Agency (BNPB). (2018). Gempabumi Sulteng (Central Sulawesi Earthquake). Accessed on 11 October 2021. https://sites.google.com/view/gempadonggala/beranda.

Kanamori, H. (1972). Mechanism of Tsunami Earthquakes, *Physics of the Earth and Planetary Interiors*, 6, 246–259.

Krautwald, C., Stolle, J., Robertson, I., Achiari, H., Mikami, T., Nakamura, R., Takabatake, T., Nishida, Y., Shibayama, T., Esteban, M., & Goseberg, N. (2021). Engineering Lessons from September 28, 2018 Indonesian Tsunami: Scouring Mechanisms and Effects on Infrastructure. *Journal of Waterway, Port, Coastal, and Ocean Engineering*, 147(2), 04020056. https://doi.org/10.1061/(ASCE) WW.1943-5460.0000620

Lagos, M. (2000). Tsunamis de origen cercano a las costas de Chile. *Revista de Geografía Norte Grande*, 27, 93–102.

Lavigne, F., Gomez, C., Giffo, M., Wassmer, P., Hoebreck, C., Mardiatno, D., Prioyono, J., & Paris, R.: Field Observations of the 17 July 2006 Tsunami in Java, *Natural Hazards and Earth System Sciences*, 7, 177–183, 2007.

LCDV. (2015). Indonesie, Anak Krakatau. https://lechaudrondevulcain.com/ blog/2019/10/21/october-21-2019-en-kamchatka-ebeko-indonesia-anak- krakatau-guatemala-pacaya-costa-rica-turrialba-poas-rincon-de-la-vieja/

Lorito, S., Romano, F., Atzori, S., Tong, X., Avallone, A., McCloskey, J., Cocco, M., Boschi, E., & Piatanesi, A. (2011). Limited Overlap between the Seismic Gap and Coseismic Slip of the Great 2010 Chile Earthquake. *Nature Geoscience*, 4(3), 173–177. https://doi.org/10.1038/ngeo1073

Mardiatno, D., Malawania, M. N., & Nisaa, M. R. (2020). The Future Tsunami Risk Potential as a Consequence of Building Development in Pangandaran Region, West Java, Indonesia, *International Journal of Disaster Risk Reduction*, 46, 101523.

Mikami, T., Shibayama, T., Matsumaru, R., Takagi, H., Latu, F., & Chanmow, I. (2011). Field Survey and Analysis of Tsunami Disaster in the Samoan Islands 2009. In *Proc. 6th International Conference on Coastal Structures*, Yokohama, Japan, pp. 1325–1336.

Mikami, T., Shibayama, T., Takewaka, S., Esteban, M., Ohira, K., Aránguiz, R., Villagrán, M., & Ayala, A. (2011). Field Survey of the Tsunami Disaster in Chile 2010. *Journal of Japan Society of Civil Engineers, Ser. B3 (Ocean Engineering)*, 67(2), 529–534.

Mikami, T., Shibayama, T., Esteban, M., & Matsumaru, R. (2012). Field Survey of the 2011 Tohoku Earthquake and Tsunami in Miyagi and Fukushima Prefectures, *Coastal Engineering Journal*, 54(1), 1250011.

Mikami, T., Shibayama, T., Esteban, M., Ohira, K., Sasaki, J., Suzuki, T., Achiari, H., & Widodo, T. (2014). Tsunami Vulnerability Evaluation in the Mentawai Islands Based on the Field Survey of the 2010 Tsunami. *Natural Hazards*, 71(1), 851–870.

Mikami, T., Shibayama, T., Esteban, M., Takabatake, T., Nakamura, R., Nishida, Y., Achiari, H., Marzuki, A. G., Marzuki, M. F. H., Stolle, J., & Krautwald, C. (2019). Field Survey of the 2018 Sulawesi Tsunami: Inundation and Run-Up Heights, and Damage to Coastal Communities. *Pure and Applied Geophysics*, 176, 3291–3304. https://doi.org/10.1007/s00024-019-02258-5

Minoura, K., Imamura, F., Sugawara, D., Kono, Y., & Iwashita, T. (2001). The 869 Jogan Tsunami Deposit and Recurrence Interval of Large-Scale Tsunami on the Pacific Coast of Northeast Japan, *Journal of Natural Disaster Science*, 23(2), 83–88.

Mori, J., Mooney, W. D., Afnimar, K. S., Anaya, A. I., & Widiyantoro, S. (2007). The 17 July 2006 Tsunami Earthquake in West Java, Indonesia, *Seismological Research Letters*, 78(2), 201–207.

Morton, R. A., Gelfenbaum, G., Buckley, M. L., & Richmond, B. M. (2011). Geological Effects and Implications of the 2010 Tsunami along the Central Coast of Chile. *Sedimentary Geology*, 242, 34–51.

Muhari, A., Imamura, F., Arikawa, T., Hakim, A. R., & Afriyanto, B. (2018). Solving the Puzzle of the September 2018 Palu, Indonesia, Tsunami Mystery: Clues from the Tsunami Waveform and the Initial Field Survey Data. *Journal of Disaster Research*, 13, sc20181108. https://doi.org/10.20965/jdr.2018.sc20181108

Murata, S., Imamura, F., Katoh, K., Kawata, Y., Takahashi, S., & Tomotsuka, T. (2009). *Tsunami. To Survive from Tsunami*. World Scientific Publishing Co Pte Ltd, Singapore.

Namegaya, Y., Koshimura, S., Nishimura, Y., Nakamura, Y., Fryer, G., Akapo, A., & Kong, L. S. L. (2010). A Rapid-Response Field Survey of the 2009 Samoa Earthquake Tsunami in American Samoa, *Journal of Japan Society of Civil Engineers, Ser. B2 (Coastal Engineering)*, 66(1), 1366–1370 (in Japanese).

National Disaster Management Agency (BNBP). (2019). Tsunami Sunda Strait (Update 14 January 2019). Accessed on 6 May 2019. https://bnpb.go.id/tsunami-selat-sunda

National Police Agency of Japan. (2012). The White Paper on Police 2012. https://www.npa.go.jp/hakusyo/h24/index.html (in Japanese).

National Police Agency of Japan. (2021). Police Countermeasures and Damage Situation Associated with 2011 Tohoku District – Off the Pacific Ocean Earthquake. Accessed on March 10, 2021. https://www.npa.go.jp/news/other/earthquake2011/pdf/higaijokyo_e.pdf

OCHA. (2020). The Facts: Indonesia Earthquakes, Tsunamis and other Natural Disasters, Reliefweb.

Okal, E. A., Fritz, H. M., Synolakis, C. E., Borrero, J. C., Weiss, R., Lynett, P. J., Titov, V. V., Foteinis, S., Jaffe, B. E., Liu, P. L. -F., & Chan, I. (2010). Field Survey of the Samoa Tsunami of September 29 2009. *Seismological Research Letters*, 81(4), 577–591.

Omira, R., Dogan, G.G., Hidayat, R., Husrin, S., Prasetya, G., Annunziato, A., Proietti, C., Probst, P., Paparo, M.A., Wronna, M., & Zaytsev, A. (2019). The September 28th, 2018, Tsunami In Palu-Sulawesi, Indonesia: A Post-Event Field Survey. *Pure and Applied Geophysics*, 176(4), 1379–1395. https://doi.org/10.1007/s00024-019-02145-z

Pararas-Carayannis, G. (2003). Near and Far-Field Effects of Tsunamis Generated by the Paroxysmal Eruptions, Explosions, Caldera Collapses and Massive Slope Failures of the Krakatau Volcano in Indonesia on August 26–27, 1883. *Science of Tsunami Hazards*, 21(4), 191–201.

Power, W., & Tolkova, E. (2013). Forecasting Tsunamis in Poverty Bay, New Zealand, with Deep-Ocean Gauges. *Ocean Dynamics*. https://doi.org/10.1007/s10236-013-0665-6

Rabinovich, A. B. (1997). Spectral Analysis of Tsunami Waves: Separation of Source and Topography Effects. *Journal of Geophysical Research, Oceans*, 102(C6), 12663–12676. https://doi.org/10.1029/97JC00479

Robertson, I., Head, M., Roueche, D., Wibowo, H., Kijewski-Correa, T., Mosalam, K., Prevatt, D. (2018): STEER – Sunda Strait Tsunami (Indonesia): Preliminary Virtual Assessment Team (P-VAT) Report, DesignSafe-CI [publisher], Dataset. https://doi.org/10.17603/DS2Q98T

Saillard, M., Audin, L., Rousset, B., Avouac, J. P., Chlieh, M., Hall, S. R., Husson, L., & Farber, D. L. (2017). From the Seismic Cycle to Long-Term Deformation: Linking Seismic Coupling and Quaternary Coastal Geomorphology along the Andean Megathrust. *Tectonics*, 36(2), 241–256. https://doi.org/10.1002/2016TC004156

Sassa, S., & Takagawa, T. (2019). Liquefied Gravity Flow-induced Tsunami: First Evidence and Comparison from the 2018 Indonesia Sulawesi Earthquake and Tsunami Disasters. *Landslides*, 16(1), 195–200. https://doi.org/10.1007/s10346-018-1114-x

Satake, K., Nishimura, Y., Putra, P. S., Gusman, A. R., Sunendar, H., Fujii, Y., Tanioka, Y., Latief, H., & Yulianto, E. (2013). Tsunami Source of the 2010 Mentawai, Indonesia Earthquake Inferred from Tsunami Field Survey and Waveform Modeling. *Pure and Applied Geophysics*, 170(9), 1567–1582.

Shibayama, T., Sasaki, J., Takagi, H., Achiari, H., & Wurjanto, A. (2006). Tsunami Disaster Survey after Central Java Tsunami in 2006. In *Proceedings of the Symposium on Tsunami, Storm Surge and Other Coastal Disasters*, Sri Lanka, pp. 9–13.

Shibayama, T., Okayasu, A., Sasaki, J., Wijayaratna, N., Suzuki, T., Jayaratne, R., Masimin, Z. A., & Matsumaru, R. (2007). Disaster Survey of Indian Ocean Tsunami in South Coast of Sri Lanka and Aceh, Indonesia. In *Coastal Engineering 2006*, vol. 5, pp. 1469–1476.

Shibayama, T., Matsumaru, R., Takagi, H., Esteban, M., & Mikami, T. (2011). Field Survey of the 2011 off the Pacific Coast of Tohoku Earthquake Tsunami Disaster to the South of Miyagi Prefecture. *Journal of Japan Society of Civil Engineers, Ser. B2 (Coastal Engineering)*, 67(2), I_1301–I_1305 (in Japanese with English abstract).

Shibayama, T., Esteban, M., Nistor, I., Takagi, H., Thao, N. D., Matsumaru, R., Mikami, T., Aranguiz, R., Jayaratne, R., & Ohira, K. (2013): Classification of Tsunami and Evacuation Areas, *Natural Hazards*, 67(2), 365–386.

Sobarzo, M., Garcés-Vargas, J., Bravo, L., Tassara, A., & Quiñones, R. A. (2012). Observing Sea Level and Current Anomalies Driven by a Megathrust Slope-Shelf Tsunami: The Event on February 27, 2010 in Central Chile. *Continental Shelf Research*, 49, 44–55. https://doi.org/10.1016/j.csr.2012.09.001

Stolle, J., Krautwald, C., Robertson, I., Achiari, H., Mikami, T., Nakamura, R., Takabatake, T., Nishida, Y., Shibayama, T., Esteban, M., & Nistor, I. (2020). Engineering Lessons from the 28 September 2018 Indonesian Tsunami: Debris Loading. *Canadian Journal of Civil Engineering*, 47(1), 1–12. https://doi.org/10.1139/cjce-2019-0049

Stolle, J., Takabatake, T., Mikami, T., Shibayama, T., Goseberg, N., Nistor, I., & Petriu, E. (2017). Experimental Investigation of Debris-induced Loading in Tsunami-like Flood Events. *Geosciences*, 7, 74. https://doi.org/10.3390/geosciences7030074

Stolle, J., Takabatake, T., Nistor, I., Mikami, T., Nishizaki, S., Hamano, G., Ishii, H., Shibayama, T., Goseberg, N., & Petriu, E. (2018). Experimental Investigation of Debris Damming Loads under Transient Supercritical Flow Conditions. *Coastal Engineering*, 139, 16–31. https://doi.org/10.1016/j.coastaleng.2018.04.026

Takabatake, T., Shibayama, T., Esteban, M., Ishii, H., & Hamano, G. (2017). Simulated Tsunami Evacuation Behavior of Local Residents and Visitors in Kamakura, Japan. *International Journal of Disaster Risk Reduction*, 23, 1–14. https://doi.org/10.1016/j.ijdrr.2017.04.003

Takabatake, T., Shibayama, T., Esteban, M., & Ishii, H. (2018). Advanced Casualty Estimation Based on Tsunami Evacuation Intended Behavior: Case Study at Yuigahama Beach, Kamakura, Japan. *Natural Hazards*, 92(3), 1763–1788, https://doi.org/10.1007/s11069-018-3277-0

Takabatake, T., Shibayama, T., Esteban, M., Achiari, H., Nurisman, N., Gelfi, M., Tarigan, T.A., Kencana, E.R., Fauzi, M.A.R., Panalaran, S., & Harnantyari, A.S. (2019). Field Survey and Evacuation Behaviour during the 2018 Sunda Strait Tsunami. *Coastal Engineering Journal*, 61(4), 423–443. https://doi.org/10.1080/21664250.2019.1647963

Takabatake, T., Esteban, M., Nistor, I., Shibayama, T., & Nishizaki, S. (2020). Effectiveness of Hard and Soft Tsunami Countermeasures on Loss of Life under Different Population Scenarios. *International Journal of Disaster Risk Reduction*, 45, 101491. https://doi.org/10.1016/j.ijdrr.2020.101491

Takabatake, T., Fujisawa, K., Esteban, M., & Shibayama, T. (2020). Simulated Effectiveness of a Car Evacuation from a Tsunami. *International Journal of Disaster Risk Reduction*, 47, 101532. https://doi.org/10.1016/j.ijdrr.2020.101532

Takabatake, T., Mäll, M., Han, D. C., Inagaki, N., Kisizaki, D., Esteban, M., & Shibayama, T. (2020). Physical Modeling of Tsunamis Generated by Subaerial, Partially Submerged, and Submarine Landslides. *Coastal Engineering Journal*, 62, 582–601. https://doi.org/10.1080/21664250.2020.1824329

Takabatake, T., Nistor, I., & St-Germain, P. (2020). Tsunami Evacuation Simulation for the District of Tofino, Vancouver Island, Canada. *International Journal of Disaster Risk Reduction*, 48, 101573. https://doi.org/10.1016/j.ijdrr.2020.101573

Takabatake, T., & Shibayama, T. (2020). Improving the Evacuation Plan of Coastal Communities Using Tsunami Evacuation Simulations: Case Study from Tagajyo, Japan. *International Journal Sustainable Future for Human Security (J-SustaiN)*, 7(2), 24–34. https://doi.org/10.1080/21664250.2019.1647963

Takabatake, T., Chenxi, D. H., Esteban, M., & Shibayama, T. (2021). Influence of Road Blockage on Tsunami Evacuation: A Comparative Study of Three Different Coastal Cities in Japan. *International Journal of Disaster Risk Reduction*, 68, 102684.

Takabatake, T., Han, D. C., Valdez, J. J., Inagaki, N., Mäll, M., Esteban, M., & Shibayama, T. (2021). Three-Dimensional Physical Modelling of Tsunamis Generated by Partially Submerged Landslides. *Journal of Geophysical Research, Oceans*, 127(1), e2021JC017826.

Takabatake, T., Mäll, M., Esteban, M., Nakamura, R., Kyaw, T. O., Ishii, H., Valdez, J. J., Nishida, Y., Noya, F., & Shibayama, T. (2018). Field Survey of 2018 Typhoon Jebi in Japan: Lessons for Disaster Risk Management. *Geosciences*, 8(11), 412. https://doi.org/10.3390/geosciences8110412

Takagi, H., Bintang, M., Kurobe, S., Esteban, M., Aranguiz, R., & Ke, B. (2019): Analysis of Generation and Arrival Time of Landslide Tsunami to Palu City due to the 2018 Sulawesi Earthquake. *Landslides*, 16, 983–991. https://doi.org/10.1007/s10346-019-01166-y

Takagi, H., Pratama, M. B., Kurobe, S., Esteban, M., Aránguiz, R., & Ke, B. (2019). Analysis of Generation and Arrival Time of Landslide Tsunami to Palu City due to the 2018 Sulawesi Earthquake. *Landslides*, 16(5), 983–991. https://doi.org/10.1007/s10346-019-01166-y

Takahashi, R. (1961). A Summary Report on the Chilean Tsunami of May 1960. In C. for the F. I. of the C. T. of 1960. (Ed.), Report on the Chilean Tsunami of May 24, 1960, as observed along the Coast of Japan.

Takaku, J., & Tadono, T. (2017). Quality updates of 'AW3D' global DSM generated from ALOS PRISM. In *Proceeding of IGARSS 2017*, IEEE, Fort Worth, TX, pp. 5666–5669. https://ieeexplore.ieee.org/document/8128293/

TDMRC (Tsunami and Disaster Mitigation Research Centre, Syiah Kuala University). (2019). The Latest Update from Post Sunda Strait Tsunami Survey. Accessed on 11 February 2019. http://tdmrc.unsyiah.ac.id/the-latestupdate-from-post-sunda-strait-tsunami-survey/

The Tohoku Earthquake Tsunami Joint Survey Group. (2011a). Nationwide Field Survey of the 2011 Off the Pacific Coast of Tohoku Earthquake Tsunami. *Journal of Japan Society of Civil Engineers, Ser. B2 (Coastal Engineering)*, 67(1), 63–66.

The Tohoku Earthquake Tsunami Joint Survey Group. (2011b). Accessed on 23 March 2022. https://coastal.jp/tsunami2011/

The Committee for Field Investigation of the Chilean Tsunami of 1960. (1961). Report on the Chilean Tsunami of May 24, 1960, as Observed along the Coast of Japan, Maruzen, 397.

Tomita, T., Arikawa, T., Kumagai, K., Matsutomi, H., Harada, K., & Diposaptono, S. (2011). Field Survey of 2010 Mentawai Earthquake Tsunami. *Journal of Japan Society of Civil Engineers, Ser. B2 (Coastal Engineering)*, 67(2), I_1281–I_1285.

Tsuji, Y., Ohtoshi, K., Nakano, S., Nishimura, Y., Fujima, K., Imamura, F., Kakinuma, T., Nakamura, Y., Imai, K., Goto, K., & Namegaya, Y. (2010). Field Investigation on the 2010 Chilean Earthquake Tsunami along the Comprehensive Coastal Region in Japan. *Journal of Japan Society of Civil Engineers, Ser. B2 (Coastal Engineering)*, 66(1), 1346–1350 (in Japanese with English abstract).

Tsuji, Y., Namegaya, Y., & Ito, J. (2005). Astronomical Tide Levels along the Coasts of the Indian Ocean. Accessed on 17 February 2005. http://www.eri.u-tokyo.ac.jp/namegaya/sumatera/tide/index.jtm

United States Geological Survey (USGS). (2018). M7.5–72 km N of Palu, Indonesia. Accessed on 11 October 2021. https://earthquake.usgs.gov/earthquakes/eventpage/us1000h3p4/executive

USGS. (2009). Magnitude 8.1 – Samoa Islands Region. http://earthquake.usgs.gov/earthquakes/eqinthenews/2009/us2009mdbi/

Vargas, G., Farías, M., Carretier, S., Tassara, A., Baize, S., & Melnick, D. (2011). Coastal Uplift and Tsunami Effects Associated to the 2010 Mw 8.8 Maule Earthquake in Central Chile *Andean Geology* 28(1), 219–238.

Watanabe. S., Mikami, T., & Shibayama, T. (2015). Laboratory Study on Tsunami Reduction Effect of Teizan Canal. *Journal of Japan Society of Civil Engineers, Ser. B2 (Coastal Engineering)*, 71(2), I_301–I_306 (in Japanese with English abstract).

Yamazaki, Y., & Cheung, K. F. (2011). Shelf Resonance and Impact of Near-Field Tsunami Generated by the 2010 Chile Earthquake. *Geophysical Research Letters*, 38(12). https://doi.org/10.1029/2011GL047508

Chapter 3

Field surveys of storm surge disasters around the world

CONTENTS

DOI: 10.1201/9781003156161-3

3.1 HURRICANE KATRINA IN 2005

Tomoya Shibayama and Miguel Esteban

Waseda University, Tokyo, Japan

3.1.1 Introduction

In 2005 Hurricane Katrina struck the Gulf coast of the USA, generating a large storm surge and high waves which affected the shorelines of New Orleans city and Louisiana and Alabama states. As a result of this there were over 1,800 casualties and a total of over $100 billion of capital stock being destroyed, with the indirect costs adding a further $23 billion (Hallegatte, 2008). The traumatic images broadcasted by the media of the country highlighted the risks posed by storm surges throughout the planet (see also Section 5.2), which could get worse in the future due to the potential intensification of these events due to climate change (see Section 7.2).

Hurricane Katrina started to form over the Bahamas on the 23rd of August 2005, intensifying as it moved towards Florida and becoming a hurricane two hours before making landfall in this state on the 25th of August. The storm lost strength as it crossed land, but once again regained hurricane-level strength after it re-entered the Gulf of Mexico, reaching Category 5 on the 28th of August, with a minimum central pressure of 902 mb and maximum sustained winds of 280 km/h. The storm continued towards

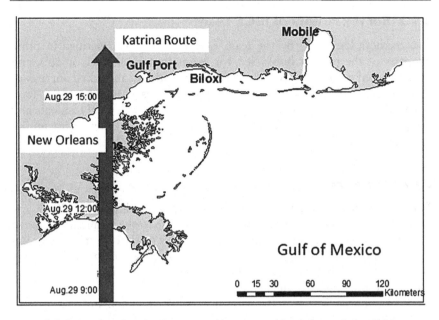

Figure 3.1.1 Route taken by hurricane Katrina and locations of the field survey (See Shibayama et al., 2006).

Louisiana, where it made a second landfall as a Category 3 hurricane, with a central pressure of 920 mbar at landfall and sustained with speeds of 190 km/h (see Figure 3.1.1). The hurricane generated a large storm surge that caused catastrophic damage to many of the levees that protected New Orleans, which resulted in the city being flooded, over 700 of its inhabitants perishing and large damage to housing taking place (Pistrika and Jonkman, 2009). The fact that the city was at risk of suffering a major storm surge was already known well before Katrina struck, as during the 20th century several hurricanes had caused substantial damage, including events in 1915, 1947 and 1965 (Jonkman et al., 2009). In this last event, known as Hurricane Betsy, approximately 40 people died and thousands had to be rescued from the flooding (Jonkman et al., 2009)

In the aftermath of hurricane Katrina a disaster survey was carried out by a team from the Japan Society of Civil Engineers (JSCE) that was headed by the chief editor of this book, Tomoya Shibayama (see Shibayama et al., 2006). The field survey was conducted along the coastal areas that were worst affected, covering 300 km from the lower part of the Mississippi River and the Borgne Lake to Pascagoula in Alabama State. The team also visited Louisiana State University and FEMA (US Federal Emergency Management Agency) to gain further understanding regarding the emergency situation and the response by local, state and national authorities. The results and conclusions of such field work will be detailed in this chapter (for a more complete description see also Shibayama, 2015).

3.1.2 Survey results of JSCE team

The methodology used by the team is similar to that described in other chapters of the present book (see, for example, Sections 2.1 or 2.2), and encompassed a variety of different types of methods, including surveys of storm surge levels through the observation and measurements of high-level watermarks along the areas flooded, and interviews with local residents and governmental organizations involved in the disaster risk management and relief operations.

3.1.2.1 Waveland

Waveland, a town with a population of 6,000 (according to the 2000 census) is situated 75 km east of New Orleans, close to the boundary between Mississippi and Louisiana states. It has a history of hurricane damage, and prior to Katrina the town had greatly suffered from the impact of hurricane Camille in 1969, highlighting the high vulnerability of many settlements in the area to the impact of hurricanes.

The town was directly in the path of hurricane Katrina, and as a result it was devastated by a combination of the storm surge and high winds. The storm surge came from the south first, through the section of the shoreline that faced the ocean (which had a low elevation that only gradually increased

Figure 3.1.2 View from Route 90 in the direction of Lake Pontchartrain.

to around 2–4 m above sea level). Government offices, firefight and police stations were all situated in this area. The highest watermark measured in this area was 8.0 m at the Civil and Cultural Center. However, the water also entered the bay from the east of the town, which then resulted in flooding to the north. As a consequence of this, most of the city was flooded, and around 50 people died.

The members of the survey team analysed a video recording taken by a local resident in a house situated 2.5–3 miles away from the coastline, around the centre of the town, showing a time series of the event:

1) The storm surge arrived to Waveland at around 9:40 in the morning.
2) In a period of around 20 minutes the water level rose very rapidly and reached the level of the roof of the 1st floor of the house at 10:00.
3) At 13:30 the water level height inside the house was reduced to less than 1 m.
4) At 15:15 there was no water inside the house, though outside the water level was still at a height roughly approximately to the middle of the doors of the cars in the street.
5) At 16:00 the street was no longer flooded.

The flood waters remained in the city for around 5–6 hours, although the major component that caused the disaster took place during the first 3 hours. The storm surge seemed like a flood, accompanied by small waves with a height of less than 10 cm. The risk to human life and property during storm surges typically depends on the flow level and velocity, and according to the video recording the effect of the waves appears to have been minor, which can help to explain the relatively low number of casualties at this location.

3.1.2.2 Northeast part of New Orleans

New Orleans was established by French colonists in the early 18th century and has a population of over 340,000, which increases to over 1,000,000 if the greater New Orleans Metropolitan area is included (US Census Bureau, 2012). It sits on the banks of the Mississippi River, and through time has gradually expanded from higher areas to lower ground, much of which is currently below the level of the river. Due to this, the city relies on a series of natural river levees to prevent flooding, which in recent times was been enhanced by an engineered concrete levees and other protection structures. Prior to Katrina the risk of a hurricane breaching these defences was well-known, though despite this authorities had failed to adequately maintain and upgrade them.

To understand the damage to the river levees the survey team visited a point (N30°04′31.10″, W89°50′44.00″) between Lakes Pontchartrain and

Borgne (see Figure 3.1.2). The storm surge waters carried with them large boats, which were left stranded on the slope of the bank. From the traces left by the boat movements and the location of the vessels themselves it was clear that the water flowed over Route 90 from the Gulf of Mexico into the lake and eventually overcame the banks.

The team then moved to a nearby location between St. Catherine's Island and Petites Coquilles (N30°08′19.5″, W89°45′17.0″), at the centre of a low coastal road which separates Lake Pontchartrain from the open sea. This area was heavily damaged by the storm surge, with the majority of houses being washed away and leaving only the pillars standing (see Figure 3.1.2).

3.1.2.3 Lower parts of the Mississippi delta

The survey team visited parts of the Mississippi delta to ascertain inundation patterns as the storm surge travelled inland. At one of the locations surveyed (N29°25′00.55″, W89°36′55.57), garbage was left hung at a height of 5.5 m between the Mississippi River bank and the coastal embankment. In the southwest shore embankment (Tropical Bend, N29°23′51.09″ W89°36′45.02″), many houses were damaged along the road situated between the west of the Mississippi River bank and coastal defences. Part of a trailer house was pushed by the storm surge into a waterway that was located close to the coastal embankment (Figure 3.1.3).

Figure 3.1.3 A trailer house was pushed onto the waterway.

3.1.2.4 Gulfport

Gulfport is located on a sand bar next to Biloxi, and constitutes the second-largest city in Mississippi, having a total population of around 67,000 (US Census Bureau, 2012). The south shore was attacked by the storm surge from the Gulf of Mexico side, whereas the north side was flooded from the Big Lake (lagoon). Due to this, houses that were located less than 250 m from the coastline suffered devastating damage, with interviews with residents indicating the storm surge was very powerful. In contrast to more sheltered areas around New Orleans and the Mississippi delta, in this city sea waves could attack the coastline directly from the open ocean, and this heavily contributed to the damage observed (see also Sections 5.2 and 7.3. related to wind waves due to hurricanes). As a result, watermarks up to a level of 10.52 m were measured (including both the storm surge and the surface wind waves superimposed on it), with Figure 3.1.4 showing a diagrammatic representation of a typical shoreline in the area.

3.1.2.5 Gautier and Graveline Bay

At Gautier (N30°21'38.25", W88°38'44.76") a high watermark of 8.5 m was observed at a location 46.1 m from the coastline. Local scour could also be seen along a wooden boat pier, as shown in Figure 3.1.5. At Graveline Bay a watermark was found at a height of 9.3 m at a point where the ground level was 4.4 m high (situated 99.2 m from the coastline, at N30°20'45.84", W88°41'49.53"). Large-scale erosion coastal erosion could be observed at this point (in that sense, the sediment transport by storms is an important problem for the US coastlines, which is described in some more detail in Section 9.4).

3.1.3 Conclusions

The devastation brought about by hurricane Katrina will be forever ingrained in the minds of coastal risk managers in the United States and other countries. The storm surge manifested itself as a flood, which was

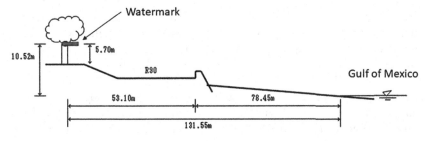

Figure 3.1.4 Survey results in Gulfport (Shibayama et al., 2006) (N30°22'12.3", W89°04'48.5").

Figure 3.1.5 Erosion area in Gautier (Shibayama et al., 2006).

accompanied by high waves in settlements that were directly exposed to the open ocean (such as Gulfport). The casualties and economic damage reported in each settlement clearly depended on the flow velocity at the point (with more severe damage being observed in houses and structures located in areas where the momentum was high), and geographical and social conditions, highlighting the importance of considering such factors when designing reliable disaster prevention countermeasures.

Major differences exist between disaster management practices in the United States and Japan. One of these differences relates to land use and population density patterns in coastal areas and the policies followed by either central or local governments. In the case of Japan, the most widespread type of counter-measures against storm surges were historically "hard" protection structures, since the population density in coastal areas is high due to land constraints. As a result, defence strategies and the overall disaster prevention philosophy are directly controlled by the central government, with local governments having limited freedom to take autonomous decisions. This results in a uniform disaster prevention management policy being established throughout the country, in contrast to the wider differences seen between the various states that form the USA (Shibayama, 2015).

3.2 CYCLONE SIDR IN 2007

Tomoya Shibayama
Waseda University, Tokyo, Japan

Tasnim Khandker Masuma
Weathernews, Chiba, Japan

Miguel Esteban
Waseda University, Tokyo, Japan

3.2.1 Introduction

Bangladesh has been ranked as being one of the most vulnerable countries to tropical cyclones (UNDP, 2004, Haque and Jahan, 2016). Between 1877 and 1995 approximately 1% of the tropical cyclones generated around the world hit the coastal areas of Bangladesh, though they resulted in the deaths of around 53% of those killed by such events (Ali, 1999). Such vulnerability is not only due to its geomorphological features, but the demographic and socioeconomic characteristics of the country also create unique challenges to ensure the sustainable development of coastal areas.

Nevertheless, over time Bangladesh has become more capable of managing disaster risks due to floods and cyclones, especially since the 1991 cyclone event that caused the loss of 140,000 lives. Long-term strategies for disaster preparedness and risk mitigation, as well as adaptation to climate change, have proven quite successful for the country to reduce human losses during natural disasters (Government of Bangladesh, 2008). However, recent low-frequency high-magnitude cyclone-induced flooding is still having devastating impacts on livelihoods and the economy in general. Furthermore, it is expected that global warming will increase the risk of large storm surge events, both due to sea-level rise and the intensification of typhoons. According to the Intergovernmental Panel on Climate Change Fifth Assessment Report (IPCC 5AR) there is high probability of major increases in tropical cyclone activity across various ocean basins (see also Section 7.2). Due to this, in the future coastal hazards are likely to be intensified by climate change and sea-level rise, increasing the risks faced by coastal communities due to tropical cyclones in the coastal region of Bangladesh (Tasnim et al., 2015).

The present chapter summarises the damage and impact caused by cyclone Sidr, with a focus on the field survey performed in the southwest coastal region of Bangladesh to understand the threats such events pose to coastal communities.

3.2.2 Cyclone Sidr: Damage and impact

Cyclone Sidr was the most powerful cyclone to impact Bangladesh since 1991, with 27 million people being affected and 1.7 billion dollars of reported damage (about 3% of the GDP of the country, see Paul, 2012; Government of Bangladesh, 2008). The storm was first detected on 9 November 2007 near the southeast of the Andaman Islands, and turned into a tropical cyclone shortly after on the 11th November. By the time it made landfall in Bangladesh on the 15th of November the cyclone had intensified to reach peak wind speeds of 59 m/s with a minimum low pressure of 944 hPa (according to the Indian Meteorological Department [IMD] observations). As a result of this and high winds, the coastal cities of Patuakhali, Barguna and Jhalokathi District were hit by a storm surge of over 5 m, which flooded numerous coastal settlements and caused 3,406 deaths. Most of the coastal residents lived in poorly constructed houses which were not strong enough to withstand the forces brought onto them by the storm surges (Chaudhury et al., 1993, and also Section 6.2). As a result, damage and losses were concentrated on the housing sector (49% of the total), the productive sector including public infrastructure (16%) and the remaining on transport and water control infrastructure, among others. Table 3.2.1 summarizes the damage and losses by sectors, estimated by the Government of Bangladesh (2008) together with international experts.

The impact of cyclone Sidr was highly concentrated in coastal districts which were already suffering from higher rates of population density and poverty, and therefore its impact was primarily borne by the poor. Approximately two million people lost their jobs and a further one million people were seriously affected by Sidr (Government of Bangladesh, 2008). Even though the Bangladesh Meteorological Department (BMD) issued a storm surge warning that alerted people to the fact that this could be over 4 m in height, many residents did not fully understand the consequences of a storm surge of this magnitude (Government of Bangladesh, 2008).

3.2.3 Post cyclone field surveys

The Japan Society of Civil Engineers (JSCE) organized a field survey in Bangladesh from 26 to 28 December 2007 to investigate the mechanisms of the disaster due to the storm surge and map the inundation heights throughout the affected areas (see Shibayama et al., 2009). The lead author of the present chapter served as the team leader of the coastal survey teams, which covered the riverside areas of Baleshwar and Burishar Rivers and the coastal area of Kuakata. The methodology used is similar to that detailed in other chapters of the present book (see Sections 2.1, 2.2 or 3.3).

Figure 3.2.1 shows the path of the cyclone, the survey route, and the distribution of measured storm surge heights (in meters). Along the Baleshwar

Table 3.2.1 Overall summary of damage and loss due to Cyclone Sidr

Sector	Sub-sector	Damage (million USD)	Losses (million USD)	Total (million USD)
Infrastructure		**1,029.90**	**30.9**	**1,060.80**
	Housing	839.3		839.3
	Transport	116	25	141
	Electricity	8.3	5.2	13.6
	Water and Sanitation	2.3	0.7	2.9
	Urban and Municipal	24.6		24.6
	Water Resources Control	71.3		71.3
Social Sectors		**65**	**21.1**	**86**
	Health and Nutrition	2.4	15	17.5
	Education	62.5	6	68.5
Productive Sectors		**25.1**	**465**	**490.1**
	Agriculture	21.3	416.3	437.6
	Industry	3.8	29.5	33.3
	Commerce		18.2	18.2
	Tourism		0.9	0.9
Cross-Cutting Issues		**6.1**	**0**	**6.1**
	Environment	6.1		6.1
Total		**1,158.00**	**516.90**	**1,674.90**

Source: Government of Bangladesh, 2008. Note that damage refers to the replacement value of totally or partially destroyed physical assets that must be included in a reconstruction program. Losses estimate the flows in the economy that arise from the temporary absence of the damaged assets.

River the inundation height was measured at Solombaria, Royenda Bazar near Sarankhola, Southkhali and Baraikhali. Inundation heights along the Baleshwar and the Burishwar rivers (although far from the coast) were relatively high compared to those observed along the coast of Kuakata, possibly because the cyclone made landfall very close to the Baleshwar River.

At Solombaria the highest inundation level with respect to the surface of the river water on the east bank of Baleshwar River (N22 28 1.2, E89 51 37.2) was 3.5 m (at a horizontal distance of 21.4 m from the river). On the east bank of the river, the ground elevation was lower and had a mild slope, resulting in a relatively higher damage in that area. The loss of several lives was reported and many houses were washed away. On the west bank of the river the ground elevation was higher, and due to this the damage to houses was comparatively smaller.

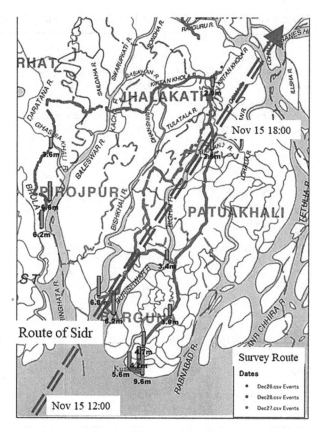

Figure 3.2.1 Route of Sidr, field survey route and distribution of measured storm surge heights (in metres) at different locations (after Shibayama et al., 2009).

At Royenda Bazar (near Sarankhola) the highest inundation level observed was about 6 m (on a palm tree situated at a distance of 68.2 m from the river). At Southkhali the highest water level observed was about 6.5 m (also on a palm tree, at a distance of 35.3 m from the river). Considerable erosion was observed in what had been (at the time) a newly constructed river bank (see Figure 3.2.2). An inundation height of 5.8 m was measured at the side of a cyclone shelter situated 100 m from the riverbank. According to interviews with local people, a first wave came to the side of the river bank (which was constructed 150 m away from the river), followed by a second wave that overtopped it and a third wave that finally reached the shelter. The wave period of the bore-like wave was estimated to be around one minute, with high water levels continuing for about 15 minutes. The water colour was red-brown, indicating a high concentration of suspended mud. The highest water level reached by the storm surge took place at around 20:00 local time, 4 hours before high tide at 24:00. The flood waters

Figure 3.2.2 Left: Partially destroyed river bank at Southkhali. Right: A primary school that also serves as a cyclone shelter at Southkhali (Shibayama et al., 2009).

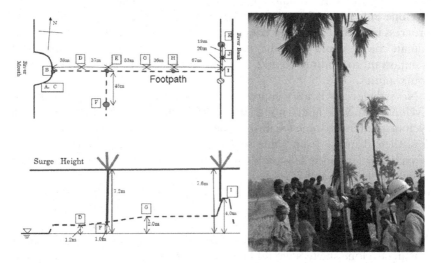

Figure 3.2.3 Surge height at Sombonia. Left: Cross-sectional and plan view. Right: Watermark on a tree (Shibayama et al., 2009).

reached many people before they could evacuate to the shelter, resulting in the deaths of around 300 of them (though many residents did manage to arrive to the shelter, the only option for evacuation in the area). The storm surge partially inundated the second floor of the building (to a level of 0.59 m from the floor), indicating that the evacuation shelter was not adequate and that there is the risk that storm surges could again flood it in the future.

At Baraikhali (N22 27 53.5, E89 51 23.1) on the east bank of the Baleshwar River, the highest inundation level observed was 5.8 m (from the river water level). Detailed field surveys were carried out at Somboniya, Naltona and Amtali Ferry Terminal, along the Burishwar river. As shown in Figure 3.2.3, at Somboniya one side of the embankment access road had collapsed at various points, with many palm trees that stood on the slope of the embankment being completely uprooted, falling down in the direction downstream of the storm surge. Again, at this location severe river bank erosion was recorded. According to interviews with local residents the highest inundation level at Somboniya and Naltona was around 6–7 m. At the right bank side of the Amtali ferry Terminal the highest inundation level observed was 3.4 m (measuring from the water surface of Burishwar River).

3.2.4 Disaster risk management

At present, economic losses due to the strong cyclones are unavoidable in Bangladesh due to the fact that structural defences against storm surges are inadequate due to lack of sufficient construction budget. Despite this, the loss of human lives, which is the main concern in the case of severe tropical cyclones, is gradually becoming avoidable. During the passage of Cyclone Sidr in 2007 it was clear that Bangladesh had made remarkable progress in its disaster preparedness and prevention skills, though given that climate change could cause the intensification of cyclones (see Section 7.2) it is important to continue to take measure to increase the resilience of coastal communities.

Nevertheless, it is also important to note that the storm surge due to Cyclone Sidr was high enough to overtop the embankments and reach cyclone shelters at some locations. The inadequate number of cyclone shelters, their distance from some settlements and lack of maintenance have often been mentioned as a problem (Tasnim et al., 2015). Under the impact of climate change and sea-level rise existing warning and evacuation plans might be insufficient if even larger scale inundation occurs, as in the case of typhoon Haiyan in the Philippines in 2013 (see Section 3.5). Extreme cyclone events in the future will likely be much stronger than at present and may remain strong longer after landfall, raising questions as to whether existing cyclone shelters would be able to withstand very high wind speeds for a prolonged period of time. Most cyclone shelters in Bangladesh were constructed decades ago, and might not be adequately designed for future climate-change induced stronger cyclones (Tasnim et al., 2015, see also Section 9.1). Thus, the construction of modern shelters that can sustain stronger winds in coastal and riverine areas, and the renovation and strengthening of existing shelters should clearly be addressed to improve the resilience of coastal communities against future cyclone events.

3.2.5 Conclusions

The present chapter reports on the field survey conducted along the southwest of the coasts of Bangladesh following the passage of Cyclone Sidr in 2007, which generated a bore-like storm surge that inundated coastal and riverside areas. The inundation heights along river sides (within the range of 5.5–7 m) were found to be relatively higher than along coastal areas (as high as 5.5 m), as the landfall location was close to the mouth of several rivers and run-up flow along their branches and associated waterways caused large scale inundation. In Somboniya, embankments played a significant role to minimize damage to life and property, and cyclone shelters were able to save a significant number of lives during Cyclone Sidr (Shibayama et al., 2009), highlighting the importance of such countermeasures in modern disaster prevention systems.

3.3 CYCLONE NARGIS IN 2008

Hiroshi Takagi

Tokyo Institute of Technology, Tokyo, Japan

3.3.1 Introduction

Cyclone "Nargis" made landfall along the southern coastline of Myanmar from the evening of May 2nd to the early morning of the 3rd 2008, representing the worst storm surge disaster since the beginning of the 21st century (see also Section 9.1). The number of deaths exceeded 138,000, at least 2.4 million people were severely affected, and more than one million people were left homeless. The damage was exacerbated by the lack of 24-hour television coverage, storm shelters, and emergency response networks throughout the country (NY Times, 2019). The government also made limited efforts to evacuate people to safety or issue warnings to the public about the gravity of the impending storm (Frontiers Myanmar, 2015). The approach of the cyclone had been reported to Yangon City and neighbouring areas through weather forecasts. However, people in the area lacked awareness about the magnitude and severity of the storm and its consequences, as they had not experienced any major cyclones in the recent past (Shibayama et al., 2008). The cyclone caused enormous damage to paddy fields and coastal forests, particularly in the downstream areas of the Irrawaddy and Yangon Rivers. As a result, the decline in vegetated area due to the cyclone was as large as 19% of its original extent (Omori et al., 2020), and the southern shoreline of the country retreated by an average of 47 m (Basset et al., 2009).

3.3.2 Field survey

Due to the military regime in power in the country at the time the exact extent of the damage and flooding was not fully investigated by international researchers, as the government refused to accept full-scale international aid, delaying aid deliveries and the issuance of visas to foreign relief workers (Frontiers Myanmar, 2015). A few exceptions were the field surveys conducted by some international teams, including Shibayama et al. (2008) around the Yangon region between 11 and 15 May 2008, and Hiraishi (2009), who surveyed the damage to Yangon ports for one week from 26 May 2008. As for the Ayeyarwady delta region, Okayasu et al. (2009)

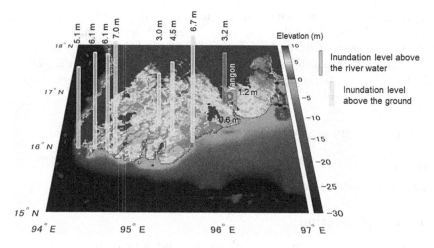

Figure 3.3.1 Inundation map, complied from the survey data by Shibayama et al.
(2008), Hiraishi (2009), and Okayasu et al. (2009). The background elevation map was created using the NOAA ETOPO1 Global Relief Model.

conducted field surveys around five townships (Kungyangon, Dedaye,
Pyapon, Bogale and Labutta) between 3 and 7 July 2008, while Fritz et al.
(2009) surveyed coastal and inland villages encompassing the Bogale and
Ayeyarwady rivers between 9 and 23 August 2008. Figure 3.3.1 shows an
inundation map aggregating the results of the field surveys outlined above.[1]
There is no doubt that the maximum storm surge height was greater than 3
m, as calculated in the numerical analysis of Tasnim et al. (2015), and given
that this height can be considered the approximate lower limit that causes
extensive damage (Takagi et al., 2017). Kyaw et al. (2021) estimated that
waves with a maximum significant height H_s of 7.3 m were generated offshore of the deltaic coast. In the following section, the situation in Yangon
(the capital at that time) and its surroundings in the aftermath of the storm
will be described, based on the survey by Shibayama et al. (2008).

3.3.2.1 Yangon City

According to a worker who stayed up all night to watch the situation at
a ferry terminal in Yangon City, the water level rose rapidly during the
cyclone and overtopped the embankment, resulting in water levels up to
1.2 m above the ground (Figure 3.3.2(a)) (the height of the point was measured to be 3.2 m above the average river water level). Many boats were
washed ashore and others sank in the river. The high water level lasted for
almost four days, after which it returned to normal. Being nearly 8 km wide,
the mouth of Yangon River resembles more an inlet or a bay than a river,

---

[1] It should be acknowledged that measurement criteria can be different among each team,
and in the coastal zone it is impossible to distinguish between the effects of storm surge
and wind waves.

though the width of the river narrows down to 3 km around Yangon, about 40 km upstream. Furthermore, the Yangon River splits into two rivers in this area, each of which narrows down to about 1 km in width. These rapid changes in river widths may have contributed to increasing the height of the storm surge near Yangon City.

3.3.2.2 Yangon suburbs

The outskirts of Yangon are characterized as being rural, with a number of villages scattered around the area. Shibayama et al. (2008) visited several villages beyond the Yangon River and proceeded southward until they were finally stopped by a neighbourhood village watch. As a result, the survey team could not reach the most severely affected areas, where the storm surge likely resulted in the land being inundated to a depth of several metres. However, they could confirm that a storm surge of at least one meter propagated over the rice fields and caused flooding damage to a large area (Figure 3.3.2(b)). Some residents also witnessed water flowing through agricultural canals. Early May is the harvest season for the "summer paddy" and mid-May is the planting season for "monsoon paddy". When the cyclone hit, the rice fields were harvested and the rice seedlings for the next harvest had just been planted. According to an official of the Ministry of Agriculture and Irrigation who accompanied the team, rice production was expected to decline by at least 20% that year. As a result, the price of this staple food began to rise immediately after the cyclone hit.

3.3.3 Unusualness of the cyclone track

The central pressure of Cyclone Nargis was estimated to have been at its lowest just before landfall. The minimum centre pressure estimated by the US Navy Joint Typhoon Warning Center (JTWC) was 937 hPa, while the Regional Specialized Meteorological Center at New Delhi estimated its intensity as 962 hPa (Kuroda et al., 2010). Figure 3.3.3 shows the tracks

(a)   (b)

Figure 3.3.2 Photos taken during the survey by Shibayama et al. (2008): (a) Yangon City (N 16° 46' 5.2", E 96° 9' 43.5"), and (b) Rakhine Chaing Village (N 16° 39' 37.5", E 96° 11' 11.6").

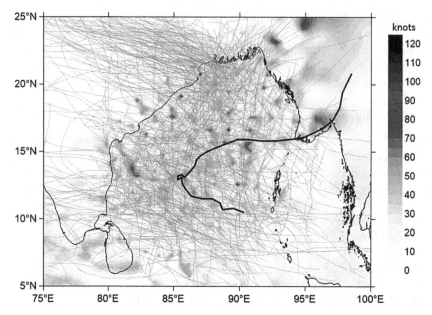

Figure 3.3.3 Historical cyclone tracks over the Bay of Bengal from 1945 to 2019. The thick line indicates the path of Nargis in 2008. The contour plot shows the maximum wind speed distribution over the Bay of Bengal and the Andaman Sea.

for Nargis and other historical cyclones (1945–2019) in the area, according to the JTWC best track data. Track analysis reveals that a small proportion of cyclones (roughly 1 or 2 every 10 years) have hit the southern coast of Myanmar (see also Section 9.1). Thus, in a sense the track that Nargis followed was rather unusual, at least amongst those of cyclones in the late 20th and early 21st centuries.

3.3.4 Influence of awareness on disaster risk management

In countries where the risk of tropical cyclone making landfall is very high, the preparedness (in terms of formulation of various countermeasures and the capacity for disaster response) typically increases gradually with time. This usually follows as a consequence of development, with countries gradually possessing more financial resources, expertise and improved governance that can improve disaster risk management in general. For example, the coast of Bangladesh is frequently hit by cyclones, but people are highly aware of the danger these weather events pose, and cyclone shelters have been built in various parts of the coastline over the past few decades (resulting in improved community-level evacuation guidance during cyclones, see Section 9.2). Cyclone Sidr in November 2007 (see Section 3.2) was as severe in magnitude as Cyclone Bhola in 1970, but the death ratio in Bangladesh became one-hundredth of what it had been (see Section 9.2). In contrast, the high death toll from Cyclone Nargis clearly highlights the danger that a severe tropical cyclone poses when it hits an area where people are less aware of the danger these weather systems represent (due to their relative smaller frequency). In addition, many residents did not evacuate during Nargis due to the misperception associated with past flooding experiences (as the areas are typically flooded several times a year due to heavy precipitation). Therefore, in addition to the rarity of the cyclone path and intensity, the lack of awareness and experience with such events likely contributed to the increase in human suffering due to this event.

3.3.5 Conclusions

Cyclone Nargis in 2008 killed nearly 140,000 people, though given restrictions by the military regime in Myanmar at the time the true scale of the disaster was not well understood by the foreign community. In this chapter, an attempt was made to describe, albeit in a limited way, the storm surge generated by Cyclone Nargis, with reference to the results of some field surveys by international researchers. Storm surges of up to 5 m or more were observed over a wide area of the Irrawaddy Delta. Yangon, the capital of the country at the time, was also affected by the storm surge propagating through the Yangon River. Compared to its neighbour Bangladesh, Myanmar has historically been hit less frequently by cyclones, which may have contributed to a lack of awareness about such hazards. The case of Cyclone Nargis demonstrates that although the probability of a cyclone event taking place is low, when it does occur the consequences can be significant.

3.4 HURRICANE SANDY IN 2012

Miguel Esteban
Waseda University, Tokyo, Japan

Takahito Mikami
Tokyo City University, Tokyo, Japan

Tomoya Shibayama
Waseda University, Tokyo, Japan

3.4.1 Introduction

On October 29, 2012, Hurricane Sandy made landfall along the eastern shorelines of the USA. The day before the storm arrived a mandatory evacuation order to the residents in Zone A (the area which was believed to be most likely to be flooded, including 375,000 residents), and a suspension of the rapid transit system (encompassing busses, subways, and trains) were carried out. The mayor and transportation authority repeatedly asked residents to recall the experience of Hurricane Irene in 2011 (New York City Government, 2014; MTA, 2014), as the same measures implemented during Sandy (a mandatory evacuation order and a suspension of the rapid transit system) were carried out during the passage of Irene (Mikami et al., 2015). This experience during Irene could have partially contributed to raising awareness and improving the response by residents, though there were still a number of problems.

As Hurricane Sandy struck New York it had a wind speed of 70 kt (36 m/s) and a minimum central pressure of 945 hPa at landfall (National Hurricane Center, 2013), generating a storm surge that struck at high tide and submerged large areas of the coastlines of New York and New Jersey. As a result of this there were 44 deaths in New York City (Bureau of Vital Statistics, New York City Department of Health and Mental Hygiene, 2014), with the storm surge also causing extensive damage to the infrastructure of the city, especially to electricity distribution and rapid transit systems. The flooding caused the failure of an electric substation in Manhattan, and a number of subway lines in Lower Manhattan remained closed for weeks after the passage of the hurricane (Mikami et al., 2015). Aerts et al. (2013) estimated that the total damage to New York City was around $28.5 billion.

3.4.2 Storm surge field survey

A field survey led by Prof. Tomoya Shibayama (the lead editor of this book) was conducted in the period between the 9th and 12th of November 2012, with the aim to measure the storm surge heights along the urban coastline and to assess the situation of the affected areas. The methodology used is similar to that described in other chapters in this book (see Sections 2.1, 3.1 and 3.7). Essentially, the storm surge heights and their locations were recorded by using a hand-held GPS instrument, a laser ranging surveying instrument and staffs. All watermarks were corrected to the heights above the estimated tide level at the time the storm made landfall.

The areas surveyed included the northwestern coast of Manhattan (along the Hudson River), the southern coast of Manhattan (Lower Manhattan), the southeastern coast of Manhattan (along the East River), and Staten Island, as shown in Figure 3.4.1.

3.4.2.1 Southern Manhattan

Lower Manhattan has a number of subway lines that converge into it, with the storm surge flooding them as well as the buildings and shops located above ground.

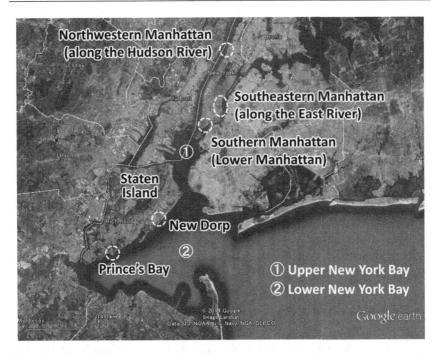

Figure 3.4.1 Location of Surveyed areas (Mikami et al., 2015).

Figure 3.4.2 Storm surge inundation in the southern coast of Manhattan. Left: Shops were flooded, though little water flow damage could be observed. Right: Entrance to South Ferry station.

At Fulton Street and Wall Street, sea water reached around 200–300 m from the shoreline, and the surveyed inundation heights were 2.5–3 m. Shops along the street were flooded, though no clear violent damage due to waves or water flow could be observed (see Figure 3.4.2 (left)). In the vicinity of Battery Park, many stations were flooded from the entrances (see Figure 3.4.2 (right)) and possibly the ventilation shafts at ground level. All electric systems inside the stations were damaged (Mikami et al., 2015).

Bowling Green station, located on relatively higher ground, was not affected by the storm surge. The entrance to Brooklyn Battery Tunnel, which connects Manhattan and Brooklyn, was also flooded during the passage of the hurricane.

3.4.2.2 Southeastern Coast of Manhattan

Around this part of the city the survey started from the E 20th Street and gradually moved north along the East River until the E 37th Street was reached. The hurricane affected a number of buildings and facilities, including a petrol station (see Figure 3.4.3 (left)), a hospital, and a car park, with inundation levels being between 1.2 and 2.57 m.

An electric substation at the east side of the E 13th Street and the E 14th Street was flooded to a level of 1.32 m during the hurricane and caused a blackout of the southern part of Manhattan. No flooding countermeasures could be observed anywhere throughout the perimeter of this substation (see Figure 3.4.3 (right)).

3.4.2.3 Northwestern Coast of Manhattan

The northwestern part of Manhattan, from W 116th Street to the W 125th Street along the Hudson River side (known as Riverside Park), has a rather high elevation and hence no clear indications of damage could be found around it. The only information about the storm surge that could be obtained was that the sea water reached a point 70 m inland from the coastline on the street (St Clair Pl) at the northern side of Riverside Park.

3.4.2.4 Staten Island

New Dorp was one of the worst affected areas in Staten Island, with many houses and vehicles being damaged (see Figure 3.4.4 (left)). The surveyed inundation height in the vicinity of the coastline was 4.03 m, and the flooded

Figure 3.4.3 Affected facilities along the southeastern coast of Manhattan. Left: Petrol station. Right: Electric substation.

Figure 3.4.4 Damage in Staten Island. Left: Local scouring at New Dorp. Right: Destroyed house at Prince's Bay.

area was large, as the gradient of the land as it moves away from the coastline was low. Watermarks were found at a house located 650 m inland from the coastline (inundation depth of 0.76 m). At Prince's Port, south of New Dorp, the measured inundation height was 4.22 m, though given that the slope of the ground was steeper, a narrower band of land was affected (see Figure 3.4.4 (right)).

3.4.3 Conclusions

The storm surge by Hurricane Sandy resulted in storm surge inundation heights of 2.5–3 m in southern Manhattan, rising to 4m along the southern coast of Staten Island. There was heavy damage due to waves or water flow in Staten Island, though no violent damage was observed in Manhattan, indicating a clear difference in the speed and power of the water flows between these two areas. Essentially, Staten Island appears to have shielded Manhattan against the worst of the storm surge. However, the southern coast of Manhattan suffered serious damage to electricity distribution and underground facilities, as there was a general lack of preparedness against flooding in the area.

Since New York experienced Hurricane Irene a year before Hurricane Sandy the city had been reasonably well prepared at the institutional level, though it is clearly necessary to continue to increase awareness at the citizen level, and improve the resilience of the infrastructure – particularly the subway and road tunnel systems – against such events. A holistic understanding of how tropical cyclones can influence the various complex systems necessary to sustain modern cities is imperative in order to ensure the long-term sustainability of human societies. Future climate change is likely to intensify the strength of such storms (see Section 7.2) and further compound the potential threats posed by them, making it imperative for the institutions to begin to expand their capacity to anticipate potential impacts and minimize vulnerabilities.

3.5 TYPHOON HAIYAN IN 2013

Takahito Mikami

Tokyo City University, Tokyo, Japan

3.5.1 Overview of Typhoon Haiyan

On the 8th November 2013, Typhoon Haiyan (known as Yolanda in the Philippines) struck the central part of the Philippines, bringing severe damage to coastal areas due to the storm surge and wind waves it generated, as well as heavy rainfall and strong winds. The National Disaster Risk Reduction and Management Council (NDRRMC) of the Philippines reported that the total number of dead and missing were 6,300 and 1,062, respectively, with most casualties being due to drowning and trauma (National Disaster Risk Reduction and Management Council, 2021), indicating that this event was one of the deadliest to have affected the country in its history.

Figure 3.5.1 shows the track of Typhoon Haiyan, together with its central pressure, every 6 hours based on the best track data from the Japan Meteorological Agency. The typhoon approached the Philippines from the east and struck Leyte and Samar Islands in the morning of the 8th November 2013 (local time, UTC+8) at almost the peak of its intensity. Most of the damage was reported in the coastal areas of Leyte Gulf, which is surrounded by the eastern coast of Leyte Island and the southern coast of Samar Island. In particular, Tacloban City, the most populated area in this region, suffered the greatest damage.

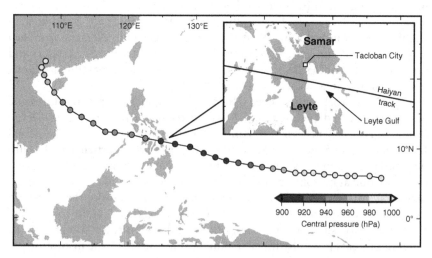

Figure 3.5.1 Track of Typhoon Haiyan and its central pressure every 6 hours, based on the best track data from the Japan Meteorological Agency.

According to Takagi and Esteban (2016), who analyzed the tropical cyclones that made landfall in the Philippines in the period between 1945 and 2013, the wind speed at the time of Haiyan's landfall (165.8 knots) was the strongest on record, and the forward speed at that moment (41 km/h) was nearly twice as fast as the average speed of other storms. These facts indicate that Typhoon Haiyan can be characterized as one of the most powerful typhoons to have landfall in the country, obviously having the potential to generate a large storm surge and high wind waves along its coast (see Section 5.2). It should also be noted that global warming may contribute to intensifying the strength of future typhoons and increasing the height of their storm surges, as indicated by numerical models (see Section 7.2 and Nakamura et al., 2016), further highlighting the importance of studying the consequences that such events can have on a given region.

3.5.2 Storm surge heights and damage

The author of this chapter and other colleagues, lead by the first editor of this book, Prof. Tomoya Shibayama, conducted field surveys after the event aiming to clarify the distribution of storm surge heights and damage along the affected coastlines (Mikami, et al., 2016; Takagi et al., 2016; Takagi et al., 2017). The first survey was conducted in December 2013 to record inundation and run-up heights, with additional surveys being conducted in May 2014 and October 2014 to complement the first survey and observe the beginning of reconstruction in the area. The methodologies used are similar to those detailed in other chapters of this book

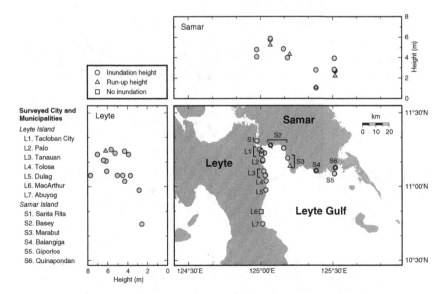

Figure 3.5.2 Distribution of storm surge heights measured during field surveys.

(see, for example, Section 2.1). Figure 3.5.2 shows the distribution of storm surge heights measured during the surveys (details of this dataset are available in Mikami et al. (2016))

As shown in Figure 3.5.2, large storm surge heights (greater than 4 m) were measured along the coastline of the innermost part of Leyte Gulf, known as San Pedro Bay. Many houses were completely or partially damaged due to the combination of the storm surge and high winds, as shown in Figure 3.5.3 (see also Section 9.1 for a more detailed discussion on the forces exerted by such phenomena on structures in the city). In particular, informal settlements along the coastline of Tacloban City, which were largely composed of densely packed wooden houses or shacks that had been erected just next to the coast, suffered severe damage (see Figure 3.5.3(a)). Even the downtown area of the city was situated in low-lying ground, and the flood water covered a wide extent of this region (see Figure 3.5.3(b)). Along the southern and eastern parts of the coast of Leyte Gulf the storm surge heights were smaller than those measured along the San Pedro Bay coast, and the damage due to the storm surge was mostly limited to houses located close to the coast.

Various residents who were interviewed during the field surveys described how the water level of the sea changed, and that this led to flooding over

Figure 3.5.3 Storm surge damage observed in the aftermath of the typhoon: (a) informal settlement in Tacloban City, (b) person indicating the storm surge water level height in the downtown area of Tacloban City, (c) damaged houses in Tanauan, and (d) damaged houses in Basey.

land. The sea water first receded towards the sea, and then the storm surge manifested itself quickly. The water flooding inland carried many sediments and other debris, and its turbulent nature made it impossible to swim in it. One resident recorded a movie, which showed that the flooding water over land had a bore-like front, making it somewhat similar to that of a violent tsunami. The storm surge withdrew as quickly as it manifested itself, with the entire episode lasting between 30 minutes to a few hours. This fast change in water level can be explained by the rapid shift in wind direction over Leyte Gulf. Figure 3.5.4 shows the hourly wind fields over Leyte and Samar Islands from 4:00 to 9:00 on 8 November 2013 (local time) based on the ERA5 reanalysis dataset (Hersbach et al., 2018). Before Typhoon Haiyan made landfall in Leyte Island a strong wind blew from the north, which made the sea water recede, and after landfall it started blowing from the south, pushing the water towards the inner part of Leyte Gulf.

3.5.3 People's awareness of storm surges

During the field surveys, questionnaire surveys and interviews with local residents and officials were also conducted to understand their preparations prior to the event, evacuation behaviour during the typhoon, and general levels of storm surge awareness (Esteban, et al., 2015; Esteban et al., 2016). Questionnaire sheets containing 16 questions were distributed to individuals at the locations surveyed, and a total of 172 valid questionnaires were obtained (Figure 3.5.5). Focus group discussions with three different groups of people who completed the questionnaires and non-structured interviews with key informants (government officers and disaster risk managers) were also conducted.

One of the problems identified by these questionnaire surveys and interviews was the poor level of understanding regarding the term "storm surge" prior to the event, as only 47% of respondents said that they understood what a storm surge was and that it could result in widespread flooding. However, it is not even clear whether this 47% of people accurately understood the threat posed by a storm surge. Indeed, during the focus group discussions, many people expressed the view that it would have been better for authorities and media to describe it as a "tsunami". Moreover, the key informant interviewee revealed that local officials relayed information regarding the storm surge to their communities, though their knowledge of what could happen was limited to strong waves (they described the storm surge as "*dagko nga balod*" which essentially translates as huge waves). These results highlight that people were not able to clearly conceptualize the phenomenon of a storm surge. Although this event has contributed to greatly increasing awareness about such phenomena in the Philippines, it is still clearly important to further disseminate the knowledge and lessons learned to other coastal communities at risk in this and other countries.

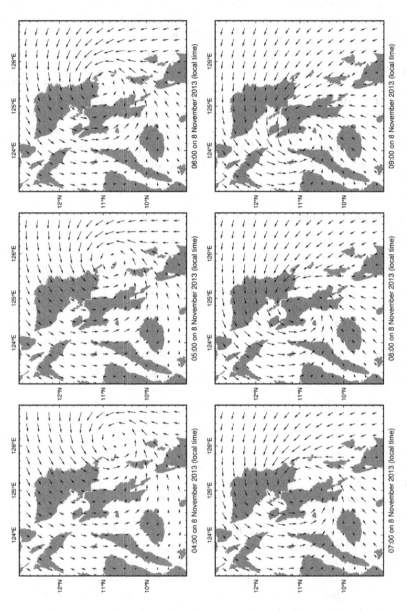

Figure 3.5.4 Hourly wind fields over Leyte and Samar Islands from 4:00 to 9:00 on 8 November 2013 (local time) based on the ERA5 reanalysis dataset (Hersbach et al., 2018).

Figure 3.5.5 Questionnaire survey: (a) questionnaire sheet distributed during field surveys and (b) respondents of questionnaires.

3.5.4 Conclusions

Typhoon Haiyan in 2013 was one of the most powerful typhoons to have made landfall in the Philippines, and severe damage due to the large storm surge it generated could be observed along the coastline of Leyte Gulf. In particular, Tacloban City and other coastal communities located along its inner part suffered the most, as the storm surge in the area was higher than 4 m. One of the characteristics of this storm surge was that the sea water first receded and then the surge manifested itself quickly, which was caused by a rapid change in wind direction over Leyte Gulf. The results of questionnaire surveys and interviews with local people showed that even though information about the storm surge reached coastal communities, people were not able to clearly conceptualize this phenomenon. Thus, in order to improve the resilience of coastal communities to such events in the future, it is clearly necessary to further disseminate information and raise awareness about how to protect against such phenomena.

3.6 STORM SURGE IN NEMURO, 2014

Ryota Nakamura
Niigata University, Niigata, Japan

3.6.1 Introduction

Storm surge hazards generated by extra-tropical cyclones (ETCs) have been frequently recorded in Europe (see Section 9.6), but have rarely taken place in Japan. However, a storm surge generated by a "bomb cyclone" (intensified ETC) in the middle of December 2014 resulted in the inundation of the town of Nemuro, situated in a peninsula along the northeast of Hokkaido island, Japan. Although no fatalities were recorded as a consequence of this event, given its rarity several Japanese research groups carried out post-hazard field surveys to record the inundation height by measuring the water level marks and interviewing and collecting photos from local residents (Nakamura et al., 2015, 2019; Bricker et al., 2015; Saruwatari et al., 2015; Kumagai et al., 2015). These studies also carried out numerical simulations to clarify the mechanism of storm surges associated with ETCs and identify the vulnerability of settlements in the area to such hazards.

This chapter will explain the fieldwork carried out by one of the survey groups, detailing also simulations that were carried out to better explain these phenomena.

3.6.2 Post-disaster field survey

A field survey was carried out by a team of researchers led by Prof. Shibayama, the lead editor of this book, on 19th December 2014 (2 days after the storm surge inundation took place), meaning that the watermarks left by the storm surge could still be clearly identified (Nakamura et al. (2015). The methodology used in these field surveys is similar to that employed in other similar exercises reported in this book (see Sections 2.1–3.7).

The watermarks surveyed showed that the storm surge flooded the area to a height of 1.8–2.0 m above mean water level (Figure 3.6.1), which was

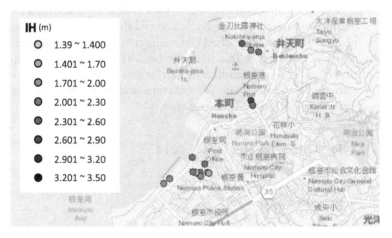

Figure 3.6.1 Maximum inundation height (IH, above mean sea water level) of the area of Nemuro affected by the storm surge (Esri Japan. Sources: Esri, HERE, Garmin, Intermap, increment P Corp, GEBCO, USGS, FAO, NPS, NRCAN, GeoBase, IGN, Kadaster NL, Ordnance Survey, Esri Japan, METI, Esri China (Hong Kong), sisstopo, OpenStreetMap contributors and the GIS User Community).

close to the maximum water level of 1.8 m recorded at Nemuro tidal station. The inundation depth (inundation height minus land height) was generally less than 1.0 m, depending on the topographical height of the land point surveyed.

The inundated area can be divided into two areas: Nemuro port and Yayoi town. In the Nemuro port area inundation took place in the apron of the port, though the extent of this flooding did not increase further as the level of the ground rapidly increases immediately behind the harbour. However, in Yayoi town the inundation extent was larger than at the port area, as the level of the land gradually decreased from the coastline to the residential area. This characteristic of Yayoi town makes the area especially vulnerable to storm surges, as if the water levels become higher than the level of the coastline then they can easily flood downwards towards the settlement.

As a result, many residential houses in Yayoi were flooded, though severe damage to structures (such as broken walls or columns), was seldom observed (see Figure 3.6.2). Nevertheless, a fishing boat was carried by the storm surge and was left stranded inland (Saruwatari et al., 2015). This indicated that floating objects carried by a storm surge can represent a hazard to structures, with the impact forces resulting having the potential for being much higher than hydrodynamic forces alone (see Section 6.2). In fact,

Figure 3.6.2 Field surveys at Nemuro, showing how despite the existence of water marks comparatively little damage to structures and houses took place.

field reports from the aftermath of Typhoon Jebi (see Section 3.7) indicate how, for example, tankers carried away by the storm surge damaged bridges between Kansai airport and the mainland (Takabatake et al., 2018), requiring repairs that took about a month to complete.

3.6.3 Numerical simulations

High-resolution numerical simulations were carried out to reproduce the coastal flood that took place at Nemuro. A first meso-scale simulation was conducted to verify the mechanism behind the increase in water levels associated with low-pressure systems, which can provide insights into the risk of future flooding to the area. A second simulation aimed to understand the hydrodynamic properties associated with coastal flooding at the local scale, which can provide a spatial and temporal distribution of flood risk at the street level, helping to explain why the damage to humans and structures was not as severe as that of other storm surges reported in this book (given also that flood levels of around 2.0 m did result in severe damage and casualties in Samar Island, Philippines, during the passage of Typhoon Haiyan of 2013 (see Mikami et al., 2016 and Section 3.5).

The storm surge simulations were carried out by using a combination of GPV-MSM for the atmospheric analysis, and the FVCOM model (Chen et al., 2003) for simulating the meso-scale and street-scale scale inundation.

3.6.3.1 Meso-scale simulations

The spatial distribution of surface wind velocity and sea-level pressure are shown in Figure 3.6.3. At first, the water piled up and the sea level increased in the Nemuro Gulf due to the westward wind associated with the approaching low-pressure system. The maximum wind velocity was over 25 m/s and the minimum sea-level pressure reached 945 hPa, which is equivalent to some of the stronger typhoons that make landfall in Japanese archipelago. Then the low-pressure system moved slightly to the north, changing the wind direction to the south, which pushed the water to the northern part of the Nemuro peninsula, increasing the water level. As a result, the height of the storm surge reached almost 1.8 m in Nemuro Gulf.

According to the simulation results, it is clear that Nemuro Gulf is vulnerable to storm surge hazards associated with low-pressure systems that approach the area in a similar manner. In fact, water levels significantly increased again around the peninsula due to typhoon Choi-Wan (2015) and another low-pressure system that took place in 2015. The storm surge caused by typhoon Choi-Wan (2015) caused inland flooding to Yayoi town due to water intruding into the sewage system that normally discharges into the sea.

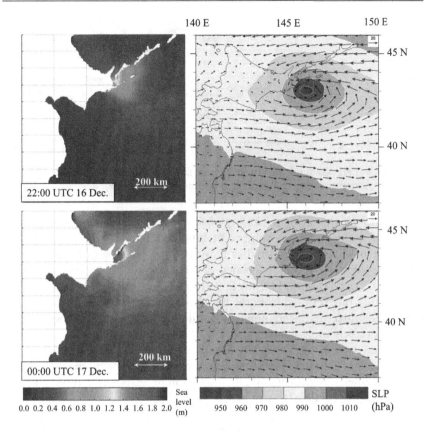

Figure 3.6.3 Spatial distribution of storm surge height (left), and sea-level pressure and surface wind velocity (right). The time of the upper and lower figures are 22:00 UTC 16th and 00:00 UTC 17th Dec. 2014, respectively.

3.6.3.2 Street-scale storm surge simulations

Street-scale simulations were carried out by inputting the results of the meso-scale simulation as boundary conditions. As explained earlier, the field surveys showed that the inundation depth was under 1.0 m in Yayoi town up to a distance of 300 m from the coastline. The right side of Figure 3.6.4. shows a digital elevation map created using data from the Geospatial Information Authority of Japan and the residential house distribution obtained from OpenStreetMap, highlighting how the centre of Yayoi town is lower than the coastline and much of the surrounding area (Figure 3.6.4). Due to this, the town is highly vulnerable to any storm surge that generates water levels higher than the ground level of the coastal zone, as this would cause a rush of water towards the lower-lying parts of the centre of the town.

The street-scale simulation results indicate that the water depth was under 1.0, which was in good agreement with observations (see also Nakamura

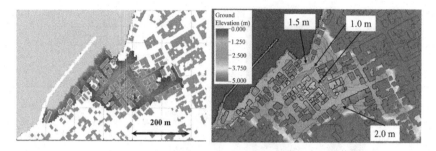

Figure 3.6.4 Left: Ground elevation of the town of Nemuro. Right. Inundation depth above the ground (Nakamura et al., 2019).

et al., 2019), highlighting the ability of simulations to predict the extent and height of floods, and the number of buildings likely to be affected. The height of inundation was smaller than in other more severe storm surge disasters, such as those associated with Cyclone Nargis in 2018 (see Section 3.3) or Typhoon Haiyan in 2013 (see Section 3.5). In such cases a high fatality rate resulted from the high hydrodynamic forces, though in the case of Nemuro there were none, indicating that the energy of the flooding water was likely much lower.

Damage to structures and people in flooded areas is typically a function of the speed of the current and the water depth, and thus momentum fluxes determine the risk of a person falling down in flooded water and dying.

The momentum flux (I_F) can be calculated by:

$$I_F = U_s^2 h_w \tag{3.6.1}$$

where, U_s and h_w denote current velocity and water depth from the simulation results, respectively.

The human instability model was proposed by previous studies based on experiments using real subjects (e.g., Abt et al., 1989; Karvonen et al., 2000). The following equation is Karvonen et al. (2000)'s human instability equation:

$$h_w U_c = 0.004HM + 0.2 \tag{3.6.2}$$

where H and M are height and weight of an individual human, respectively.

Finally, the hazard ratio (HR), considering the critical current velocity can be calculated by:

$$HR = \min\left(1.0, U_s / U_c\right) \tag{3.6.3}$$

with a human becoming fully unstable when HR becomes 1.0.

These equations were used to calculate the impact of hydrodynamic forces and human instability in Yayoi town. The value of I_F was lower than

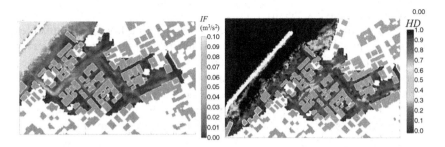

Figure 3.6.5 Spatial distribution of I_F for buildings and *HR* rate for human instability (Nakamura et al., 2019).

0.10 m³/s² in almost all the inundated area, which is much lower than in other more damaging coastal flooding events (for instance, in the case of the *2011 Tohoku Earthquake and Tsunami*, I_F reached over 100). Therefore, it can be said that the hydrodynamic forces associated with the inundation due to the storm surge in Nemuro were not strong enough to cause severe damage to the structures in this town (Figure 3.6.5).

The calculated results show that the maximum *HR* reached 0.5 across the town, and was lower than 0.3 in almost all the inundated areas. This means that even amongst 5-year-old children the fatality rate would be low, so it would be highly unlikely for adults to be harmed.

3.6.4 Conclusions

A storm surge generated by a strong extra-tropical cyclone resulted in the inundation of the town of Nemuro in the middle of December 2014. The present chapter detailed a field survey that was carried out to clarify the inundation patterns that took place, showing how the low-lying ground close to the coastline is highly vulnerable to storm surges. However, computer simulations showed that the hydrodynamic forces acting on structures and humans were low, explaining why the damage caused by this event was not as significant as that in other recent events around the planet.

3.7 TYPHOON JEBI IN 2018

Martin Mäll

Yokohama National University, Yokohama, Japan

3.7.1 Introduction

Extreme tropical storms in Japan and in the Northwest Pacific area are known as typhoons, generally affecting the region during the warmer half of the year (May to November). These weather disturbances can become hazardous and have a negative effect on human socio-economic systems, as they can cause heavy rainfall, coastal and riverine flooding, landslides, high winds and dangerous nearshore conditions (rip currents, high waves, etc.). Since Japan has a long history in disaster risk management, it is generally considered to be well prepared against typhoons. However, occasionally there are particularly impactful typhoon events that can provide new lessons on how disaster risk management in the coastal zone should be improved. This was the case with Typhoon Jebi, which struck Japan on the 4th of September 2018 (Figure 3.7.1).

Typhoon Jebi (2018), better known in Japan as Typhoon Nr. 21 of 2018, started out as a tropical depression and, according to the Japan Meteorological Agency, developed into a typhoon on the 28th of August (JMA1 n.d.), continuing to gather strength as it moved towards central Japan. The minimum central pressure reached as low as 915 hPa and the maximum one-minute sustained winds were 55 m/s (198 km/h) 2 days prior to making landfall on the 2nd September 2018 at 09:00 UTC (JMA2 2018), making it a Category 3 typhoon on the Saffir–Simpson scale. Prior to making landfall on Shikoku Island, the intensity of the typhoon was sustained by the high sea surface temperatures in the seas around Japan (28°C or more; JMA1 n.d.), and during landfall on the 4th September the sustained winds were above 44 m/s (158 km/h; JMA2 2018). These meteorological characteristics made Typhoon Jebi (2018) the strongest storm to hit mainland Japan in 25 years (since Typhoon Yancy in 1993; Le et al., 2019). The typhoon track crossed the western edge of the Osaka Bay, providing favourable conditions for the generation of a high storm surge.

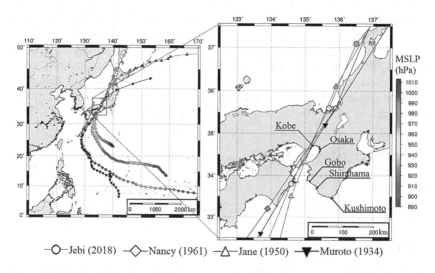

—○—Jebi (2018) —◇—Nancy (1961) —△—Jane (1950) —▼—Muroto (1934)

Figure 3.7.1 Notable historical typhoons to have hit the Osaka Bay area, together with their tracks with minimum sea-level pressure (MLSP) (after Takabatake et al., 2018).

3.7.2 Storm surge and preparedness

The high meteorological conditions and favourable track resulted in large sea-level oscillations in Osaka Bay. The Kobe and Osaka tidal stations recorded the surge heights to be +2.33 m and +3.29 m above Tokyo Peil (T.P.; datum corresponding to mean sea level in Tokyo Bay). However, the design heights for coastal defences in Kobe and Osaka are T.P. +2.8 m and T.P. +3.9 m, respectively. Emergency flood warnings were issued prior to landfall of the typhoon (approximately 6 hours before), which prompted the closure of flood gates in the city. However, several areas were reported to have been flooded during the typhoon, and images from social media sites showed that this was the case even in some areas inside the flood gates (e.g., in Higashi-Kawasaki town). Furthermore, several artificial islands in Osaka Bay were reported to have been flooded. This included Kansai Airport, which was constructed on an artificial island in 1994, and had to stop its service for around ten days due to the flooding. Another area that was flooded was Suzukaze town, built of newly reclaimed land in 1997. These newly reclaimed areas had no previous experience of suffering flooding due to storm surges or high waves due to strong typhoons coming to Osaka Bay. In the aftermath of the disaster various research groups (e.g., Mori et al., 2018) conducted field surveys to ascertain the impact of the typhoon. In this chapter the main findings from the Shibayama Lab at Waseda University are presented (after Takabatake et al., 2018).

3.7.3 Field surveys

A field survey was conducted on September 6th, two days after Typhoon Jebi (2018) made landfall. During the passage of the typhoon many photos and videos were posted on various social media sites (e.g., Twitter and Instagram) that showcased and helped the group to better understand which areas were likely to have been affected by the storm surge (i.e., the extent of inundation, if any). Four locations in two different cities were surveyed in detail: Higashi-Kawasaki town, Mikageishi town, and Fukaehama town in Kobe City, and Suzukaze town in Ashiya City (Figure 3.7.2).

The survey team had three main aims overall: to determine the extent of the inundation and its height, establish patterns of damage to infrastructure and ascertain the behaviour of local citizens before and after the typhoon. To measure inundation heights the team used a variety of instruments, similar to that used in other surveys detailed in this book (see Sections 3.1–3.6). This included GPS devices (eTrex 20XJ, Garmin, Ltd., Olathe, KS, USA) to record measurement locations, laser ranging instruments (IMPULSE 200LR; minimum reading: 0.01 m; Laser Technology Inc., Centennial, CO, USA), prisms, staffs and a salinity meter (EN-901, DRETEC Co., Ltd., Saitama, Japan). More details on the survey techniques can be found in Esteban et al. (2017). These instruments were used to determine the inundation height

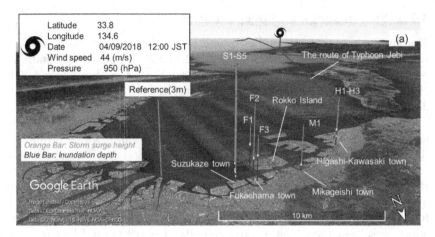

Figure 3.7.2 Locations visited by the Shibayama survey team in Osaka Bay. The vertical bars show inundation depth and height at the various survey locations. Numbers show specific measurements at any given location (e.g., HI indicates Higashi-Kawasaki town, with the I denominating that this was the 1st of 3 measurements taken in that town). Figure after Takabatake et al. (2018).

from the sea surface level by relying on water marks left behind (Figure 3.7.3). The reliability of water marks was confirmed through eyewitness accounts and photos (from social media sites) when and where possible. In some locations where access to the seafront was difficult, digital elevation data (DEM; 5 m resolution) was used to estimate the inundation height from the water marks. Using DEM data is a useful approach where access is limited, though it is important to note that this method has limitations in accuracy (+/–30 cm). In addition, GPS was used to collect the precise locations of drift cargo containers, found along the coastline in Suzukaze Town.

A questionnaire survey was handed to local citizens in residential areas (Higashi-Kawasaki and Suzukaze town). The structured questionnaire was developed based on prior ones that were conducted to study the storm surge awareness in the Philippines after the 2013 super typhoon Haiyan (Esteban et al., 2016, see also Section 3.5). The questionnaire consisted of a total of 18 questions (taking 5–10 minutes to answer) and valid answers were provided by 17 individuals (out of 21). For statistical analysis of such questionnaires a larger sample is needed, though due to limitations and circumstances it was difficult to reach more respondents, so the results can only be considered as being indicative of how and why the larger population might have acted.

The survey results showed that in some of the areas surveyed coastal inundation had taken place. However, the flood countermeasures had also worked and stopped most of the surge water from intruding into the protected zones behind them. For instance, Higashi-Kawasaki town (Figure 3.7.2)

Figure 3.7.3 Example of water marks and points of interest during the survey at Fukaehama town. (a) Coastal dike at Fukaehama town. (b) Survey Point F1 indicates that the direction of debris was towards inland. (c–d) Presence of debris at Fukaehama town. (e) Survey Point F2 (watermark at school gate). (f) Survey Point F3 (watermark at a model house). After Takabatake et al. (2018).

has a flood gate which was closed before the peak of the typhoon, yet water marks up to 68 cm were found inside the gate. Through eyewitness accounts it became clear that the high level was caused by a multitude of factors, including water seepage from manholes, rain water and local topographic features where precipitation tends to pool into one area. Another example is the residential area of Suzukaze town, which sits on a reclaimed artificial island. This area was protected by a T.P. +5.6 m embankment (which was not destroyed) while the Kobe and Osaka tidal stations measured maximum water level to be +2.33 m and +3.29 m, respectively. However, significant inundation still occurred (maximum inundation depths of up to +0.64 m), with survey observations and eyewitness accounts suggesting that the sea water overtopped the embankment for some period of time, which was most likely due to a combination of both the storm surge and wave overtopping.

3.7.4 Disaster risk management lessons learnt

The field survey highlighted the need to consider the potential impacts from drift objects that are carried from one place to another during such extreme events. In Suzukaze town several port containers were found on the embankments, which could cause potentially serious damage to structures, coastal defences and threaten the safety of residents. The questionnaire survey indicated that residents in the newly reclaimed area of Suzukaze town did not have a high level of awareness about the danger posed by storm surges, which seemed to stem from a lack of past experiences and the failure of local authorities to issue evacuation orders well in advance of the event (all respondents stayed in their houses). The impact from Typhoon Jebi (2018) was not as devastating as it might have otherwise been due to relatively good levels of preparedness and solid coastal defences.

The field study results highlighted some of the areas that should be further investigated in the future. For instance, the hazard maps that were in place indicated that Suzukaze town should have been safe even in the case of typhoons stronger than Jebi (2018). These hazard maps did not include the effects from wave overtopping, which was considered to be the main cause of flooding in the Suzukaze town. In addition, the existing hazard maps did not take into account the potential of fluvial floods (e.g., in Mikageishi town). Another risk management consideration was the through capacity of drainage system, as in Higashi-Kawasai town the flood gates stopped the intrusion of seawater, though some flooding still took place (water infiltrating through manholes and rainfall). Therefore, the hazard maps should be revised to take into account the effects of wave overtopping (especially for the case of reclaimed islands in the Osaka Bay) as well as considering compound flooding from rivers and heavy rainfall. Unless existing defences are improved it is possible that they might not be as effective in the future if they are challenged by a stronger typhoon (which could come about as a result of climate change, see Section 7.2).

3.7.5 Conclusions

Typhoon Jebi (2018) was the strongest typhoon to hit mainland Japan in 25 years, affecting many coastal areas along the Osaka Bay. Existing coastal defences and the high level of preparedness kept storm damage to a minimum, despite several areas being flooded. The Shibayama lab survey team conducted a field survey to ascertain the level of impact and behaviour of residents during the event, providing new insights for disaster risk managers on lessons that could be learned from this disaster. The main conclusions are: (1) it is advisable to determine the location of potential large drift objects and evaluate their movement during extreme events, (2) hazard maps should be reassessed to include combined surge and wave overtopping effects (especially in the case of the more exposed artificial islands), as well as fluvial and rainfall effects, and (3) it is necessary to improve the disaster risk awareness of residents in higher risk areas (e.g., in Suzukaze town).

REFERENCES

Aerts, J.C.J.H., Botzen, W.J.W., De Moel, H., Bowman, M. (2013). Cost estimates for flood resilience and protection strategies in New York City, *Annals of the New York Academy of Sciences*, 1294, 1–104.

Abt, S.R., Wittler, R.J., Taylor, A., Love, D.J. (1989). Human stability in a high flood hazard zone. *Journal of the American Water Resources Association* 25(4), 881–890.

Ali, A. (1999). Climate change impacts and adaptation assessment in Bangladesh. *Climate Research*, 12(2–3), 109–116.

Basset, M., Anthony, E.J., Dussouillez, P., Goichot, M. (2009). The impact of Cyclone Nargis on the Ayeyarwady (Irrawaddy) River delta shoreline and nearshore zone (Myanmar): Towards degraded delta resilience? *Comptes Rendus Geoscience*, 349, 238–247.

Bricker, J.D., Volker, R., Fukutani, Y., Kure, S. (2015). Simulation of the December 2014 Nemuro storm surge and incident waves. *Journal of Japan Society of Civil Engineers, Ser. B2 (Coastal Engineering)*, 71(2), I_1543–I_1548. https://doi.org/10.2208/kaigan.71.I_1543

Bureau of Vital Statistics, New York City Department of Health and Mental Hygiene. (2014). Summary of vital statistics 2012 the City of New York: Executive summary with a special section on deaths due to hurricane sandy.

Chen, C., Liu, H., Beardsley, R.C. (2003). An unstructured, finite-volume, three-dimensional, primitive equation ocean model: Application to coastal ocean and estuaries. *Journal of Atmospheric and Oceanic Technology*, 20, 159–186.

Esteban, M., Takagi, H., Mikami, T., Bahbouh, L., Becker, A., Nurse, L., Shibayama, T., Nagdee, M. (2017). How to carry out bathymetric and elevation surveys on a tight budget: Basic surveying techniques for sustainability scientists. *International Journal of Sustainable Future for Human Security*, 5(2), 86–91. https://doi.org/10.24910/jsustain/5.2/8691

Esteban, M., Valenzuela, V.P., Yun, N.Y., Mikami, T., Shibayama, T., Matsumaru, R., Takagi, H., Thao, N.D., De Leon, M., Oyama, T., Nakamura, R. (2015). Typhoon Haiyan 2013 evacuation preparations and awareness *International Journal of Sustainable Future for Human Security*, 3(1), 37–45.

Esteban, M., Valenzuela, V.P., Matsumaru, R., Mikami, T., Shibayama, T., Takagi, H., Thao, N.D., Leon, M.D. (2016). Storm surge awareness in the Philippines prior to typhoon Haiyan: A comparative analysis with Tsunami awareness in recent times. *Coastal Engineering Journal*, 58(1), 1640009-1–1640009-28. https://doi.org/10.1142/S057856341640009X

Fritz, H., Blount, C., Thwin, S., Thu, M.K., Chan, N. (2009). Cyclone Nargis storm surge in Myanmar. *Nature Geoscience*, 2, 448–449.

Frontiers Myanmar. (2015). Controlling catastrophe. https://www.frontiermyanmar.net/en/controlling-catastrophe/

Government of Bangladesh. (2008). *Cyclone Sidr in Bangladesh: Damage, Loss and Needs Assessment for Disaster Recovery and Reconstruction*. Government of Bangladesh, Dhaka.

Hallegatte, S. (2008). An adaptive regional input-output model and its application to the assessment of the economic cost of Katrina. *Risk Analysis*, 28(3), 779–799.

Haque, A., Jahan, S. (2016). Regional impact of cyclone Sidr in Bangladesh: A multi-sector analysis, *International Journal of Disaster Risk Science*. https://doi.org/10.1007/s13753-016-0100-y

Hersbach, H., Bell, B., Berrisford, P., Biavati, G., Horányi, A., Muñoz Sabater, J., Nicolas, J., Peubey, C., Radu, R., Rozum, I., Schepers, D., Simmons, A., Soci, C., Dee, D., Thépaut, J.-N. (2018). ERA5 hourly data on single levels from 1979 to present, Copernicus climate change service (C3S) climate data store (CDS). Accessed on 26 December 2021. https://doi.org/10.24381/cds.adbb2d47

Hiraishi, T. (2009). Field survey on facility damage in Yangon harbour due to 2008 storm surge. Technical Note of the Port and Airport Research Institute, No. 1192.

JMA1. (n.d.). Jebi (T1821). RMSC Tokyo – Typhoon center japan meteorological agency. Accessed on 23 September 2021. https://www.mri-jma.go.jp/Dep/typ/typ1/TY_VER/T1821_JEBI.pdf

JMA2. Japan Meteorological Agency. (2018) Wind storm and storm surges by Typhoon No. 21. 2018. Accessed on 14 September 2018. https://www.data.jma.go.jp/obd/stats/data/bosai/report/2018/20180911/20180911.html (in Japanese)

Jonkman, S.N., Kok, M., van Ledden, M., Vrijling, J.K. (2009). Risk-based design of flood defence systems: A preliminary analysis of the optimal protection level for the New Orleans metropolitan area. *Journal of Flood Risk Management*, 2(3), 170–181.

Jonkman, S.N., Maaskant, B., Boyd, E., Levitan, M.L. (2009). Loss of life caused by the flooding of New Orleans after hurricane Katrina: Analysis of the relationship between flood characteristics and mortality. *Risk Analysis*, 29(5), 676–698; Best paper Award 2009; Risk Analysis - an international journal.

Karvonen, R.A., Hepojoki, H.K., Huhta, H.K., Louhio, A. (2000). The use of physical models in dam-break flood analysis, development of Rescue Actions Based on Dam-Break Flood Analysis (RESCDAM). Final report of Helsinki University of Technology, Finnish Environment Institute. Accessed on 5 August 2019. http://ec.europa.eu/echo/files/civil_protection/civil/act_prog_rep/rescdam_rapportfin.pdf

Kumagai, K., Seki, K., Fujiki, T., Tomita, T., Tsuruta, N., Sakai, K., Yamamoto, Y., Kakizaki, E. (2015). Damage of Nemuro port and its surrounding areas due to the storm-surge in 17 December 2014. Technical Note of National Institute for Land and Infrastructure Management 854. Accessed on 5 August 2019 http://www.nilim.go.jp/lab/bcg/siryou/tnn/tnn0854pdf/ks0854.pdf (in Japanese with English Abstract)

Kuroda, T., Saito, K., Kunii, M., Kohno, N. (2010). Numerical simulations of Myanmar cyclone Nargis and the associated storm surge part I: Forecast experiment with a nonhydrostatic model and simulation of storm surge. *Journal of the Meteorological Society of Japan*, 88(3), 521–545.

Kyaw, T.O., Esteban, M., Mall, M., Shibayama, T. (2021). Extreme waves induced by cyclone Nargis at Myanmar coast: Numerical modeling versus satellite observations. *Natural Hazards*, 106, 1797–1818.

Le, T.A., Takagi, H., Heidarzadei, M., Takata, Y., Takahashi, A. (2019). Field surveys and numerical simulation of the 2018 Typhoon Jebi: Impact of high waves and storm surge in semi-enclosed Osaka Bay, Japan. *Pure and Applied Geophysics*, 176, 4139–4160.

Mikami, T., Esteban, M., Shibayama, T. (2015). Storm Surge in New York City Caused by Hurricane Sandy in 2012. In *Handbook of Coastal Disaster Mitigation for Engineers and Planners*. Esteban, M., Takagi, H., Shibayama, T. (eds.). Butterworth-Heinemann (Elsevier), Oxford, UK.

Mikami, T., Shibayama, T., Takagi, H., Matsumaru, R., Esteban, M., Thao, N.D., De Leon, M., Valenzuela, V.P., Oyama, T., Nakamura, R., Kumagai, K., Li, S. (2016). Storm surge heights and damage caused by the 2013 Typhoon Haiyan along the Leyte Gulf Coast. *Coastal Engineering Journal*, 58(1), 1640005.

Mori, N., Yasuda, T., Arikawa, T., Kataoka, T., Nakajo, S., Suzuki, K., Yamanaka, Y., Webb, A. (2018). 2018 Typhoon Jebi post-event survey of coastal damage in the

Kansai region, Japan. *Coastal Engineering Journal*, 61(3), 278–294. https://doi.org/10.1080/21664250.2019.1619253

MTA. (2014). Accessed on 10 October 2014. http://www.mta.info/

Nakamura, R., Iwamoto, T., Shibayama, T., Mikami, T., Matsuba, S., Mäll, M., Tatekoji, A., Tanokura, Y. (2015). Field survey and mechanism of storm surge generation invoked by the low pressure with rapid development in Nemuro Hokkaido in December 2014. *Journal of Japan Society of Civil Engineers, Ser. B3 (Coastal Engineering)*, 71(2), I_31–I_36. https://doi.org/10.2208/jscejoe.71.i_31 (in Japanese with English Abstract).

Nakamura, R., Mäll, M., Shibayama, T. (2019). Street-scale storm surge load impact assessment using fine-resolution numerical modelling: A case study from Nemuro. *Japan Natural Hazards*, 99, 391–422.

Nakamura, R., Shibayama, T., Esteban, M., Iwamoto, T. (2016). Future typhoon and storm surges under different global warming scenarios: Case study of Typhoon Haiyan (2013). *Natural Hazards*, 82(3), 1645–1681.

National Disaster Risk Reduction and Management Council. (2021). Final report re effects of typhoon "YOLANDA" (HAIYAN). Accessed on 22 December 2021. https://ndrrmc.gov.ph/21-disaster-events/1329-situational-report-re-effects-of-typhoon-yolanda-haiyan

National Hurricane Center. (2013). Tropical cyclone report: Hurricane Sandy (AL182012) 22-29 October 2012, http://www.nhc.noaa.gov/data/tcr/AL182012_Sandy.pdf

New York City Government. (2014). Accessed on 10 October 2014. http://www.nyc.gov/

NY Times. (2019). "U Tun Lwin," who warned of deadly Myanmar cyclone, Dies at 71.

Okayasu, A., Shimozono, T., Thein, M.M., Aung, T.T. (2009). Survey of storm surge induced by cyclone Nargis in Ayeyarwaddy. *Japan Society of Civil Engineers*, B2-65(1), 1386–1390.

Omori, K., Sakai, T., Miyamoto, J., Itou, A., Oo, A.N., Hirano, A. (2020). Assessment of paddy fields' damage caused by Cyclone Nargis using MODIS time-series images (2004–2013). *Paddy and Water Environment*, 19, 271–281.

Paul, B.K. (2012). Human injuries caused by Bangladesh's cyclone Sidr: An empirical study. *Natural Hazards*, 54, 483–495. https://doi.org/10.1007/s11069-009-9480-2

Pistrika, A.K., Jonkman, S.N. (2009). Damage to residential buildings due to flooding of New Orleans after hurricane Katrina. *Natural Hazards*, 54(2), 413–434.

Shibayama, T., Tajima, Y., Kakinuma, T., Nobuoka, H., Yasuda, T., Ahsan, R., Rahman, M., Shariful Islam, M. (2009). Field survey of storm surge disaster due to cyclone sidr in Bangladesh. In *Proc. of Coastal Dynamics 2009, No. 129*.

Shibayama, T., Yasuda, T., Kojima, H., Tajima, Y., Kato, H., Nobuoka, H., Yasuda, T., Tamagawa, K. (2006). A field survey of storm surge caused by Hurricane Katrina. *Annual Journal of Coastal Engineering*, 53(1), 401–405, JSCE (in Japanese).

Shibayama, T., Takagi, H., Hnu, N. (2008). Report of the field investigation after the Cyclone Nargis in 2008. *Japan Society for Natural Disaster Science*, 27(3), 331–339.

Shibayama, T. (2015). 2005 Storm Surge by Hurricane Katrina. In *Handbook of Coastal Disaster Mitigation for Engineers and Planners*. Esteban, M., Takagi, H., Shibayama, T. (eds.). Butterworth-Heinemann (Elsevier), Oxford, UK.

Takabatake, T., Mäll, M., Esteban, M., Nakamura, R., Kyaw, T.O., Ishii, H., Valdez, J.J., Nishida, Y., Noya, F., Shibayama, T. (2018). Field survey of 2018 Typhoon Jebi in Japan: Lessons for disaster risk management. *Geosciences*, 8, 412.

Takagi, H., Esteban, M., Shibayama, T., Mikami, T., Matsumaru, R., De Leon, M., Thao, N.D., Oyama, T., Nakamura, R. (2017). Track analysis, simulation and field survey of the 2013 Typhoon Haiyan storm surge. *Journal of Flood Risk Management*, 10(1), 42–52.

Takagi, H., Li, S., de Leon, M., Esteban, M., Mikami, T., Matsumaru, R., Shibayama, T., Nakamura, R. (2016). Storm surge and evacuation in urban areas during the peak of a storm. *Coastal Engineering*, 108, 1–9.

Tasnim, K.M. Esteban, T., Shibayama, T., Takagi, H. (2015). Observations and Numerical Simulations of Storm Surge due to Cyclone Sidr 2007 in Bangladesh. In *Handbook of Coastal Disaster Mitigation for Engineers and Planners*. Esteban, M., Takagi, H., Shibayama, T. (eds.). Butterworth-Heinemann (Elsevier), Oxford, UK.

U.S. Census Bureau. (2012). Largest U.S. metropolitan areas by population, e 1990–2010, world almanac and book of facts 2012.

UNDP (United Nations Development Programme). (2004). A global report: Reducing disaster risk: A challenge for development. Accessed on 15 June 2014. http://www.undp.org/bcpr

Chapter 4

High wave attacks and coastal flooding

Case studies from Japan

CONTENTS

4.1 TYPHOON JONGDARI IN 2018

Tomoya Shibayama

Waseda University, Tokyo, Japan

4.1.1 Introduction

Typhoons approaching Japan usually move from west to east under westerly winds, with events moving from east to west along the Pacific coast being very rare. Nevertheless, such storms do take place from time to time, such as the typhoon that approached the Kanto region of Japan in August 1945. As the event took place before the surrender of the country during WW2, and thus before the US military moved in, detailed weather information (such as the exact path of the typhoon or its central pressure) were not collected. Typhoon Louise in 1963 also took a westward course and caused heavy damage to the country, mainly due to heavy rains. Typhoon Jongdari in 2018 was the first significant typhoon to take a westward course since that time, generating high waves along the western side of Sagami Bay.

Although damage due to this event was not particularly high, as the weather system was not particularly strong at the time of landfall, there is concern that future typhoons could become intensified as a consequence of climate change (see Section 7.2), and that the paths they follow might also change. This highlights the importance of studying such rare events, and attempts to draw lessons on how the population at risk could be affected in the future, in order to improve risk management countermeasures that enhance the resilience of coastal settlements (see Section 8.3)

4.1.2 Typhoon Jongdari

The weather system originated southeast of Guam on the 19th of July and intensified into a typhoon on July 26th over the Ogasawara islands, then eventually turning westwards and making landfall over Ise, in Mie prefecture in Japan. Normally, when typhoons approach the southeastern coastline of Honshu island in Japan waves arrive from the west. However, as shown in Figure 4.1.1, the typhoon approached the Kanto region from a southeastern direction at a slow speed, and moved from east to west along the Pacific coast of Japan, passing over Sagami Bay.

As the typhoon approached, the swell generated offshore by the prolonged strong winds (given the slow movement and persistence of the typhoon such swell waves had ample time to propagate) caused the significant wave height to surge from about 30 cm to over 3 m. Wave overtopping of breakwaters and coastal revetments took place, with significant overflowing of embankments also being reported. Overtopping by wind waves is usually tolerated to a certain extent, as structures are rarely damaged when the quantity of overflowing water is limited. However, in the case of typhoon Jaongdari the run-up at the steep slopes of hills along the coastline of the Izu Peninsula was unexpectedly large. Since the main road connecting the peninsula is constructed on the hillside of this slope, seawater reached the tarmac.

Due to the unusual route of the typhoon, the storm surge and high waves came as a surprise to local residents, with the surge seeping away an emergency vehicle while rescuing a car on the road. Coastal roads at many other places were also flooded, causing delays to transport, and cars were swept away in Odawara City, Manazuru Town and Yugawara Town. At some points the high waves reached over 8 meters above sea level.

4.1.3 Recent changes in behaviour of typhoons around the Japanese archipelago

In recent years, the position of the westerlies has been unstable, and there have been increasing numbers of cases of typhoons stagnating around the

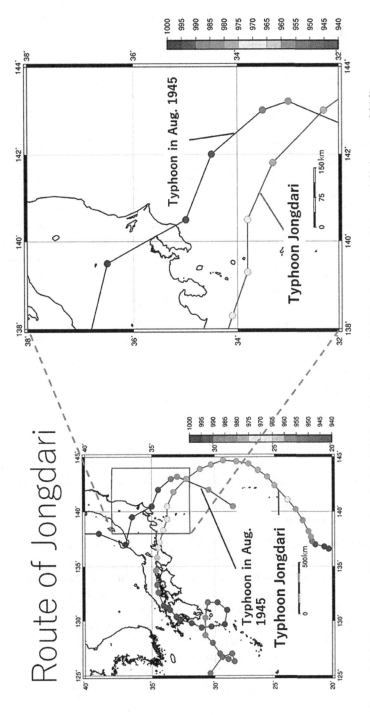

Figure 4.1.1 Trajectory of Typhoon Jongdari in 2018 and the August 1945 Typhoon (see Noya and Shibayama, 2019).

Japanese archipelago or moving westward. Some even made landfall directly on the coast of the Tohoku region in the north part of Japan (e.g. Typhoon Lionrock in 2016).

In the case of Typhoon Fitow in 2007, the slow speed of the typhoon and the long duration of strong winds in the sea around Japan caused the generation of long-period waves. As a result, the Seisho Bypass collapsed between the Oiso-Nishi and Tachibana Interchanges. Typhoons cause both storm surges and wind waves, and in many cases, they occur simultaneously, increasing the risk of a disaster taking place, so both need to be considered together (see also Section 9.1. for further discussion of disasters around the Japanese islands).

In the Japanese disaster prevention system it is the role of the local government at the prefectural level to predict the maximum storm surge heights. The author of the present chapter serves as the chairperson for the committees of storm surge prediction (Sagami and Tokyo bays, separately) in Kanagawa Prefectural Government. In the simulations that were performed under such committees four different directions, including typhoon Jondari's direction, 9–12 parallel courses for each direction and two different speeds were set as scenarios to simulate storm surges and calculate all possible typhoons that can generate storm surges and high waves (Kanagawa Prefectural Government, 2021). From the simulation results the maximum surge height is selected for each location along the coastline of Kanagawa prefecture. Based on these simulations, the extent of expected flooded areas, water depth and flood duration time can be calculated, with the results being open to the public. For example, in the Tokyo Bay side of the prefecture, the maximum height was estimated to be around 5 m in the inner bay area and 2 m in the mouth of the bay. The maximum duration of flooding, in case of a lack of drainage capacity due to local topography or the availability of pumps, would be around 3 days. Using these calculation results, the local government of each city or town can formulate appropriate evacuation plans for the residents of the areas that are expected to be flooded.

4.1.4 Conclusions

Typhoon movements in the vicinity of the Japanese islands are changing rapidly, possibly as a consequence of the increase in sea surface temperatures of the oceans around the archipelago. Examples of this behaviour include the route of typhoons moving from east to west, a low velocity of the centre of the storm, meandering of the typhoon course, or storms making landfall on the north part of the country. Given such changes to the usual typhoon routes residents can be caught unprepared, with high waves hitting coastal areas in different patterns from past experiences. In order to improve the resilience of coastal settlements and ensure the long-term sustainability of Japanese society it is thus necessary to continue to perform research on how to simulate future changes in high wave attacks and formulate appropriate countermeasures.

4.2 TYPHOON FAXAI, 2019

Naoto Inagaki

Waseda University, Tokyo, Japan

4.2.1 Typhoon synopsis and high wave attack over Fukuura Coast

Typhoon Faxai approached the Tokyo Metropolitan Area and made landfall in the Miura peninsula at around 02:30 am on September 9, 2019 (Japan Standard Time: UTC+9:00), before moving into Tokyo Bay and making landfall again on the north-east end of the bay at around 5 am. Figure 4.2.1 provides the best-track location indicating the path of the storm, together with its maximum wind speed and minimum central pressure. Although Faxai was notable for its low central pressure, it also had an exceptionally small wind field for a typhoon of its intensity, and this steep pressure gradient generated strong winds that caused great damage around Tokyo Bay. Chiba City, located on the east side of Tokyo Bay, reported a record-breaking maximum instant wind speed of 57.5 m/s at 5 am, just as the typhoon made landfall. Given the enclosed nature of Tokyo Bay, with only a narrow opening to the open sea to the south, it is rare that high wind waves propagate inside it, although in the west side of the bay severe wind waves and flood damage were recorded around the Port of Yokohama at 2 am.

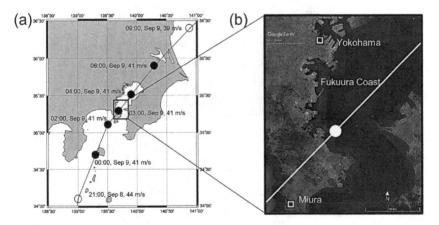

Figure 4.2.1 Track of Typhoon Faxai in 2019 over Tokyo Bay, according to observations by the Japanese Meteorological Association (JMA). (a) Track of the typhoon, together with its maximum wind speed. The typhoon had a central pressure of 960 hPa in the vicinity of the coast (represented as a black marker). (b) West side of Tokyo Bay, showing the main locations of interest. The rectangles indicate the wind stations of Yokohama and Miura. Fukuura Coast is located between these stations, as shown on the map (see also Inagaki et al., 2022).

A unique feature of this event is that, while wind damage was reported at many locations, significant flood damage was spatially limited to only the area around Yokohama city (Suzuki et al. 2020). Although countermeasures against potential storm surge hazards have been constructed to protect industrial areas around Tokyo Bay, the flooding that took place caused severe damage to industrial machinery and resulted in the suspension of operations by a number of companies.

4.2.2 Field survey of Fukuura Coast and wave hindcasting

A field survey was conducted along Fukuura Coast on the 15th of September 2019, which was the most seriously affected area inside of Tokyo Bay (Figure 4.2.1(b)). The methodology used in the field surveys was similar to that detailed in other chapters of the present book (see Sections 3.1–3.7), and included the measuring of inundation heights and run-up using surveying equipment. Fukuura District has been, like most of the shorelines of Tokyo Bay, artificially reclaimed from the sea, and the coastline runs in a straight line from north to south at this point.

The average height of the ground at this location is T.P. + 3.0 m (Tokyo Peil[1]). Partial destruction of the coastal defences was observed over a stretch of 600 m, as shown in Figure 4.2.2(a, b). According to the watermarks, the

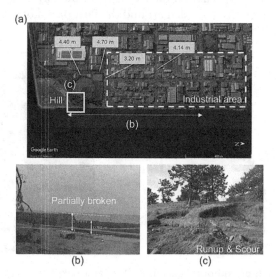

Figure 4.2.2 Summary of the field survey at Fukuura Coast. (a) Aerial view of the coast, showing the measured height (in reference to Tokyo Peil, or T.P.) of the water marks surveyed. (b) Damaged coastal defences. (c) Local scour at the bottom of a small hill (see Inagaki et al., 2022).

---

[1] The ground elevation in Japan is measured with reference to T.P. (Tokyo Peil), which was determined by the tidal records at Tokyo Bay during the period between 1873 and 1879, and corresponds to the mean water level in the bay.

waves overtopped the defences and reached over 300 m inland. There is a small hill with a maximum height of around T.P. + 10.7 m at the southern end of the coast (see the rectangle with the solid line in Figure 4.2.2(a)), with Figure 4.2.2(c) showing the local scour at its bottom, indicating also that the water did not overtop the hill.

The tidal level at Yokohama and Yokosuka stations (10 km away from Fukuura Coast, see Figure 4.2.1(b)) at 2 am was around T.P. + 0.8 m (including astronomical tide and pressure surge), which was less than the highest record of around T.P. + 1.5 m. Since the recorded water level was low (indicating a very modest storm surge), the destruction of the coastal defences can almost certainly be attributed to high wind waves. The coastal defences around this area had all been designed to the same height, up to a level +1.5 m above land. Figure 4.2.2(b) shows a section in which the upper 0.51 m part of some of these defences sheared and was carried inland. Considering that a recorded water level of T.P. + 0.8 m was measured just north of the study area, the wave set-up and wave height reached over 3.7 m (the difference between the top of the defence (3.0 + 1.5 m T.P.) and the tidal level (+0.8 m T.P.)). The inundation depth varied along the coastline, as summarized in Figure 4.2.2(a). Considering the height of the defences that were destroyed and the track of the storm, wind waves with a height of at least 3.2 m should have been generated during the passage of the typhoon between 2 and 4 am.

Wave hindcasting was conducted utilizing the SWAN wave model and MSM meteorological model. Although the simulated significant wave height Hs could reproduce to some extent the observed wave patterns (see the double peak in Figure 4.2.3(d)), the peak wave height (2.7 m according to Figure 4.2.3(c)) was generally less than the observed values.

Figure 4.2.4 illustrates the distribution of wind and wave fields over Tokyo Bay at 2 am, when the flooding was considered to have happened. According to Figure 4.2.4(a), a strong wind blew from the east. Given the wind profile and Hs distribution (Figure 4.2.4(b)), the wind could have been blowing the full diagonal distance across the bay from the northeast towards the southwest (Fukuura Coast), essentially representing the maximum fetch length that could be possible in Tokyo Bay. Thus, it appears that at the time when the flooding happened, Fukuura Coast was likely being affected by waves propagating from the north-east end of the bay, and others that entered through the mouth of the bay and refracted towards the northwest.

4.2.3 Generation of strong gust-winds and their potential effect on overtopping

One of the reasons why Hs was underestimated in the SWAM simulations could be due to the effect of the strong gust-winds that were generated by the typhoon, though such phenomenon is not well captured by the time-averaged meteorological data of MSM. Figure 4.2.5 shows a wind rose at two locations (Miura and Yokohama, see Figure 4.2.1(b)) on the west side of Tokyo Bay from 0 am to 6 am on September 9th (Fukuura is located

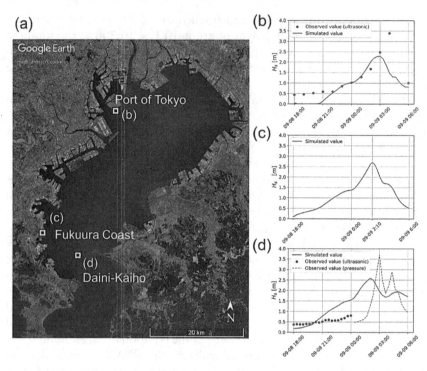

Figure 4.2.3 Time variation of *Hs* at selected locations along Tokyo Bay. For the case of (b) Port of Tokyo and (d) Daini-Kaiho, the observed records of *Hs* are also provided (recorded every 1 hour in (b), and every 20 minutes in (d). The data of the pressure gauge was retrieved from Suzuki et al. (2020).

between these two points). The data includes sustained wind speed and gust-winds every 10 min, together with the wind direction, with Figure 4.2.5(a, d) providing simulated data from MSM (Figure 4.2.4), and Figure 4.2.5(b, c, e, f) plotting the observed values retrieved from the AMeDAS-JMA data-base. The observed data shows that wind from the east (E or NE) was pre-dominant, which agrees well with Figure 4.2.5(a, d). The second dominating wind direction was from W or SW, but this was due to the blowback after passage of the typhoon around 4–6 am. According to the data of sustained wind (Figure 4.2.5(b, e)), the wind speed in the dominating direction was 10–20 m/s, while gust-winds (Figure 4.2.5(c, f)) were around 30-41 m/s. The maximum wind speed in the MSM simulation was around 23 m/s, which suggests it was closer to the recorded sustained wind than the instant gust-wind. Hence, the MSM-SWAN model is likely to have ignored the effect of strong gust-winds on wind waves in the nearshore area. Given that the SWAM model considers energy transfer from wind to waves in order to calculate *Hs*, evaluating the wave-run up and overtopping using this model would result in an underestimation of the effects of the typhoon.

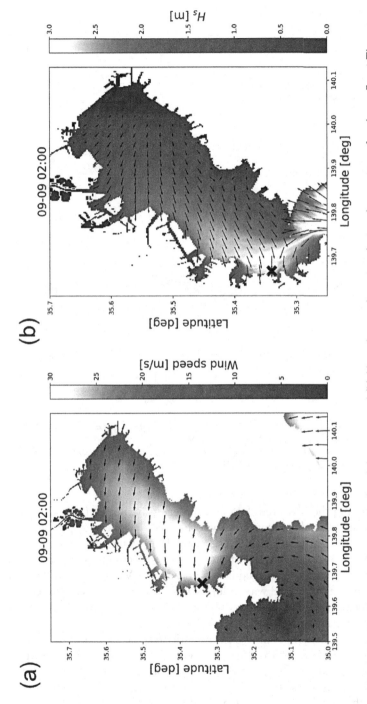

Figure 4.2.4 (a) Distribution of wind speed and direction and (b) *Hs* and wave direction during the passage of typhoon Faxai. The cross mark indicates the location of Fukuura Coast (see Inagaki et al., 2022).

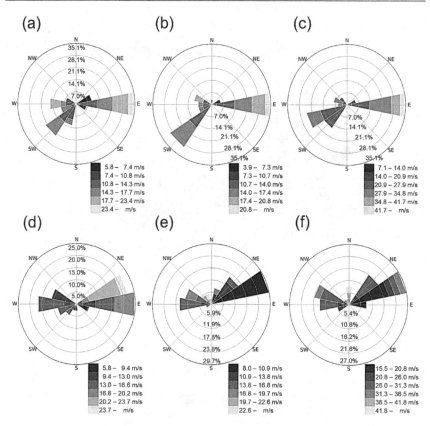

Figure 4.2.5 Wind rose at Miura and Yokohama (see Figure 4.2.1(b)); (a) wind from the MSM, (b) sustained wind, (c) gust-wind at Miura, (d) wind from the MSM, (e) sustained wind and (f) gust-wind at Yokohama (see Inagaki et al., 2022).

Considering the observed data and underestimation of *Hs* by the MSM-SWAN coupled model, it could be hypothesized that the strong gust-wind in the nearshore area helped to exacerbate the large wave run-up and overtopping rate. However, such interaction is out of the scope of the energy transfer and of the action balance equation on which SWAN is based. Therefore, to correctly evaluate wind waves during this event it is necessary to consider the transportation of water mass by momentum rather than as the process of energy transfer (Inagaki et al., 2022).

Accordingly, Inagaki et al. (2022) investigated the effect of wind on wave overtopping numerically by utilizing a hydrodynamic model (the FVM-VOF scheme). The major finding of the research was that even such advanced hydrodynamic models still had difficulty simulating highly turbulent behaviour and the interaction between air and water flows.

4.2.4 Conclusion

This chapter attempted to clarify the damage that took place along the Fukuura Coast (in Tokyo Bay, Japan) during the passage of Typhoon Faxai in 2019. Field surveys revealed the high waves caused structural damage to coastal defences, resulting in local flooding and scouring to a hill facing the sea. The hindcasted wave height using the SWAN model (a third-generation wave model) was not high enough to explain the damage observed. Considering that recorded data indicated that the typhoon had a strong wind field which was particularly intense along Fukuura Coast, the underestimation was likely due to the third-generation model failing to take into account the effect of strong gust-winds.

The effect that instantaneous winds can have on dynamic water behaviour remains a poorly understood phenomenon, which should be further investigated in the future. Considering that as a consequence of climate change typhoons could increase in intensity in the future (see Section 7.2), such strong gust-winds could lead to similar hazards taking place more frequently during the passage of other storms.

REFERENCES

Inagaki, N., Shibayama, T., Takabatake, T., Esteban, M., Maell, M., Kyaw, T.O. (2022). Increase in Overtopping Rate Caused by Local Gust-Winds during the Passage of a Typhoon. *Coastal Engineering Journal*, 64, 116–134.

Kanagawa Prefectural Government. (2021). Storm Surge Evaluation in Kanagawa Prefecture. Accessed on 23 March 2022. http://www.pref.kanagawa.jp/docs/jy2/takashio.html

Noya, H., Shibayama, T. (2019). Reproduction of Waves by Typhoon Joangdari 2018 and Analysis of High Waves at Seisho Coast. In *2019 Proc. of Annual Conference*, Japan Society for Coastal Zone Studies, I-site Namba (in Japanese).

Suzuki, T., Tajima, Y., Watanabe, M. N., Tsuruta N., Takagi, H., Takabatake, T., Suzuki, K., Shimozono, T., Shigihara, Y., Shibayama, T., Kawaguchi, S., Arikawa, T. (2020). Post-event Survey of Locally Concentrated Disaster Due to 2019 Typhoon Faxai along the Western Shore of Tokyo Bay, Japan. *Coastal Engineering Journal* 62(2), 146–158.

Chapter 5

Mechanics of tsunamis and storm surges

The driving forces and hydrodynamics

CONTENTS

DOI: 10.1201/9781003156161-5

5.1 TSUNAMIS

Takahito Mikami

Tokyo City University, Tokyo, Japan

5.1.1 Causes and generation mechanism of tsunami

A tsunami is a wave generated by the sudden vertical deformation of the boundary of a body of water, and can be generated both in the outer ocean or in enclosed water areas such as bays and lakes (see Sections 2.7 and 5.3). There are several potential causes of the deformation of land masses which can generate a tsunami. Figure 5.1.1 shows the causes of tsunami events that claimed at least one death from the year 1900 to 2020, based on data from the NCEI/WDS Global Historical Tsunami Database (National Geophysical Data Center/World Data Service n.d.). Earthquakes, landslides,

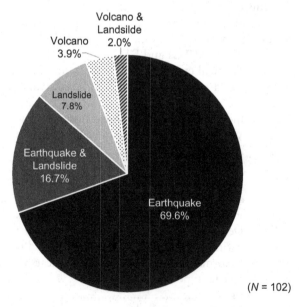

Figure 5.1.1 Causes of tsunami events 1900–2020, based on data from the NCEI/ WDS Global Historical Tsunami Database.

volcanic eruptions, or the combination of these geological activities, have all caused tsunamis in the last 120 years, though these events were predominantly caused by earthquakes.

Figure 5.1.2 shows the locations of recent major tsunami events, together with their causes. Generally speaking, most areas affected by tsunamis can be found along the coastlines facing plate boundaries, which have a potential to generate major earthquakes and also typically suffer from volcanic activity. The details of several major events are described in other chapters of this book (see Sections 2.1–2.8).

As mentioned earlier, most tsunami events in recent history were caused by earthquakes (particularly those of a submarine nature), with Figure 5.1.3

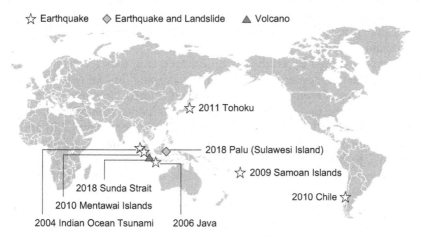

Figure 5.1.2 Locations of recent major tsunami events with their causes.

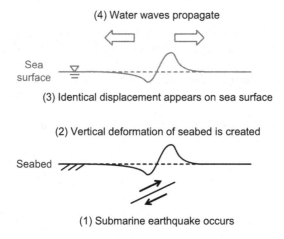

Figure 5.1.3 Generation mechanism of a tsunami caused by a submarine earthquake.

showing the generation mechanism of such events. When a submarine earthquake occurs, it sometimes results in a vertical deformation of the seabed, which pushes the water column above it and creates an identical displacement of the surface of the sea, which becomes the wave source. The waves then propagate from the source, eventually reaching coastal areas as a tsunami.

Although there are different types of earthquakes, those caused by a normal or reverse fault in a relatively shallow area (several tens of kilometres in depth) are particularly prone to generate tsunamis. An earthquake that is too deep inside the crust of the Earth is unlikely to result in a significant vertical deformation of the seabed, and is thus less likely to generate a tsunami. An earthquake resulting from a strike-slip fault is also unlikely to result in a large vertical deformation, though it has been reported that if these take place along a steep slope they can contribute to the generation of a tsunami (Tanioka & Satake, 1996).

Tsunamis caused by landslides or volcanic eruptions are relatively rare compared to those generated by an earthquake, and thus knowledge regarding these types of events is still comparatively limited. However, recent studies based on the findings from the *2018 Palu Tsunami* in Sulawesi Island (caused by an earthquake and multiple landslides, see Section 2.7) and the *2018 Sunda Strait Tsunami* (caused by the flank collapse of a volcano, see Section 2.8) have revealed some important characteristics of such events. Following these events research has intensified, and for example Takabatake et al. (2020) conducted laboratory experiments to clarify the characteristics of three different types of landslide-generated tsunamis (subaerial, partially submerged, and submarine), as observed in the *2018 Palu Tsunami*.

5.1.2 Heights of tsunami

When evaluating the impact of a tsunami to coastal areas it is necessary to know its height, which can be measured in two different ways, namely inundation and run-up heights (Figure 5.1.4). In general, the height of a tsunami is defined as the vertical distance from the estimated tide level when the wave arrived to any trace left by it. When a trace of the tsunami is found between the shoreline and the most inland point that the tsunami reached

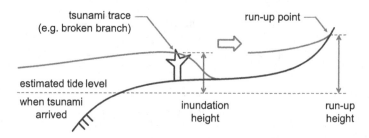

Figure 5.1.4 Definition of inundation and run-up heights due to a tsunami.

(run-up point), the height of this trace is considered the inundation height at that point (i.e., the vertical distance from the ground level below a trace is called the inundation depth). When a run-up point is identified (generally on the slope of a hill), the height of this point is considered as the run-up height at that point. By looking at the distribution of inundation and run-up heights along a coastline it is possible to understand the process in which a given tsunami penetrated inland and gradually lost its energy (for example, see the distribution of tsunami inundation and run-up heights during the *2011 Tohoku Tsunami*, as described in Section 2.6). The height of a storm surge (see Section 5.2) is also described in the same way.

There are various types of tsunami traces that can be found in affected coastal areas, such as a broken branch of a tree, a mudline left on the surface of a structure, debris left on the slope of a hill, and vegetation wilted by salty water (Figure 5.1.5). When it is difficult to find such traces (mainly due to rain after an event or when rapid recovery works erase traces), the descriptions by residents who witnessed the event can also be considered (though it should be noted that this is considered to have lower accuracy than those identified by physical traces). If a trace is significantly higher at one point than others nearby it can be judged to have taken place due to splash, and can be excluded from the records of tsunami heights at the discretion of the survey team. In many cases, measurement surveys of tsunami heights are

(a) (b) (c) (d)

Figure 5.1.5 Typical tsunami traces: (a) broken branch of a tree, (b) a mudline left on the surface of a structure, (c) debris left on the slope of a hill, and (d) vegetation wilted by salty water.

carried out using a hand-held laser ranging instrument and a surveying staff, as described in other chapters of this book (see Sections 2.1–2.3, for example) and in Esteban et al. (2017).

5.1.3 Physical characteristics of tsunami waves

During the propagation of a tsunami wave its physical characteristics, such as speed, height, and direction, can change according to the bathymetry and topography of the area. In order to adequately formulate countermeasures against a tsunami and increase the resilience of coastal settlements it is important to understand these features, including its propagation speed, shoaling and breaking, reflection, refraction, and diffraction. These are described in the following section, together with some examples regarding recent events.

5.1.3.1 Propagation speed

In the field of coastal engineering water waves are classified into three types, namely deep water, shallow water, and long waves, according to their relative water depth (h/L, the ratio of water depth to wavelength). The wavelength of a tsunami is much larger than the water depth over which it propagates, and it is therefore classified as a long wave. The propagation speed of a long wave (c) is given by the following equation:

$$c = \sqrt{gh} \qquad\qquad (5.1.1)$$

where g is gravitational acceleration and h is water depth. Essentially, this equation indicates that the propagation speed of a tsunami depends only on the water depth, which will be much higher in deeper areas of the sea. For example, if a tsunami propagates over the Pacific Ocean, where the average depth is around 4,000 m, the average propagation speed can be calculated to be around 710 km/h.

5.1.3.2 Wave shoaling and breaking

As waves arrive to shallower areas their height gradually increases, a phenomenon known as wave shoaling. As most of the times tsunamis are generated offshore, when they approach the shoreline their height will also increase due to wave shoaling. If a tsunami enters a bay (especially a v-shaped bay), this type of coastline will result in a further amplification of energy, as they become funnelled towards the inner part of the bay. During the *2011 Tohoku Tsunami* offshore GPS buoys recorded the waveforms of the incoming tsunami to be around 2.6–6.7 m (Kawai et al., 2011), though the tsunami heights measured in coastal areas (see Section 2.6) were much larger due to wave shoaling and energy concentration in bays.

There is a limit to the height a wave can reach before becoming unstable and breaking, thereby dissipating part of its energy. When the slope of the seafloor is steep waves break at a point close to the shoreline, though when the slopes are milder waves start breaking far from the coastline and the energy dissipation process takes place over a longer distance. Generally speaking, the energy dissipation due to wave breaking is not significant for the case of a tsunami, though when there is a long shallow water area in front of the coastlines (such as in the case of a wide coral reef flat), the effects of breaking and turbulence should be carefully considered. The Samoan Islands have wide coral reef flats along their shorelines, and breaking of the tsunami on top of the reef was witnessed by residents during the 2009 event (see Section 2.3). Some residents mentioned that they saw the wave breaking over the coral reef and this made them aware of the approaching danger (Shibayama et al. 2010). Numerical analysis conducted by Mikami and Shibayama (2012) showed that, if the reef flat is wide enough, the energy of the tsunami will gradually dissipate due to wave breaking, resulting in the wave height gradually decreasing toward the shoreline.

5.1.3.3 Wave reflection

When a wave arrives at a beach or hits an obstacle some amount of its energy is reflected. In particular, a vertical seawall or revetment can reflect back to the sea almost all incoming wave energy. Shibayama and Ohira (2010) numerically simulated several tsunami scenarios in Tokyo Bay and pointed out that protection structures constructed along the coast reflected the energy of the tsunamis, resulting in the wave energy being trapped inside the bay and multiple waves reaching the coast in succession.

5.1.3.4 Wave refraction

When a wave arrives at an angle to the boundary where the water depth changes, its direction will be altered in a process known as wave refraction. If the water depth gradually changes, then the direction of the wave will also gradually change as it propagates. For the case of a tsunami, wave refraction due to the bathymetry of an area becoming shallower can cause the concentration of the energy towards a specific area of the coast. For example, when a tsunami comes to a cape, the wave energy is concentrated around the head of the cape due to wave refraction, and thus higher tsunami heights can be observed there.

Refraction can also take place when the water depth suddenly drops. However, in this case, if the angle of incidence is larger than a certain limit (the critical angle), the wave energy is completely reflected at the boundary. This phenomenon is called total internal reflection, and a typical example of this can be found at the edge of a continental shelf. Along the coast of Chile there is a wide continental shelf. When a tsunami is generated on the shelf, some amount of the wave energy is reflected at the edge of the shelf, as well as along the coast. Hence, much of the wave energy can be trapped within the shelf, meaning that tsunamis can travel for a long distance along the coast of Chile. Mikami et al. (2011) reported that a high tsunami arrived even four hours after the earthquake in some of the coastal areas affected by the *2010 Chile Tsunami* (see Section 2.4).

5.1.3.5 Wave diffraction

When a wave encounters an obstacle that does not completely block it, some amount of its incident energy will be transmitted to the area behind the obstacle, in what is known as wave diffraction. For the case of a tsunami, this phenomenon can be observed when a tsunami travels around an island. In the *2009 Samoan Islands Tsunami*, a tsunami came to the islands from the south, and thus the areas worst hit were those that were in this direction. However, due to wave diffraction around the sides of the islands, the tsunami was also observed along the northern coast, though its height was smaller than along the southern coast.

5.1.3.6 Tsunami waveforms and physical characteristics

Figure 5.1.6 shows two different waveforms recorded in the *2009 Samoan Islands Tsunami* and the *2010 Chile Tsunami*. For the case of the Samoan

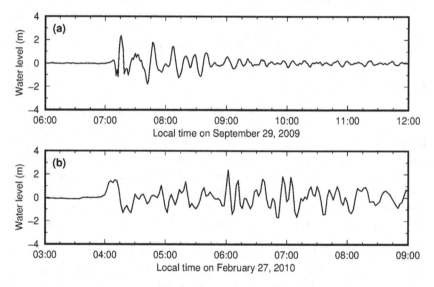

Figure 5.1.6 Waveforms recorded during tsunami events (tide levels were removed from the recorded water levels): (a) At Pago Pago, American Samoa during the 2009 tsunami, (b) at Valparaiso, Chile during the 2010 tsunami (data source: National Tsunami Warning Center (2009, 2010)).

Islands, the tsunami came from the south, and wave reflection, refraction, and diffraction near or at the coasts resulted in several waves being able to reach coastal areas. However, several hours later, once most of the wave energy travelled around the islands, water level fluctuations became very small. In the case of Chile, the tsunami was generated on the continental shelf, and much of the wave energy was trapped on the shelf because of reflection and refraction both at the coast and the edge of the shelf. The tsunami travelled back and forth on the shelf, and thus even 3–4 hours after the earthquake relatively high waves were reaching coastal areas. These examples clearly show that tsunami waveforms vary from place to place depending on the bathymetry and topography of each area.

5.1.4 CONCLUSIONS

Tsunamis can be caused by earthquakes, landslides, volcanic eruptions, or the combination of these geological activities, though those generated by earthquakes remain the most common. The main parameters used to record the height of a tsunami as it floods over land are the inundation and run-up heights. Measuring these heights in the field can provide disaster risk managers and academics with a better understanding of the behaviour of the waves in each area, allowing for the formulation of disaster countermeasures. In addition, a basic understanding of the physical characteristics of a tsunami, including its propagation speed, shoaling and breaking, reflection, refraction, and diffraction, is important to analyze where its effects might be magnified, how long it will be felt along a coast, and what type of waveform could be observed at each location.

5.2 STORM SURGES AND WIND WAVES

Ryota Nakamura
Niigata University, Niigata, Japan

5.2.1 Introduction

Storm surges and high wind waves can be generated by tropical cyclones (TCs, also known as typhoons, hurricanes, or just cyclones, depending on the region) and extra-tropical cyclones (ETCs, also known as winter storms, gales or "bomb cyclones", depended on their intensity). As shown in Sections 3.1 to 3.7, the characteristics of storm surges and high waves can be attributed to a combination of the intensity and trajectories of cyclones, together with the topographical features of the region in which they take place. This chapter will first briefly describe the characteristics of TCs and ETCs and then discuss the mechanisms of storm surge hazards. Finally, some types of unusual phenomena such as rogue waves and freak waves will also be explained.

5.2.2 Characteristics of TCs and ETCs

The major difference between TCs and ETCs can be found in the region of the planet in which they are generated, the shape and intensity of the low-pressure systems and their particular thermodynamic characteristics. The area where the cyclone forms, together with its intensity, will in turn greatly influence the size of the storm surge. The intensity of TCs and ETCs is often evaluated using their minimum sea-level pressure and maximum sustained wind velocity. In terms of shape, TCs are typically symmetrical in their physical and thermodynamic features, while ETCs show an asymmetry in horizontal wind distribution due to the existence of front structures that cause severe rainfall. However, it is worth noting that some powerful ETCs, such as "bomb cyclones", show an almost symmetrical distribution of wind velocity and sea-level pressure.

5.2.2.1 Tropical cyclones

While some recent studies employ numerical meteorological simulation models to hindcast historical TCs, older empirical and analytical models (such as Myers and Malkin, 1961; Holland, 1980) can provide useful insights into their physical structures. The spatial distribution of maximum wind velocity and sea-level central pressure are described in the analytical studies of Holland (1980), where sea-level pressure p and gradient wind V_g follow the equations given below:

$$p = p_c + (p_n - p_c)\exp(-A / r^B) \tag{5.2.2.1}$$

$$V_g = \sqrt{AB(p_n - p_c)\exp\left(-\frac{A}{r^B}\right) / \rho r^B + \frac{r^2 f^2}{4}} - rf / 2 \tag{5.2.2.2}$$

here, p_c is the central (minimum) sea-level pressure of a cyclone, p_n is the standard atmospheric pressure at the sea surface of the Earth (1013.25 hPa), A is maximum sustained surface wind velocity (m/s), r is the distance from the centre of a cyclone (m), f is the Coriolis parameter, ρ is the density of air, and B is a parameter related to the distribution of sea-level pressure.

The forward speed and wind circulation pattern inside a TC influence the surface wind velocity inside it. For example, in the northern hemisphere, wind velocity to the right side of the forward direction of the TC is much stronger than that to the left (given that the wind circulation inside TC is in the counterclockwise direction).

5.2.2.2 Extra-tropical cyclones

ETCs, sometimes called mid-latitude cyclones or winter storms, can be generated in areas where the difference between upper and lower air temperature is significant, or air pressure turbulence is amplified due to mountains and the confluence of air extremes. In Europe ETCs have a long history of generating storm surge inundation in the winter seasons (see Section 9.7). For example, in 1959, severe coastal flooding took place due to a storm surge with a height of over 3 m along the coasts of the British islands and the Netherlands, and in 2005 ETC Gudrun also generated a significant storm surge in Estonia (see Section 9.6). Examples from such weather systems in Asia include the storm surge in Nemuro described in Section 3.6.

5.2.3 Storm surge mechanisms

The fundamental physical mechanism that generates storm surges relates to the level of seawater rising due to reduced atmospheric pressure and the forcing of wind on the surface of the ocean. There are four main components of storm surges: low atmospheric pressure, wind-induced set-up, astronomical tides and wave set-up due to radiation stress associated with high waves (see Figure 5.2.1). Among the four components, the wind-induced set-up can often be the major contributor to the storm surge.

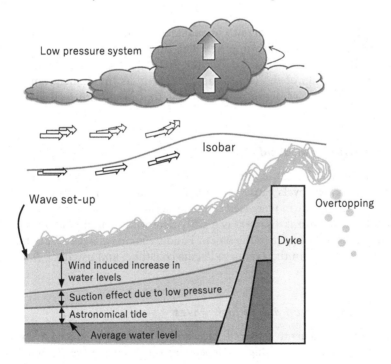

Figure 5.2.1 Storm surge components.

5.2.3.1 Increased water level due to low pressure

A low atmospheric pressure will exert less force on the surface of the water, letting the water level rise. For every 1 hPa decrease in atmospheric pressure the water level of the sea will rise by 0.99 cm:

$$\Delta h = 0.99 \left(p_n - p \right) \qquad\qquad (5.2.2.3)$$

here, Δh is the increase in water level (cm), p_n is the normal atmospheric pressure at sea level (usually, 1013.25 hPa) and p is the actual atmospheric pressure at sea level.

5.2.3.2 Wind-induced increase in water levels

Strong winds push the mass body of water towards the coastline, increasing water levels. This rise in the surface of the ocean is linearly proportional to the square of surface wind velocity:

$$\Delta h = \frac{K}{D} U^2 \qquad\qquad (5.2.2.4)$$

Here K is a parameter related to several factors: stratification of water body, bottom stress and the direction in which the wind blows, D is the depth of water and U is the wind speed (WMO, 2011). There is uncertainty regarding the estimation of the wind drag coefficient, as the momentum transfer between the air-sea boundary is a process that is still not well understood (Powel et al. 2006).

5.2.3.3 Astronomical tide

Astronomical tides can contribute to exacerbate or mitigate the effects of abnormal water levels caused by other phenomena, and events that hit at high tide can result in increased coastal flooding. The degree to which tides can influence natural disasters is greatly determined by the tidal range. For example, the tidal range in the Bay of Fundy, Canada is over 15 m, though it is only 1.0–2.0 m in the Japanese Pacific coastline, and usually 0.5 m along the Sea of Japan.

5.2.3.4 Wave set-up due to high waves

Wave set-up is generated by radiation stress, which is enhanced by progressive wave trains, and is highly dependent on the wave height and length. Among the numerous equations that have been proposed to calculate the height of wave set-up, Kato et al. (1989) provide a simple one based on the wavelength (L_o) and height (H_o) in the offshore region:

$$\frac{\xi}{H_0} = 0.052 \left(H_0 / L_0 \right)^{-0.2}$$ (5.2.2.5)

The contribution of wave set-up is generally less significant than that of other components, and is usually only in the order of several dozen cm. However, it should be noted that wave set-up is also influenced by the slope of the seafloor beneath waves, and it should always be taken into account when determining the overall storm surge.

5.2.3.5 Other factors influencing the height of a storm surge

The height of a storm surge is also influenced by the track and speed of ETCs and TCs, the bathymetry, and the shape of the coastline where the storm makes landfall. Sometimes, storm surge resonance can be observed inside a Gulf, highlighting the need to carefully consider such characteristics.

- **Track of cyclones:** The track of a cyclone is an important factor to determine the level of the storm surge. As explained earlier, in the northern hemisphere the wind velocity to the right side of the forward direction of a cyclone is stronger than that on the left side. Therefore, if the right side of a cyclone crosses over a given gulf or bay, the height of the storm surge generated in that region is likely to increase.
- **Translation speed of cyclones:** The translation speed of a cyclone influences the height of the surge that is generated, as the increase in the water level due to low pressure and wind-induced surge could be influenced by the duration of these atmospheric variables. If the translation speeds slows then the wind will act over a certain area of the sea for longer, increasing the duration of the fetch and hence the waves. However, it should also be noted that very fast TCs can also generate higher storm surges.
- **Water depth and shape of the affected coastline:** Shallower water depth can contribute to the generation of higher storm surges. According to Tatekoji et al. (2016), part of the reason behind the high water levels during the storm surge in Tokyo Bay in 1917 was due to the land reclamation activities that had been carried out around the circumference of the bay. Also, as stated earlier, the shape of the coastline can contribute to the intensification of storm surges, with semi-enclosed water areas (such as gulfs) being particularly at risk.

5.2.4 High wind waves

Ocean waves are generated by wind propagating above them, with the stronger the wind and longer the area it acts on (known as fetch) resulting in higher waves. There are a variety of ways to predict such waves, as will be explained in the following section. Also, to estimate the size of the waves that can be generated it is important to understand their propagation and deformation processes in the nearshore region.

5.2.4.1 Definitions of wind wave heights and periods

The characteristics of wind waves are often defined statistically using four parameters, namely the average wave height $\left(\bar{H}\right)$, significant wave height (H_s), 1/10 maximum wave height $(H_{1/10})$, and maximum wave height (H_{max}). H_s and $H_{1/10}$ are calculated from the top 1/3 and 1/10 percentiles of a series of waves, and $\bar{H}$ and H_{max} are the average and maximum wave heights for a given wave train, respectively. The rough statistical relationship between these parameters is given below:

$$H_s = 1.60\bar{H},\ \ H_{1/10} = 2.03\bar{H} = 1.27H_s \tag{5.2.2.6}$$

The period of the waves is also an important parameter, as those with a long period can have a particularly high impact on the coastline (wind waves with long periods being known as swell).

5.2.4.2 Estimation of oceanic wave height and period

Numerous equations have been proposed to estimate the wave heights and periods of oceanic waves. Among these equations, the SMB (Sverdruv Munk Bretschneider) method was frequently employed until numerical models became available. Three numerical schemes have been widely employed: the wave action balance equation model, mild slope equation model, and Boussinesq-type wave model. Wave action balance equation models can predict the wave height at the mesoscale level, covering a wide ocean area that is under the influence of a low-pressure system. Therefore, in this section the SMB method and the wave action balance spectrum model will be introduced.

5.2.4.2.1 SMB method

The SMB method was proposed by Sverdrup and Munk (1947), who first attempted to understand the relations between wind, sea waves and swells. The equation was later modified by Wilson (1965) in order to incorporate the observed wind velocity and wave height. The SMB equations can be used to estimate the significant wave height (H_s) and period (T_s) by using the wind velocity:

$$\frac{gH_s}{U^2} = 0.30\left\{1 - \frac{1}{\left[1 + 0.004\left(gF/U^2\right)^{1/2}\right]^2}\right\} \qquad (5.2.2.7)$$

$$\frac{gT_s}{2\pi U} = 1.37\left\{1 - \frac{1}{\left[1 + 0.008\left(gF/U^2\right)^{1/3}\right]^5}\right\} \qquad (5.2.2.8)$$

here, g is the gravitational acceleration, U is surface wind velocity (m/s), and F is the wind fetch (km). The wind fetch can also be made time-dependent to increase the accuracy of wave forecasting.

5.2.4.2.2 Wave action balance equation

The wave action balance equation can be employed to numerically simulate oceanic wave fields at the mesoscale level and capture the development of oceanic waves. Due to these merits, it is widely employed to estimate the waves generated by low-pressure systems and forms the basis of the popular third-generation nearshore wave model SWAN (developed by Delft University Technology; Booji et al. 1999).

The wave action equation is based on the relationship between changes in wave energy:

$$\frac{\partial N}{\partial t} + \nabla\left[\left(\overrightarrow{C_g} + \vec{U}\right)N\right] + \frac{\partial c_\sigma N}{\partial \sigma} + \frac{\partial c_\theta N}{\partial \theta} = \frac{S_{tot}}{\sigma} \qquad (5.2.2.9)$$

$\overrightarrow{C_g}$ is wave group velocity, $\vec{U}$ is current velocity, C_σ and C_θ are the wave velocity (σ and θ denote the relative frequency and direction of the waves, respectively). S_{tot} represents the change in wave energy in the nearshore region. Note that the wave action density is defined as $N = E(\sigma,\theta)/\sigma$ and, $\sigma^2 = gk \tanh(kd)$. k is wave number and g is gravitational acceleration.

In addition, the energy source to generate the wind-induced surface waves is given by:

$$S_{tot} = S_{input} + S_{nl3} + S_{nl4} + S_{dsp,wc} + S_{dsp,btm} + S_{dsp,brk} \qquad (5.2.2.10)$$

here S_{input} represents the increase in wave energy due to changes in wind velocity and surface pressure, $S_{dsp,\,wc}$ denotes the decrease in wave energy due to white capping, $S_{dsp,\,btm}$ is the wave height decrease due to bottom friction, $S_{dsp,\,brk}$ is the energy decrease due to wave breaking and S_{nl3}, S_{nl4} are the third-order and fourth-order non-linear interactions between surface waves.

Several methods have been proposed to consider the wave diffraction processes (e.g., Holthuijsen et al. 2003), as the wave action balance equation theoretically cannot capture this phenomenon. This means that it is difficult to simulate the wave fields behind breakwaters using this equation, and its application is restricted to meso-scale simulations.

5.2.4.3 Deformation of wind waves in the nearshore area

As waves approach the shoreline they start to feel the effect of the seabed and undergo a series of transformation processes, including refraction, diffraction, shoaling and breaking.

Refraction occurs when waves travelling towards the shallow water are propagating at an angle to the shoreline. Such a phenomenon causes the wave crest line to change its alignment as it moves over the shallow water, attempting to keep a right angle towards the shoreline. This change in the wave crest line is caused by the change in wave phase velocity, which is greatly determined by the change in water depth.

Diffraction can be observed when ocean waves propagate around islands or coastal structures such as breakwaters. The height of a diffracted wave is often smaller than the original, though it can nevertheless still cause problems inside areas that should be protected, such as ports.

Shoaling occurs when ocean waves approach a shallow area and start to suffer the effect of the coastal sea bed. As a result, their phase velocity decreases, wavelength decreases and height gradually increase. Eventually, the shoaling effect will cause the wave to become large enough to break. From a hydrodynamics point of view, breaking is caused by the water particles at the top of waves travelling faster than the speed of the rest of the wave. There are many formulas to calculate this, though in Japan that of Goda (1975) is often employed.

5.2.4.4 Special types of waves

There are several types of special waves that have the potential to cause particularly significant damage to coastal structures, coastal settlements and ships, including what are known as freak waves, infra-gravity waves and meteo-tsunamis. For instance, freak waves can be much larger than normal waves, and have been known to sink vessels. Infra-gravity waves generated a "wave bore" that caused severe damage to houses that were inundated by the storm surge due to Typhoon Haiyan (2013). Also, meteo-tsunamis generated by the interaction between a low-pressure system and long waves can flood coastal areas.

Freak wave: Freak waves (also called rogue waves) are statistically rare events generated by wave-wave interactions, though they can be much higher than the significant wave height. For example, an observed freak wave was reported to have reached seven times the height of the significant wave height in the Black Sea (Lopatoukhin and Boukhanovsky, 2004). According to the European Space Agency (ESA), satellite imagery shows that freak waves of over 30 m in height are not rare events (ESA, 2004).

Infra-gravity wave: Infra-gravity waves have periods of 25–250 s, which are longer than swells but lower than storm surges. They can lead to the rapid inundation of coastal areas and damage houses. For example, in the case of Typhoon Haiyan, a wave bore was observed along the Pacific Ocean coast of Samar Island (see Section 3.5). This wave could have been formed due to the wave-wave interaction of small-amplitude surface waves, eventually forming a distinctive bore.

Meteo-tsunami: Meteo-tsunamis induce abnormal variations in the sea surface. In the southern coasts of Kyushu in Japan, a meteo-tsunami can sometimes be over 3 m in height along coastal areas, even when strong surface winds are not observed. Although several generation mechanisms are possible, one of the major ones includes Proudman resonance, which occurs when the transient speed of a low-pressure system and an oceanic surface are almost identical and in a similar direction. This situation amplifies surface waves due to the interaction with the suction effect of the low-pressure system (as shown in Figure 5.2.2).

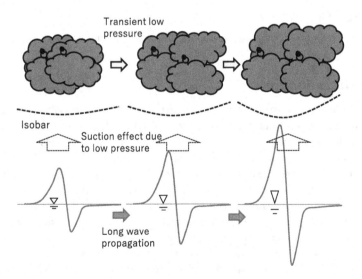

Figure 5.2.2 Mechanism behind Proudman resonance.

Proudman resonance can be calculated by using the equation below, which estimates the amplified rate α by resonance:

$$\alpha = \frac{1}{1 - V^2 / gh} \qquad\qquad (5.2.2.11)$$

where V is the velocity of the low-pressure system and h is the water depth.

5.2.5 Conclusions

This chapter briefly introduced the phenomena of ocean waves and storm surges. Storm surges are generated by a combination of several components, including low atmospheric pressure, wind-induced set-up, astronomical tides and wave set-up. High waves can be generated by the wind acting over the surface of the ocean, and their parameters can be predicted using a variety of equations. The mechanisms of wave deformation were also described in order to understand the different transformations that they undergo as they approach the nearshore area. Finally, some other types of unusual phenomena that have the potential to generate hazardous waves were also explained at the end of this chapter.

5.3 SLOSHING OF CLOSED WATER-NUMERICAL ANALYSIS OF SEISMIC WATER LEVEL OSCILLATIONS

Koichiro Ohira

Chubu Electric Power Company, Incorporated, Nagoya, Japan

5.3.1 Introduction

When ground vibrations due to an earthquake expand over an enclosed body of water (such as lakes, bays or canals) sloshing can take place, creating a flood risk that can have consequences for any vessels in it and/or anybody close to the shore (Ohira et al. 2020a). In a sense, this sloshing phenomenon is somewhat similar to that of a tsunami in that both result from an earthquake, though the former can take place in enclosed water bodies (such as lakes in a mountainous area) and the mechanism behind them is different. Figure 5.3.1 shows the location of some major sloshing events that have taken place throughout the world, with the waves generated by such events being generally limited to a maximum wave height of 1 m (Table 5.3.1). Recently, with the widespread proliferation of smartphones and cameras, some of these events have been recorded. For example, one

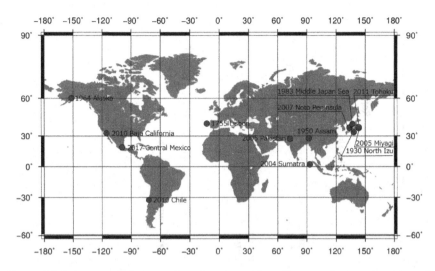

Figure 5.3.1 Location of the earthquakes that induced known sloshing in enclosed water bodies.

Figure 5.3.2 Left: Aerial photograph of Lake Saiko (taken in May 2015 by the Geospatial Information Authority of Japan). Right: Photograph of wave traces at Point A (taken in April 2011).

month after the *2011 Tohoku Earthquake and Tsunami* (see Section 2.6), Ohira and Shibayama (2012) conducted a field survey in Lake Saiko, with Figure 5.3.2 (left) showing that watermark traces indicated that a run-up height of about 0.8 m occurred at Points A and B on the lakeside. The red line in Figure 5.3.2 (right) shows a line of branches that were swept by the waves, which indicates the run-up height. Moreover, at Point C, reports indicated that a boat was washed up on the shore immediately after the *2011 Tohoku Earthquake*, and interviews with residents provided testimonies that waves with heights of up to 1 m were generated immediately after the earthquake (with those interviews indicated that the "water level of the lake slowly raised and fell").

Table 5.3.1 List of seismic water level oscillations in enclosed bodies of water reported in the past

Earthquake event	M_W	Location Country	Place	Observed water level fluctuations	Reference
1755 Lisbon	8.5–9.0 (est.)	UK Germany Norway	Loch Lomond River Trave Several lakes	0.7 m 1.2–1.5 m possibly ≥ 1 m	Kvale (1955)
1920 Gansu 1930 North Izu	7.8 (M_L) 7.8 (M_S)	Norway Japan	Several localities Lake Ashi	0.15–0.75 m (wave amplitude) 0.18 m (total amplitude)	Kvale (1955) Imamura and Kodaira (1932)
1950 Assam	8.6	UK Norway	Water reservoirs 29 Fjords	0.05 m 0.05–1 m (wave amplitude)	Kvale (1955) and McGarr and Vorhis (1968)
1964 Alaska	9.2	USA Australia	Coastal regions, lakes, reservoirs (US 850, Austalia 4 gauging points)	<0.4 m (total amplitude)	McGarr and Vorhis (1968)
1983 Middle Japan Sea	7.8	Japan	Hukaura tidal station	0.44 m	Omachi and Inoue (2001)
2002 Denali Alaska	7.9	USA	Portage Bay, Lake Union	0.08–0.16 m (estimated water wave) 0.6–3.0 m (observed run-up)	Barberopoulou et al. (2004)
2004 Sumatra	9.1–9.3	India Nepal Thailand Bangladesh	Ponds	— (eyewitness)	Manimaran (2009)
2005 Pakistan	7.6	India Bangladesh	Dam lakes, reservoirs	0.6 m	USGS (2005)
2005 Miyagi 2007 Noto Pensinsula	7.2 6.7	Japan Japan	Kugi Bay Toyama Bay	<0.1 m <0.2 m	Nagai et al. (2007) Inoue et al. (2008)

Event	Magnitude	Country	Location	Measurement	Reference
2010 Chile	8.8	USA	Lake Pontchartrain	0.15 m	Erdman (2010)
2010 Baja California	7.2	USA	Private pools	— (eyewitness)	Sheets and Zilber (2019)
2011 Tohoku	9.0	Japan	Lake Saiko	0.8 m (run-up height)	Ohira and Shibayama (2012)
			Lake Ashi	0.4 m (total amplitude)	Harada et al. (2014)
			Lake Oshioko	— (eyewitness)	YouTube (2011)
			Manaduru Port	<1.0 m	Li et al. (2017)
			Ikehara Dam Lake	— (eyewitness)	Ohira and Shibayama (2012)
			Ria coasts in Tohoku	<1.0 m	Nemoto et al. (2015)
		Greece	Gulf of Corinth	0.06–0.08 m (total amplitude)	Canitano et al. (2017)
		Russia	Lake Baikal	0.15 m (total amplitude)	Granin et al. (2018)
		China	85 wells/boreholes	−2.13 to +3.09 m (majority ±0.3 m)	Yan et al. (2014)
2012 Sumatra	8.6	Norway	Several fjords	1.20 m (total amplitude)	Bondevik et al. (2013)
2016 Sumatra	7.8	Russia	Lake Baikal	0.24 m (total amplitude)	Granin et al. (2018)
2017 Central Mexico	7.1	Russia	Lake Baikal	0.025 (total amplitude)	Granin et al. (2018)
		Mexico	Xochimilco Canals	0.5 m	Pleasance (2017)

5.3.2 Wave generating mechanism

The generating mechanism for the events described above is the same as sloshing in a water tank. Sloshing, generally speaking, refers to the movement of water in a container or a tank due to an external vibration (forced vibration). In the case of large enclosed water bodies in the vicinity of an earthquake, ground vibrations from the earthquake can push the water, and this movement induces waves similar to sloshing. It is worth noting that submarine or aerial landslides can also generate long waves in enclosed water bodies (see Section 2.7), and thus it is important to consider that many simultaneous phenomena may interact.

5.3.3 Selection of analytical method

Ohira et al. (2020a) conducted a 3D sloshing numerical analysis in order to reproduce the historical events and perform a risk assessment of the danger of sloshing due to future earthquakes, recreating their generation due to the vibrations due generated by earthquakes and the complex propagation process. The 3D sloshing analysis is a method that was previously used for liquefied gas and water tanks, which was extended and then applied to large-scale water systems like bays, lakes and canals.

OpenFOAM, an open-source numerical fluid analysis tool that enables gas-liquid two-phase fluid analysis using the volume-of-fluid method, was used for the analysis. This software employs a high-accuracy method for the analysis of sloshing, which is effective for complex shapes and fluid non-linear phenomena. The interDyMFoam solver was used, which is an analytical solver in OpenFOAM that involves a dynamic meshing function being added to an interFoam solver, which analyses two-phase isothermal, immiscible and uncompressed flows. The basic equations used are the continuity equation of an incompressible fluid and the Navier–Stokes equation, solved by discretisation using a finite volume method. By using this method, a passive oscillation of water due to ground vibration can be recreated.

5.3.4 Case examples of reproducibility of calculations

Table 5.3.2 provides a number of different examples of the assessments of different historical sloshing events in lakes, bays and canals using the method described earlier (see Ohira et al., 2020a, and Ohira et al., 2020b). For lakes, there are the examples of Lake Saiko and Lake Ashi at the time of the *2011 Tohoku Earthquake* in Japan (see Section 2.6), both situated in the mountains at an altitude of 900 m and 720 m, respectively. The 3D terrain

Table 5.3.2 Simulation conditions and results for various case studies of sloshing

Analysis case	Area	Max. depth	Earthquakes	Hypocenter dist.	Max. acc. of seismic movement	Calc. result Max. height
Lake Saiko	2.1 km²	72 m	Tohoku (Mw 9.0)	appx. 470 km	34 gal	0.7 m
Lake Ashi	7.0 km²	44 m		appx. 470 km	92 gal	0.8 m
Fjord	—	appx. 1,000 m		appx. 8,300 km	0.13 gal	0.6 m
Xochimilco Canal	—	3 m	Central Mexico (Mw 7.3)	appx. 110 km	225 gal	0.5 m
Keihin Canal	—	Const. 2.2 m	Southern Tokyo Metropolitan (Mw 7.3)	appx. 3.8 km	396 gal	0.5 m

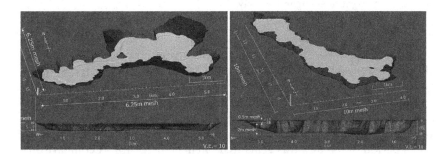

Figure 5.3.3 Terrain model for calculation. Left: Lake Saiko. Right: Lake Ashi.

models used in the calculation are shown in Figure 5.3.3. In the case of Lake Saiko, the calculated results correlate well with the reports of witnesses and the trace marks recorded (about 1 m in height above the water level). Similarly, in the case of Lake Ashi a comparison of the calculation results and observed values (Harada et al., 2014) at a sluice in the northern part of Lake Ashi is shown in Figure 5.3.4 (Left). Although there is some time lag between the observed and recorded waves, the wave heights and periods are reproduced well.

For bays, in Leikanger on the coast of the Sogne fjord, in Norway, which is about 8,300 km from the epicentre of the *2011 Tohoku Earthquake*,

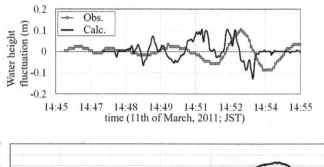

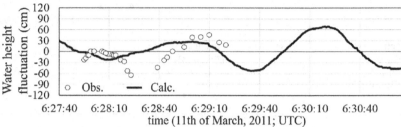

Figure 5.3.4 Comparison of the water height fluctuations between the observed and the calculated results at a specific point. Left: Lake Ashi, Japan. Right: Leikanger, Norway.

sloshing was also reported after the event (Stein et al., 2013). The abnormal water–level fluctuations associated with this phenomenon were captured on video, and the time series of the waveforms of such fluctuations could be recorded as observation values by reading the water level changes at the edge of a pier. This indicated a wave with a height of 1.2 m (total amplitude) and a period of 64–66 s was generated (Stein et al., 2013). A comparison of the observed values and the calculation results for the time-series waveform of the water level at the location where the video was recorded is shown in Figure 5.3.4 (Right) (the observed values are readings from Stein et al., 2013). The simulated waves had a height of about 0.6 m and a period of about 1–2 min, which were mostly consistent with the observed values. Additionally, as shown in the snapshot of the analysis results in Figure 5.3.5 (Left), the water surface shape during the backwash shows no indication of sharp waveforms or large breaker waves. Instead, the water level changed slowly over a long period of time.

For canals, water level oscillations were observed at the Xochimilco Canal in Mexico City during the *2017 Central Mexico Earthquake*. A video recorded there shows that standing waves with a wavelength of a few meters were generated. Figure 5.3.5 (Right) presents a snapshot of the simulated results, showing how standing waves with a wavelength of a few meters observed in the video were successfully reproduced in the simulation (Ohira et al., 2020b).

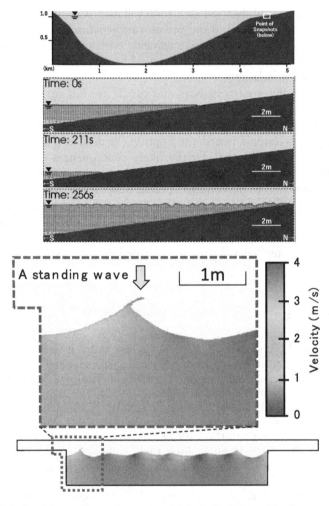

Figure 5.3.5 Snapshots of simulation results. Left: Leikanger. The figure on the top shows a general cross section of the lake, with those below snapshots at different points in time. Right: Xochimilco Canal.

5.3.5 Conclusions

The simulation analysis that was conducted indicates that, generally speaking, the waves generated by sloshing due to earthquake vibrations in enclosed water bodies are limited to a maximum wave height of 1 m. Thus, the risk of widespread casualties taking place is low, particularly when compared to normal tsunamis. However, damage from waves running ashore, drifting objects, capsizing ships, mooring line breakage, and other phenomena associated with wave run-up cannot be ignored. While no significant wave amplification due to resonance was observed in any of the simulations conducted, if the waveforms generated by sloshing and/or waves generated by an earthquake coincide with the natural frequency of the terrain and/or a canal, the waves could be amplified by resonance and produce significant damage.

Inland lakes, inner parts of bays and canals are not considered to be at risk of normal tsunamis, and hence are not normally protected against flooding, though there is some residual risk due to the sloshing phenomena described in this chapter. In order to improve the disaster prevention in these areas it is necessary to consider the likely consequences of sloshing, and design minor structures and sluice gates to take into account the contribution of this phenomenon to maximum expected water levels. Also, it is necessary to consider its effect on ships and other floating objects, which can include mooring considerations to ensure that these are not washed away or damaged. It is also important for users of such water bodies to remain alert to abnormal water level fluctuations caused by earthquakes and follow evacuation plans.

REFERENCES

Barberopoulou, A., A. Qamar, T.L. Pratt, K. Creager, and W.P. Steele. (2004). "Local amplification of seismic waves from the Denali Earthquake and damaging seiches in Lake Union, Seattle, Washington." *Geophysical Research Letters* 31: L03607. https://doi.org/10.1029/2003GL018569

Bondevik, S., B. Gjevik, and M.B. Sørensen. (2013). "Norwegian seiches from the giant 2011 Tohoku earthquake." *Geophysical Research Letters* 40(13): 3374–3378. https://doi.org/10.1002/grl.50639

Booij, N., R.C. Ris, and L.H. Holthuijsen, 1999, A third-generation wave model for coastal regions, Part I, Model description and validation, *Journal of Geophysical Research*, 104(C4): 7649–7666.

Canitano, A., P. Bernard, and S. Allgeyer. (2017). "Observation and modeling of the seismic seiches triggered in the Gulf of Corinth (Greece) by the 2011 Mw 9.0 Tohoku earthquake." *Journal of Geodynamics*, 109: 24–31. https://doi.org/10.1016/j.jog.2017.06.001

Erdman, J. (2010). "How strong & rare was quake?." The Weather Channel. Accessed on October 20, 2019. https://web.archive.org/web/20100405003158/http://www.weather.com/outlook/weather-news/news/articles/0227-quake-perspective_2010-02-27/

Esteban, M., H. Takagi, T. Mikami, L. Bahbouh, A. Becker, L. Nurse, T. Shibayama, and M. Nagdee. (2017). "How to carry out bathymetric and elevation surveys on a tight budget: Basic surveying techniques for sustainability scientists." *International Journal Sustainable Future for Human Security*, 5(2): 86–91.

European Space Agency. (2004). Ship-sinking monster waves revealed by ESA satellites. https://www.esa.int/Applications/Observing_the_Earth/Ship-sinking_monster_waves_revealed_by_ESA_satellites

Goda, Y. (1975). "Deformation of irregular waves due to depth-controlled wave breaking." *Report of Port and Harbour Research Institute*, 12(3): 31–69 (in Japanese); "Irregular wave deformation in the surf zone." *Coastal Engineering in Japan, JSCE*, 18: 13–26.

Granin, N.G., N.A. Radziminovich, M. De Batist, M.M. Makarov, V.V. Chechelnitcky, V.V. Blinov, I.A. Aslamov, R.Y. Gnatovsky, J. Poort, and S.G. Psakhie. (2018). "Lake Baikal's response to remote earthquakes: Lake-level fluctuations and near-bottom water layer temperature change." *Marine and Petroleum Geology*, 89: 604–614. https://doi.org/10.1016/j.marpetgeo.2017.10.024

Harada, M., K. Itadera, and Y. Yukutake. (2013). "Seiche at Lake Ashinoko triggered by the 2011 Off the pacific coast of Tohoku earthquake." *Bulletin of the Hot Springs Research Institute of Kanagawa Prefecture*, 46: 9–16.

Harada, M., K. Itadera, and Y. Yukutake. (2014). "Seiche at Lake Ashinoko triggered by the 2011 off the pacific coast of Tohoku earthquake." *Bulletin Hot Springs Research Institute Kanagawa Prefecture* 46: 9–16 (in Japanese).

Holland, G.J. (1980). "An analysis of wind and pressures profile in hurricane." *Monthly Weather Review*, 108(8): 1212–1218.

Holthuijsen, L.H., A. Herman, and N. Booij (2003). "Phase-decoupled refractiondiffraction for spectral wave models." *Coastal Engineering*, 49: 291–305.

Imamura, A., and T. Kodaira. (1932). "Seiches in Lake Ashinoko caused by the earthquake." *Journal of the Seismological Society of Japan*, 1(4): 57–70 (in Japanese).

Inoue, S., T. Ohmachi, and T. Takahashi. (2008). "Simulation of the 2007 Noto Hanto earthquake tsunami by using the observation data." *Proceedings of Coastal Engineering, JSCE*, 55: 341–345. https://doi.org/10.2208/proce1989.55.341 (in Japanese)

Kato, K., S. Yanagishima, T. Isogami, and H. Murakami. (1989). "Wave set-up near the shoreline -Field observation at HORF." *Report of the Port and Harbor Research Institute*, 28(1): 3–42.

Kawai, H., M. Satoh, K. Kawaguchi, and K. Seki. (2011). "The 2011 off the pacific coast of Tohoku earthquake tsunami observed by GPS buoys." *Journal of Japan Society of Civil Engineers, Ser. B2 (Coastal Engineering)*, 67(2): I_1291–I_1295 (in Japanese with English abstract).

Kvale, A. (1955). "Seismic seiches in Norway and England during the Assam earthquake of August 15, 1950." *Bulletin of the Seismological Society of America*, 45(2): 93–113.

Li, Y., K. Itadera, M. Harada, and M. Ukawa. (2017). *The Seiche Excited by the 2011 Great East Japan Earthquake at Manaduru Port, Japan, No. 52*, 181–190. Tokyo, Japan: Nihon Univ., Institute of Natural Sciences (in Japanese).

Lopatoukhin, L.J., and A.V. Boukhanovsky. (2004). "Freak wave generation and their probaliity." *International Shipbuilding Progress*, 51(2–3).

Manimaran, G. (2009). "History and geotectonic of tsunami with special reference to the Indian ocean." In *Geological Hazards: Causes, Consequences & Methods of Containment*, edited by M. Ramkumar, 37–58. New Delhi, India: New India Publishers.

McGarr, A., and R.C. Vorhis. (1968). *Seismic Seiches from the March 1964 Alaska Earthquake. U.S. Geological Survey Professional Paper No. 544-E*. Washington, DC: USGS.

Mikami, T., and T. Shibayama. (2012). "Numerical analysis of tsunami propagating on wide reef platform using turbulence model." *Journal of Japan Society of Civil Engineers, Ser. B2 (Coastal Engineering)*, 68(2): I_76–I_80 (in Japanese with English abstract).

Mikami, T., T. Shibayama, S. Takewaka, M. Esteban, K. Ohira, R. Aranguiz, M. Villagran, and A. Ayala. (2011). "Field survey of tsunami disaster in Chile 2010." *Journal of Japan Society of Civil Engineers, Ser. B3 (Ocean Engineering)*, 67(2)" I_529–I_534 (in Japanese with English abstract).

Myers, V.A., and W. Malkin. (1961). "Some properties of hurricane wind field as deduced from trajectories." NHRP Report No. 49, p. 45.

Nagai, T., A. Nozu, H.J. Lee, M. Kudaka, S. Adachi, and T. Ohmachi. (2007). "Characteristics of the observed pre-tsunami dynamic seabed pressure wave records." *J. Jpn. Soc. Civ. Eng., Ser. B* 63 (4): 368–373. https://doi.org/10.2208/jscejb.63.368 (in Japanese)

National Geophysical Data Center/World Data Service. (n.d.). "NCEI/WDS global historical tsunami database, NOAA National Centers for Environmental Information." Accessed on December 9, 2021. https://doi.org/10.7289/V5PN93H7

National Tsunami Warning Center. (2009). "Samoa Islands Region Tsunami of 29 September 2009." https://tsunami.gov/previous.events/09-29-09-Samoa/09-29-09.htm

National Tsunami Warning Center. (2010). "Offshore Maule, Chile Tsunami of 27 February 2010." https://tsunami.gov/previous.events/Chile_02-27-10/Tsunami-02-27-10.htm

Nemoto, M., T. Kito, M. Osada, and K. Hirata. (2015). "Tsunami generation by horizontal crustal deformation of the 2011 Tohoku earthquake." *Journal of Japan Society of Civil Engineers, Ser. B2 (Coastal Engineering)*, 71(2): 157–162 (in Japanese).

Ohira, K., and T. Shibayama. (2012). "The generation of waves caused by long-period ground motion at a remote location from epicenter." *Journal of Japan Society of Civil Engineers, Ser. B3 (Coastal Engineering)*, 68(2): 55–59. https://doi.org/10.2208/jscejoe.68.I_55 (in Japanese)

Ohira, K., T. Takabatake, T. Mikami, and T. Shibayama. (2020a). "Impact assessment of sloshing in bays and lakes." *Journal of JSCE*, 8(1): 13–25. https://doi.org/10.2208/journalofjsce.8.1_13

Ohira, K., T. Takabatake, M. Esteban, R. Aranguiz, and M. Mäll. (2020b). "Numerical analysis of seismic seiches in canals." *ASCE's Journal of Waterway, Port, Coastal, and Ocean Engineering*, 146(6): 04020042.

Omachi, T., and O. Inoue. (2001). "An analysis research for the feathers of sea surface fluctuations before tsunami arrivals." *Proceeding JSCE Earthquake Engineering Symposium* 26: 153–156. https://doi.org/10.11532/proee1997.26.153 (in Japanese)

Pleasance, C. (2017). "Tourist boats are tossed around as 7.1 magnitude earthquake hits Mexico leaving at least 248 people dead including 20 children after school collapsed." Mail Online. Accessed on December 29, 2019. https://www.dailymail.co.uk/news/article-4901920/Tourist-boats-tossed-Mexican-earthquake-strikes.html

Powell, M.D., P.J. Vickery, and T.A. Reinhold. (2006). "Reduced drag coefficient for high wind speeds in tropical cyclones." *Nature*, 422: 279–283.

Sheets, M., and A. Zilber. (2019). "REVEALED: Powerful 7.1 magnitude earthquake that sparked huge fires and caused multiple injuries in Southern California was 11 TIMES stronger than 6.4 tremor 32 hours earlier." Daily Mail, July 6, 2019.

Shibayama, T., and K. Ohira. (2010). "A study on tsunami behavior in Tokyo Bay and estimation of damage under earthquake of Tokyo metropolitan." *Proceedings of Civil Engineering in the Ocean, JSCE*, 26, 219–223 (in Japanese with English abstract).

Shibayama, T., T. Mikami, R. Matsumaru, H. Takagi, and F. Latu. (2010). "Field survey and analysis of tsunami disaster in the Samoan Islands." *Journal of Japan Society of Civil Engineers, Ser. B2 (Coastal Engineering)*, 66(1): 1376–1380 (in Japanese with English abstract).

Stein, B., G. Bjørn, and B. Mathilde. (2013). "Norwegian seiches from the giant 2011 Tohoku earthquake." *Geophysical Research Letters*, 40: 3374–3378.

Sverdrup, H., and W.H. Munk. (1947). *Wind, Sea, and Swell: Theory of Relations for Forecasting*. Washington, DC: U.S. Navy Hydrographic Office, No. 601 Wilson (1965).

Takabatake, T., M. Mäll, D. Chenxi, N. Inagaki, D. Kishizaki, M. Esteban, and T. Shibayama. (2020). "Physical modeling of tsunamis generated by subaerial, partially submerged, and submarine landslides." *Coastal Engineering Journal*, 62(4): 582–601.

Tanioka, Y., and K. Satake. (1996). "Tsunami generation by horizontal displacement of ocean bottom." *Geophysical Research Letters*, 23(8): 861–864.

Tatekoji, A., R. Nakamura, and T. Shibayama. (2016). "Influence of historical bathymetric changes due to urbanization on the vulnerability of storm surge in Tokyo Bay." In *Proceedings of the 35th International Conference on Coastal Engineering (ICCE)*, Antalya, Turkey.

USGS. (2005). "Today in earthquake history M7.6-Pakistan, 2005." Accessed on December 15, 2019. https://earthquake.usgs.gov/learn/today/index.php?month=10&day=8&submit=View+Date

Wilson, B.W. (1965). "Numerical prediction of Ocean waves in the North Atlantic for December 1959." *Deutsche Hydrographische Zeitschrift*, 18: 114–130.

World Meteorological Organization. (2011). "Guide to storm surge forecasting." WMO-No. 1076, p. 120. https://library.wmo.int/doc_num.php?explnum_id=7747

Yan, R., H. Woith, and R. Wang. (2014). "Groundwater level changes induced by the 2011 Tohoku earthquake in China mainland." *Geophysical Journal International*, 199(1): 533–548. https://doi.org/10.1093/gji/ggu196

YouTube. (2011). "Resonance phenomena 2011 March 11? At Lake Oshioko, Tomioka City, Gunma Prefecture." Accessed on December 13, 2019. https://www.youtube.com/watch?v=83pz7Ksb5kM (in Japanese).

Chapter 6

Mechanics of disasters and the damage to coastal settlements

CONTENTS

DOI: 10.1201/9781003156161-6

6.1 COASTAL DYKES AND FLOODING OVER LAND

Takahito Mikami

Tokyo City University, Tokyo, Japan

6.1.1 Introduction

Japan has long stretches of coastal defence structures along its shorelines to protect hinterlands against tsunamis, storm surges, high waves, and coastal erosion. The *2011 Tohoku Tsunami* caused serious damage to many of the structures along the Pacific Coast of the northeastern part of Japan. In particular, coastal dykes suffered severe damage due to the tsunami (see Section 2.6). According to the Ministry of Land, Infrastructure, Transport and Tourism (2011), 190 km out of a total of 300 km of coastal dykes in the three most affected prefectures (Iwate, Miyagi and Fukushima) were completely or partially destroyed due to the tsunami.

Figure 6.1.1 shows typical coastal dyke failures observed after the 2011 tsunami. In many cases, the damage on the landward side of the dyke was more severe than that on its seaward side. Scouring could often be seen to have taken place at the back of a dyke and this seems to have led to the subsequent collapse of the inner core and the concrete panels covering its landward side. This type of failure was mainly caused by prolonged tsunami overflow over the dykes, and hence it is important to understand the characteristics of such phenomena and the effect they can have on coastal defences.

6.1.2 Overflowing tsunami behaviour around coastal dykes

To understand how tsunamis behaved around coastal dykes during the 2011 tsunami, it is useful to analyse videos recorded during the event and the distribution of measured tsunami heights in coastal areas (The 2011

(a)   (b)

Figure 6.1.1 Typical coastal dyke failure observed after the *2011 Tohoku Tsunami* (a) in Iwanuma City, Miyagi Prefecture and (b) in Soma City, Fukushima Prefecture.

Tohoku Earthquake Tsunami Joint Survey Group, 2012). These materials are available for various locations affected during the 2011 tsunami. Here, two examples showing different tsunami behaviours will be outlined.

Figure 6.1.2 shows video snapshots that capture the tsunami flow over a coastal dyke in Oirase Town, Aomori Prefecture. This video shows that the water level gradually rose in front of the dyke, and once the water level exceeded its top the tsunami inundated the hinterland as a high-speed but shallow flow. Figure 6.1.3 shows that the measured heights behind the dyke were T.P. + 1–5 m, which were smaller than the height of the dyke (T.P. + 6.0–7.5 m). It was reported that the dyke suffered damage to its landward slope, likely caused by the high-speed flow behind it.

Figure 6.1.2 Snapshots of a video capturing the 2011 tsunami flow over a coastal dyke in Oirase Town, Aomori Prefecture.

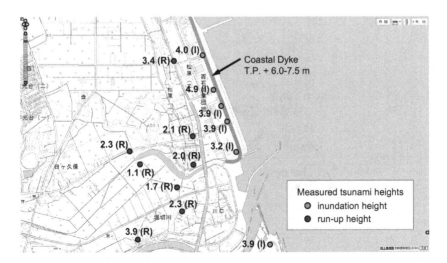

Figure 6.1.3 Distribution of measured tsunami heights around the coastal dyke in Oirase Town (base map: Geospatial Information Authority of Japan).

Figure 6.1.4 shows snapshots of another video capturing the tsunami flow over a coastal dyke in Kuji City, Iwate Prefecture. This video shows that the water overflew the dyke and accumulated in the narrow area between it and the hill, which was rapidly filled up to a level similar to that at the top of the structure. Figure 6.1.5 shows that the measured heights behind the dyke were T.P. + 11–16 m, which were almost equal to or even higher than the height of the dyke (T.P. + 12.2 m). In this area, it was reported that the dyke remained almost intact after the event, and there was no clear damage to its landward side. It can be said that the rapid accumulation of water between the dyke and the hill prevented the overflowing tsunami from becoming a high-speed flow for a prolonged time, which has the potential to cause damage to its landward side, as showcased by the case of Oirase Town.

Figure 6.1.4 Snapshots of a video capturing the 2011 tsunami flow over a coastal dyke in Kuji City, Iwate Prefecture.

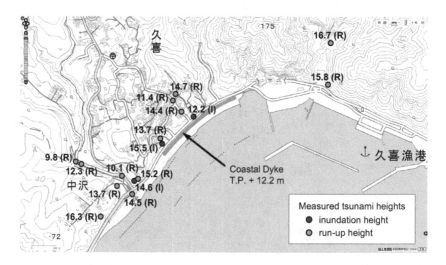

Figure 6.1.5 Distribution of measured tsunami heights around the coastal dyke in Kuji City (base map: Geospatial Information Authority of Japan).

6.1.3 Physical characteristics of overflowing tsunamis

As mentioned in the previous section, the behaviour of a tsunami around coastal dykes can be different from place to place due to the influence of the surrounding environment, which in turn can alter the patterns of coastal dyke failure. A high-speed and shallow flow behind a dyke can cause damage to the landward slope in one place, while the water gradually accumulating behind a different structure may not cause severe damage at another location. These findings show that it is necessary to consider the various possible types of flow when analysing coastal dyke failures induced by overflowing tsunamis.

Overflowing tsunamis can be classified into three types based on the classical classification of a flow over a dyke given by Hom-ma (1940), as shown in Figure 6.1.6. In type (a), the subcritical flow (slow-speed and deep flow) changes into a supercritical flow (high-speed and shallow flow) near the crown of the dyke, with this supercritical flow then running down into the hinterland. In type (b), a hydraulic jump is generated around the toe of the landward slope. In type (c), the main stream flows near the water surface and a vortex with lower velocity is generated below.

Mikami et al. (2013) investigated the physical characteristics of each type of flow by conducting laboratory experiments. Figure 6.1.7 shows the velocity fields around a dyke model for three different flow types, which were visualized through the use of a laser light. In types (a) and (b), a supercritical flow was generated along the surface of the dyke model. However, in type (c) a supercritical flow was not generated and the velocity near the surface of the dyke model was much lower than the flow around the water surface. These results indicate that, from the point of view of the velocity field, the

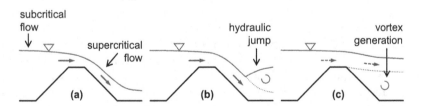

Figure 6.1.6 Three types of flow over a dyke based on the classification given by Hom-ma (1940).

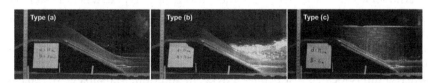

Figure 6.1.7 Velocity fields around the dyke model for three different flow types, visualized through the use of a laser light.

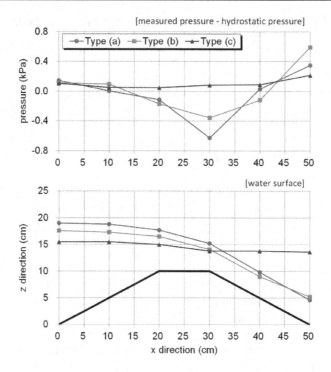

Figure 6.1.8 Distributions of pressures acting on the surface of the dyke model for three different flow types.

high-velocity flow along the surface of the dyke observed in types (a) and (b) can cause scouring failure. Figure 6.1.8 shows the distributions of pressures acting on the surface of the dyke model for the three different flow types (recorded using pressure transducers). In types (a) and (b), the centrifugal force due to the curved flow caused a different pressure profile on the landward slope than in type (c). On the top of the landward slope the pressure was lower, while at the toe of the landward slope the pressure was higher. These results indicate that the non-uniform pressure profile on the landward slope in flow types (a) and (b) can lead to the instability of concrete panels covering the core of a dyke.

The experimental results highlight that the physical characteristics of overflowing tsunamis depend on the various flow types. When a tsunami overflows a coastal dyke flow type (a) first appears. If this flow continues for a while it can cause damage to the dyke, especially to the landward side. However, when the tsunami rapidly changes to flow type (c) its potential to cause damage could be reduced. This situation can be found in an area where a hill is located just behind a dyke. In addition, Matsuba et al. (2015) conducted laboratory experiments and pointed out that a coastal forest just behind a dyke can also increase the water level and reduce the pressure exerted by the tsunami on the dyke. These findings indicate that it is necessary to consider the topography and environments around dykes and understand what types of flow can appear in order to improve their resilience against tsunamis.

6.1.4 Conclusions

During the *2011 Tohoku Tsunami* overflowing tsunamis caused severe damage to coastal dykes in many locations along the coastline of Japan. Field investigations after the event revealed that the landward sides of dykes typically suffered more severe damage than the seaward sides. To understand the characteristics of overflowing tsunamis and their physical effects on dykes, field data (videos and measured tsunami heights) were analysed and laboratory experiments were conducted. These works showed that overflowing tsunamis can be classified into three types, each with its own distinctive characteristics. If a high-speed and curved flow continues for a prolonged time it can result in the failure of the landward side of a dyke, although the type of flow can change rapidly due to the topography and environment around the structure. Thus, it is necessary to understand what types of flow can appear around dykes when assessing their potential failure mechanisms due to overflowing tsunamis.

6.2 DEBRIS LOADING

Jacob Stolle

Institut National de la Rescercher Scientifique, Québec City, Canada

Ioan Nistor

University of Ottawa, Ottawa, Canada

6.2.1 Introduction

Flood hazard assessment has generally focused on hydraulic conditions (i.e., floodplain mapping, hydraulic loading), though field investigations have shown the critical nature of secondary effects, such as erosion and debris (Ghobarah et al., 2006; Palermo et al., 2013; Robertson et al., 2007). Debris, defined as solid objects entrained within a flow, can range from small (sediments, pieces of wood, etc.) to large (ships, hydro poles, etc.), and can influence flood hazards in a variety of ways (Stolle et al., 2020a).

As a tsunami flows over land (see also Section 5.1), it accumulates within it an enormous volume of debris consisting of vegetation, building materials, automobiles, boats and ships, and in some cases entire buildings. The resulting debris flow will impact structures in its path, often leading to localized damage due to impact strikes, or overloading of structural elements or the entire structure through the effects of debris damming (see also Section 9.1 related to the structural analysis of buildings against flows).

Due to the ubiquity and range of scales in which debris can act, debris hazards can be assessed using a range of tools. This chapter will focus on a macro-scale examination of how individual debris (i.e., tracked in a Lagrangian manner) exert loads on structure. In this way, the discussion often centres around larger objects, as it is possible to identify individual debris (such as ships) and debris sources (i.e., ports). Debris hazards can also be defined by their average properties (i.e., tracked in an Eulerian frame) commonly used for debris flow by characterizing their density, though this approach will not be addressed in this chapter.

Debris loading on structures can be broadly classified into two categories: impact and damming. Debris impact is a rapid impulse force applied as the material entrained in the water flow hit a structure. Debris damming occurs due to the accumulation of debris at the face of or around structures, resulting in an increase in the drag forces exerted on them. The different time scales at which these loads act requires different considerations (i.e., structural response in the case of debris impact) from a design perspective.

The relatively rare occurrence of extreme hydrodynamic events such as tsunami-induced coastal inundation, and difficulties in determining debris sources in field investigations following such events, make determining debris motion challenging. Currently, only one detailed field investigation and subsequent analysis of debris motion following an extreme hydrodynamic event (Naito et al., 2014) has been performed. The majority of debris motion studies have been performed through experimental studies at relatively small scales. Hence, an assessment of the scale effects of these studies has yet to be addressed in order to determine the applicability of the recommendations from these smaller-scale experiments to prototype scale.

This chapter will provide several examples of debris loading taken from field investigations of tsunami events throughout the world and will then look at the theoretical background of debris impact loads and how they are currently applied within design standards. The chapter will conclude with an examination of future steps that are needed to address uncertainties in debris hazard assessment.

6.2.2 Case studies

6.2.2.1 2004 Indian Ocean Tsunami

The *Indian Ocean Earthquake* of December 26, 2004 occurred due to the rupturing of the subduction zone between the Indian plate and the Burma microplate (see also Section 2.1). The Indian plate has been moving northeast at a rate of approximately 60 mm per year, subducting under the overriding Burma microplate. The epicentre of the quake was about 155 km west of Sumatra and about 255 km south-east of Banda Aceh, Indonesia, with the focal point being at a depth of 30 km. The ruptured fault length was estimated to be 1300 km. The rupturing initiated near the south end of the fault and gradually progressed towards north, taking approximately 500–600 s. Vertical uplift reportedly ranged between 7 to 10 m at the ocean floor, displacing a huge amount of water that led to a tsunami (see Section 5.1). The earthquake was strongly felt along the south-western coast of Sumatra Island and significantly affected local communities. A team of Canadian researchers, which included one of the authors of this chapter (Nistor), conducted a 10-day forensic engineering field survey of Phuket and Phi-Phi Islands (Thailand), as well as Banda Aceh (Indonesia).

Damage along southern Phuket (Thailand) was limited to coastal erosion and partial failures of non-engineered reinforced concrete and timber frame structures along the coast. The most populated beach town along the west coast of Phuket was Patong, which suffered extensive damage to low-rise buildings. The watermark on buildings varied between 4.0 m to 6.0 m from the average sea level. The building inventory in Patong consisted of a large number of non-engineered one- to two-storey reinforced concrete hotels and timber frame shops. There were also a number of multi-storey engineered reinforced concrete buildings, where extensive damage to masonry infill walls was observed. Limited damage occurred to reinforced concrete structural elements, though significant damage was seen in timber structural elements. The entire shopping district of Patong was destroyed within an area extending approximately 2 km inland from the shore.

Nai Thon, further north of Patong on the island of Phuket, suffered extensive structural and non-structural damage to reinforced concrete frame buildings. The wave run-up was in excess of 10 m in areas between the shore and the nearby hilly terrains, which led to significant destruction. Figure 6.2.1 illustrates the extent of structural damage observed in this area.

Banda Aceh (Indonesia) is a city which had a population of about 340,000 inhabitants before the tsunami (see also Section 8.2). It was subjected to damaging forces not only due to the tsunami and its associated debris, but also the earthquake. Most casualties took place in this city itself, with coastal areas being entirely swept away by tsunami waves, leaving only piles of timber as the remains of buildings. Many non-engineered reinforced

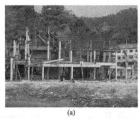

(a)　　　　　(b)　　　　　(c)

Figure 6.2.1 Structural damage in Nai Thon (Thailand) due to tsunami waves and debris impact.

Figure 6.2.2 Column failures due to debris impact in Banda Aceh (Indonesia) during the *2004 Indian Ocean Tsunami*.

concrete buildings suffered structural damage, especially to their first-floor columns. Nevertheless, it is worth noting that a large number of mosques survived the disaster, though they also suffered damage to masonry walls.

The overall extent of the destruction was impressive. It was estimated that the city lost more than half of its population as a result of the extensive tsunami flooding. Observations on column behaviour indicated many failures occurring at mid-height, especially in Banda Aceh. This was attributed to the effects of debris impact, over and above the tsunami wave pressure. Indeed, floating building remains, as well as floating large objects such as fishing boats and cars impacted on columns, causing them to fail away from the supports (Figure 6.2.2).

6.2.2.2 2011 Tohoku Tsunami

On March 11, 2011, the *Tohoku Earthquake* of Moment Magnitude (M_w) 9.0 (38.322 N, 142.369 E, depth 32 km) generated a tsunami that devastated the northeastern coast of the main island of Honshu, Japan. A tsunami reconnaissance trip by the American Society of Civil Engineers Structural Engineering Institute (ASCE SEI), in collaboration with the team led by Prof. Tomoya Shibayama (lead editor of this book) from Waseda University,

Japan, investigated and documented the performance of buildings and other structures along the Tohoku coastline of Japan, with the specific intent to apply this experience in the work to develop tsunami structural design provisions for the ASCE 7 Standard, Minimum Design Loads for Buildings and Other Structures (Chapter 6). From this event, there was a great amount of information collected that greatly enhanced tsunami design practice and its application to risk mitigation.

Floating debris was observed in virtually all of the videos capturing the incoming and outgoing tsunami flows in Japan. Often the debris concentration was higher during the outgoing flow, given the accumulation of damaged and failed structures and vegetation. Direct strikes by floating debris on structural members were hard to identify conclusively, as much of the debris had been cleared by the time we performed our survey. However, a number of sites clearly demonstrated the effects of debris strikes, even if the debris was no longer evident. For instance, the Sendai Port shipping container storage area was inundated up to a level of approximately 5–6 m above grade. All of the container stacks were dislocated and floated inland, accumulating against port buildings and other structures, with Figure 6.2.3 illustrating such impacts.

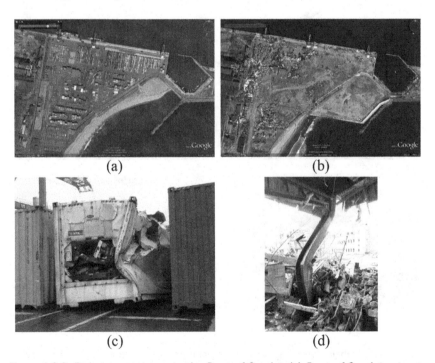

Figure 6.2.3 Debris containers in the Port of Sendai: (a) Port of Sendai prior to the 2011 tsunami; (b) Port of Sendai after the 2011 tsunami; (c) damaged container due to impact with adjacent structure; (d) damage to steel structure in Sendai Port due to container impact.

The large debris load generated by failed buildings, vegetation and other objects resulted in significant amounts of debris accumulation on the side of surviving buildings. This debris accumulation resulted in damming effects that may have increased the lateral load on the buildings, particularly if the cases where the original structure consisted of an open frame system. As buildings fail, they also became debris for other neighbouring structures.

Figure 6.2.4 show typical cases of debris damming formed around various structures. Figure 6.2.4(b) shows a common condition where structural elements from one failed building became debris load on adjacent buildings. Damage from these impacts varied from superficial to the total collapse of the impacted building. Figures 6.2.4(c) and 6.2.4(d) show examples where entire steel-framed buildings collapsed and were transported into an adjacent building. These are instances where the adjacent building was strong enough to survive this impact and damming load, but there were likely numerous conditions where this was not the case and both buildings failed.

Figure 6.2.4 Debris damming: (a) Debris damming on reinforced concrete building in Natori; (b) debris accumulation on a steel-framed building in Otsuchi; (c) debris accumulation on a surviving building in Otsuchi; (d) steel-framed building collapse leading to debris accumulation on an adjacent concrete building in Otsuchi.

6.2.2.3 2018 Indonesian Tsunami

On September 28th, 2018, an M_w 7.5 strike-slip earthquake occurred along the coast of Central Sulawesi, Indonesia (Hui et al., 2018). The earthquake's epicentre was located 95 km north from Palu City, the capital of Central Sulawesi. The cascading hazard of earthquake, tsunami, and liquefaction resulted in over 2,100 casualties and 4,500 missing, with the majority occurring within Palu City. While the specific generation events are still being debated, the tsunami waves were likely co-generated by the earthquake and associated submarine landslides (Aránguiz et al., 2020; Schambach et al., 2021). A maximum wave run-up of 10.5 m was measured on the south side of the bay (see also Section 2.7).

Figure 6.2.5(a and b) shows examples of vehicle impacts, with 6.5(a) showing the plastic crumpling of the vehicle when it impacted a structure (located outside of the figure). In the case of Figure 6.2.5(b), a truck became lodged in the concrete slab upon impact. Minor damage was induced by the vehicle to the slab and its pillar, which were built using reinforced concrete with a facade of masonry. Several cracks could be found on the masonry wall, and a concrete block in front of the building was destroyed as a result

(a) (b)

(c) (d)

Figure 6.2.5 Examples of debris loading during the *2018 Palu Tsunami*: (a) vehicle that impacted a structure (Palu City); (b) truck lodged in a non-engineered structure after impact (Mamboro); (c) boats accumulated on structural columns of a local house (Wani); (d) accumulation of debris from a local warehouse (Mamboro).

of the vehicle impact. Vehicles represent a particularly interesting case of debris impact due to safety features meant to reduce the deceleration and forces acting on the passengers. Debris impact equations (as will be discussed later) assume an elastic-plastic model, where the impact force is limited by the elastic response of the debris. However, this is not realistic in the case of vehicles, which deform plastically (Figure 6.2.5(a)) to dissipate some forces during accidents.

Figure 6.2.5(c–d) shows examples of debris damming during the tsunami event. Figure 6.2.5(c) shows a specific example in Wani Harbour, where residents noted the formation of a debris dam by ships washed in from the harbour. The debris dam had a significant impact on the damage patterns in the community. There was severe damage to structures close to the harbour due to the high inundation depths there, with the degree of damage rapidly reducing further inland. Interviews with residents indicated that the debris accumulated across the narrow roadways leading inland perpendicular to the approaching tsunami wave front, reducing the inundation depths on the leeside of the debris dams significantly.

Figure 6.2.5(d) shows another example of debris accumulation at a warehouse area in Mamboro. The light steel frame construction forming the warehouse was largely destroyed, and goods inside the warehouses were found accumulated at several points in front of structural elements. It is possible that debris dams could contribute to the progressive failure of structures throughout a flooding event, though the temporal aspect of this process is not well understood and requires further future research.

6.2.3 Theoretical background

6.2.3.1 Debris impact

The focus of debris impact modelling in hydraulic engineering has emphasized the role of individual impacts. In deriving an analytical approach for estimating the maximum impact loading, the most common methodology uses a spring-mass Single Degree-Of-Freedom (SDOF) system (Haehnel and Daly, 2004). The model is based on the assumption that the structure will be rigid and the impact duration short (i.e., damping is not important). The SDOF system of equations can be solved in a variety of manners, but most commonly, and assuming a linear relationship between penetration depth and normal force, the maximum impact force can be calculated by using

$$F = u\sqrt{km} \tag{6.2.1}$$

where u is the impact velocity, m is the mass of the debris, and k is the debris stiffness.

Another common approach is the use of the Hertzian contact model (Hertz 1882), where the deformation is assumed to occur at the face of the structure (as opposed to being distributed throughout the structure). Ikeno et al. (2013) proposed the following equation to estimate the impact force of a single tsunami-driven debris:

$$F = 0.243 \left(C_M m\right)^{\frac{3}{5}} u^{\frac{6}{5}} \left(\frac{1}{\pi \left(K_s + K_d\right)}\right)^{\frac{2}{5}}$$

(6.2.2)

where C_M is the inertia coefficient with $C_M = 1 + C_0$, C_0 is the added mass coefficient, K is the material stiffness defined as $K = 1 - \upsilon^2/\pi E$, υ is Poisson's ratio, and E is the elastic modulus. While the parameters between the two models are similar, the major difference is related to the relationship between the impact velocity and force.

These equations consider only the response of the debris to the impact; Stolle et al. (2018b) showed that the response of the structure can also have a significant influence on the impact force. Khowitar et al. (2014) looked into the response of the structure, examining the longitudinal impact by a pole on a column governed by Timoshenko's (1914) beam theory. The impacting pole is governed by the one-dimensional wave equation (Paczkowski et al., 2012), and is considered to be sufficiently short that the deformation of the beam is dominated by shear. Stolle et al. (2019b) further showed the importance of considering the structural response by extending Eq. (6.2.1) to consider the structure as a second mass–spring system, particularly in examples where the stiffness of the structure and debris are of similar magnitude.

The equations above represent idealized debris impacts on structures, though significant deviation has been observed in experimental results due to the impact geometry. Haehnel and Daly (2004) identified that the mass of the debris does not uniformly act around the impact point, resulting in the rotation or redirection of the debris. This is commonly referred to as obliqueness (β), where the debris velocity vector is at an angle from the impact vector, and eccentricity (e), where the impact point occurs out of line with the centre-of-gravity (CG) of the debris (inducing rotation). Derschum et al. (2018) showed the impact geometry was particularly relevant in transient flow conditions where 3D flow features near the structure cause significant lateral motion of the debris just before impact. Stolle et al. (2020b) investigated the influence of multiple debris impact structures as agglomerations and showed that a correction could be made to Eq. (6.2.1) for the additional mass and concentration of the debris within the flow.

6.2.3.2 Debris damming

Debris damming loads are often considered in the context of drag force, tending to act as a quasi-static load (Yeh et al., 2013). Hydraulic researchers have primarily investigated the formation of debris dams at bridge piers in steady-state conditions. However, Stolle et al. (2018a) found that a similar phenomenon also occurs in transient flow conditions. Parola et al. (2000) found that the drag coefficient was dependent on the blockage ratio (the fraction of the total unobstructed cross section that is blocked by the debris dam) and on the Froude (Fr) number. In transient flow, Stolle et al. (2018a) determined that the Reynolds (Re) also needs to be considered.

While the debris dam influences the loads exerted on a structure, the formation of the dam can also have secondary effects that must be considered in the design process. The constriction of the flow path results in backwater rise, potentially overtopping flood protection structures adjacent to those at the location of interest (Schmocker and Hager, 2013). Stolle et al. (2018a) determined that the run-up could be estimated using the Bernoulli equation by assuming the kinetic energy of the flow was transferred to potential energy (in confined flow cases). Debris dam-induced flow constrictions also cause flow accelerations underneath and downstream of the dam (Pagliara and Carnacina, 2013), which can results in significant scouring (see also Section 6.3).

6.2.4 Design standards

Several design standards address debris loading, though the concept is similar between the different methodologies. To simplify, this chapter will only look at the newest standard: the American Society of Civil Engineers Chapter 6 Tsunami Loads and Effects (ASCE, 2016). The ASCE 7 applies debris impact loads to buildings where the minimum inundation depth exceeds 0.91 m, with the forces being applied to the perimeter gravity-load carrying structural components perpendicular to the inflow and outflow directions specified elsewhere in the code. Such loads are to be applied to points critical for flexure and shear within the inundation depths. All buildings will be designed for impact by floating wooden poles, logs, vehicles, tumbling boulders and concrete debris; and buildings in proximity to a port or shipping yard should have the potential for strikes from shipping containers, ships and barges determined. The estimation of the impact loads follows Eq. (6.2.1), except for vehicles and boulders, where a constant value is assigned based on previous research.

The ASCE 7 does not explicitly address debris damming, though the influence of debris accumulation is mentioned within hydrodynamic loads. A common practice for mitigating hydrodynamic loading is to design structures with breakaway walls, which reduce the drag loads by reducing the cross-sectional area of the structure. The ASCE 7 sets a lower limit to the reduction in cross-sectional area as a result of debris accumulating and filling the sections where the walls break away, with the loading exerted on the structure being calculated using the drag force equation.

6.2.5 Discussion

Debris hazards have been increasingly identified as critical considerations in the design of tsunami resilient infrastructure, and this chapter outlines some of the fundamental considerations in assessing such loads. While the basic processes have been clarified, more research is needed to address uncertainties, particularly related to the type of debris as well as secondary effects (Nistor et al., 2017b).

As mentioned earlier, vehicles are commonly found within the debris after major flooding events. The majority of research has focused on positively buoyant, elastic debris. Vehicles are neutrally buoyant, resulting in a deeper and unequal (due to the engine being situated at the front) draft, and the added mass coefficient (caused by the deceleration of the surrounding fluid) could potentially increase the load compared to experiments that do not consider the surrounding fluid (Arrighi et al., 2015; Shafiei et al., 2016). Additionally, vehicles deform plastically upon impact, which differs significantly from the models described earlier. This deformation would result in a reduction of the impact force that is not currently captured in existing models.

Another aspect that is not captured within current design standards is the coupling between debris and flow conditions. For example in the warehouse in Mamboro, in Palu (Indonesia) described earlier, the high concentration of debris could have potentially increased the resistance to the flow, therefore reducing the wave front velocity (Stolle et al., 2020a). Flow resistance from obstructions, such as buildings, has been well documented through experimental and numerical modelling (Goseberg et al., 2009; Stolle et al., 2018a). The ASCE 7 accounts for high concentrations of suspended sediment and smaller debris in the drag force equation through a fluid density factor, which increases the density of water by a factor of 1.10. However, this is not captured in the flow resistance, which is based on the physical modelling of a built environment (which uses clear water) (Bricker et al., 2015). The other example is Wani Harbor, where debris damming resulted in blockage and subsequent rise of water levels in the port. This two-way coupling between debris and flow conditions has so far not been adequately explored in physical or numerical modelling (Nistor et al., 2017a; Stolle et al., 2019a).

6.2.6 Conclusions

The present chapter presents a brief overview of debris impact loads due to tsunamis, with a focus on existing standards which consider the presence of such loads. For extreme hydrodynamic conditions, in particular, the focus has been on the determination of the maximum debris impact loads. The methods currently used by the design guidelines/standard discussed are based on the one-degree-of-freedom model for a single debris impacting a rigid structure, with each method using simplifying assumptions to calculate the maximum loads. There are multiple solutions to the debris impact model, with each solution requiring the debris impact velocity and mass. Additionally, each solution requires a variable that is difficult to quantify: stopping distance, stopping time, or effective stiffness.

While the impact of a single debris on a rigid structure has been addressed, there are still many aspects of debris loading that are not well understood. In several extreme hydrodynamic events field observations have noted large concentration of debris (Takahashi et al., 2010). Therefore, simultaneous multiple debris impacts or impacts of agglomerations of debris may and most probably do occur. The multiple debris impacts are most certainly not well represented by the single debris impact formulas, and the momentum transfer between the debris could result in significantly larger impact forces than those provided by current models. Additionally, the agglomeration of the debris at the front face of the structure likely results in increased hydrodynamic loads. Future research is expected to address the remaining complex issues related to debris impact which have yet to be properly understood and quantified.

6.3 LOCAL AND OVERTOPPING SCOUR

Ravindra Jayaratne
University of East London, London, UK

6.3.1 Introduction

Tsunami-induced scour has been identified as one of the most critical failure modes of coastal defence structures and buildings damaged by recent tsunami events such as the *2004 Indian Ocean* and the *2011 Tohoku Earthquake and Tsunami* (Kato et al., 2012; Bricker et al., 2012; Jayaratne et al., 2013, see also Sections 2.1 and 2.6). Tsunami-induced scour can trigger failure in isolation or, in combination with overflow-induced sliding or bore impacts, by weakening the resisting force of the foundation of a given structure. Scour can occur when the plunging jet impacts the landward toe of the coastal structures, digging out a hole that can expose its foundation and eventually even the core. This hole can also cause landside slope armour to fail, leaving the core material vulnerable to erosion. Similarly, rapid-drawdown-induced liquefaction can exacerbate local scour and therefore increase the likelihood of failure in a building. Therefore, the accurate prediction of local and plunging scour depths is of paramount importance to formulate resilient structural designs against a given tsunami scenario.

In order to accurately predict scour depths, there is a need for both measured data and also the development of practical and/or numerical models to replicate/predict such laboratory experiments or field observations.

Rajaratnam (1981), Sumer et al. (1992), Stein et al. (1993), Noguchi et al. (1997), Dehghani et al. (2010), Ca et al. (2010), Yamamoto et al. (2011), Kato et al. (2012), Jayaratne et al. (2016), Link et al. (2016), Suzuki et al. (2018), Larsen et al. (2018) and Jordan et al. (2019) studied local and over-topping scour using laboratory experiments of river and coastal structures under steady, non-steady flows and dam-break conditions. While Lida et al. (2016) and Larsen et al. (2017, 2018) developed comprehensive numerical models for time-varying scour depth predictions, Noguchi et al. (1997), Tonkin et al. (2003) and Dehghani et al. (2010) established mathematical models for local and plunging-jet scour depths that could be beneficial for practising coastal engineers.

Due to the limitation in the range of applicability of the models mentioned above and the inherent complexity of some of them, Jayaratne et al. (2016) and Nicholas et al. (2016) developed two simple practical tsunami scour depth predictive models for coastal structures and buildings, respectively. Fieldwork in the aftermath of the *2011 Tohoku Tsunami* and a series of small-scale 2D laboratory experiments carried out at the hydraulic wave flume of the University of East London (UEL) were used to develop these two models, which will be detailed in this chapter.

6.3.2 Field observations

Jayaratne et al. (2013) summarize the results of two field surveys conducted in 2011 and 2012, as part of a research collaboration between the University of East London (UEL) and Waseda University. These surveys covered armoured breakwaters, earth embankments, concrete seawalls and sea dikes in Soma, Yamamoto, Watari, Iwanuma, Higashimatsushima and Ishinomaki cities. In the report of the field surveys the authors detail linear measurements (geometry) of damaged structures, scour depth and extent, recorded geographical information (GPS data), and provide digital photos and sketches of scour profiles. Additionally, sand samples were taken at the location of the scour failure at each location. Jayaratne et al. (2013) found that the most probable mode by which coastal structures failed in seven locations of the Miyagi and Fukushima prefectures was due to tsunami-induced scour at the leeward toe.

Figure 6.3.1 illustrates examples of landward toe scour failure at a sea-wall, a dyke and two buildings located along the coast during the *2011 Tohoku Tsunami*. This failure was initiated by the leeward toe scour which was caused by the massive tsunami overflow (Figure 6.3.1(a)). Successive wave action generated a large scour hole and, eventually, the leeward primary armour was scattered due to the loss of support. Overflowing wave pressure and super-critical flow conditions undermined the body and some parts of the dike were completely washed away due to hydrodynamic forces that were exerted through the open parts of the structure (Figure 6.3.1(b)). Photographic evidences of damaged buildings were gathered in the Tohoku

Figure 6.3.1 (a) Field observations and photographic evidence of tsunami-induced scour failure on coastal infrastructure following the *2011 Tohoku Tsunami*; (a) leeward of a seawall in Ishinomaki East, (b) landward of the coastal dike in Iwanuma (Jayaratne et al., 2013), (c) Seaward edge of a building in Minami-Sanriku (courtesy of Takayuki Suzuki, Yokohama National University), and (d) exposed building foundation in Natori (courtesy of Takayuki Suzuki, Yokohama National University).

region from published information (e.g., MLIT, Japan) in order to explore the tsunami-induced scour damage to the seaward side of the buildings (Figure 6.3.1(c), (d)).

6.3.3 Existing scour depth predictive models

There are analytical and empirical predictive models of plunging jet, pier and rapid-drawdown-induced liquefaction scour depths under steady, unsteady and tsunami waves, developed by several researchers. Rajaratnam (1981),

Sumer et al. (1992), Stein et al. (1993), Noguchi et al. (1997), Dehghani et al. (2010) and Yamamoto et al. (2011) performed laboratory experiments on plunging jet scour around hydraulic structures (e.g., seawalls, vertical cylinders) to establish empirical and analytical relationships. The researchers who conducted experiments on pier scour are Nadal et al. (2010), Ca et al. (2010), Link et al. (2016) and Larsen et al. (2018). Among them, the Colorado State University (CSU) scour depth model (as suggested by Nadal et al., 2010) is widely used in scour depth estimation at bridge piers. Rapid drawdown is known to enhance scour due to the reduction of inter-granular effective stress that occurs in the presence of a pore pressure gradient. In this regard, Tonkin et al. (2003) established a rapid-drawdown-induced lique-faction scour depth predictive model for a large cylindrical structure at the shoreline on a uniform slope for solitary waves.

6.3.4 Laboratory experiments at the University of East London

6.3.4.1 Case study I: Leeward toe scour (overtopping) at coastal structures

In order explore the tsunami-induced overtopping scour at laboratory-scale, Jayaratne et al. (2016) conducted a series of fully controlled experiments on four coastal dike models at the University of East London's (UEL) hydraulic flume. Two of the model geometries were fabricated to replicate the damage to coastal defences at Iwanuma and Soma cities by the *2011 Tohoku Tsunami* (see Figure 6.3.2), including a sand layer at the landward region (see Figure 6.3.3).

Based on the field and laboratory scour data and the work of Mizutani and Imamura (2002) and Matsutomi et al. (2010), Jayaratne et al. (2016) established a new scour depth predictive model by setting up non-dimensional tsunami parameters and geometry of the dike (inundation height, h, height, H_{d2} and angle, θ_2 of the leeward slope, see Figure 6.3.4) using Buckingham π theorem and regression analysis. They observed from the trend of the field and laboratory data that a decreasing exponential function seemed to exist between the relative scour depth and the inverse of the

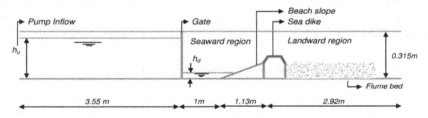

Figure 6.3.2 Schematic sketch of the dam-break scour experimental set-up at UEL.

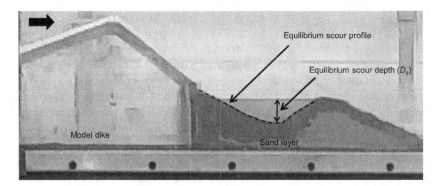

Figure 6.3.3 Example of an equilibrium scour hole at the toe of a model dike (Jayaratne et al., 2016).

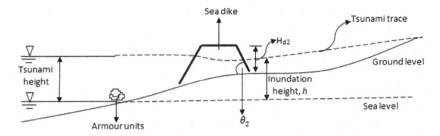

Figure 6.3.4 Diagrammatic representation of scour depth model parameters.

relative maximum overflowing pressure (Figure 6.3.5). The proposed new scour depth predictive model is given below:

$$\frac{Ds}{H_{d2}} = \lambda\left(\exp-\left[\frac{\sqrt{H_{d2}}}{2.5\lambda\sqrt{h}\,\sin\theta_2}\right]\right) h > H_{d2} \quad 10^{-4} < K < 10^{-3} \quad (6.3.1)$$

where Ds is the representative scour depth, λ is a fitted coefficient which was found to be 0.85, and K is the permeability coefficient. An interesting feature of this new model is that it includes measurable parameters in the field, such as height of structure measured on the leeward face (H_{d2}), inundation height (h) and the leeward slope angle (θ_2), making it a simple practical model for practising coastal engineers.

6.3.4.2 Case study II: Seaward toe scour (local) at building foundations

As mentioned in the introduction of this chapter, one of the dominant failure modes of buildings that were damaged by the *2011 Tohoku Tsunami* was scour. Contrary to leeward toe scour at coastal dikes, many researchers

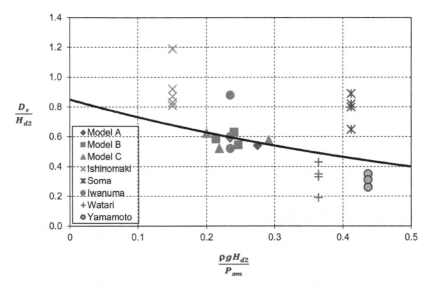

Figure 6.3.5 Relationship between relative scour depth and maximum overflowing pressure using field and laboratory scour data (Jayaratne et al., 2016).

(MLIT, Japan; EEFIT, 2011) observed that scour holes at the seaward corners of building foundations contributed to both moderate and severe damage (Figure 1(c), (d)). Nicholas et al. (2016) collected measurements of scour depths and lengths, geometry of the foundations of the damaged building, tsunami inundation heights and sediment characteristics at various locations along the Japanese coastlines from published information. They then developed a novel scour depth predictive model using the Buckingham π theorem and regression analysis, by considering relative scour depth, shear velocity, frictional properties of the soil surface and geometry of the building. Finally, they obtained a power relationship for the local scour depth (z) at the seaward corner of the building foundations in terms of measurable quantities in the field (see Figure 6.3.6).

$$\frac{Z}{B_h} = \lambda \left[\frac{gB_h}{CV^2} \right]^{-1} \tag{6.3.2}$$

where B_h is the half seaward width of the building, C is Darcy–Weisbach friction factor, V is the tsunami flow velocity [$=\sqrt{(gh)}$], h is the inundation height, λ is a fitted coefficient which is found to be 16, and g is the gravitational acceleration.

The 2011 Tohoku post-tsunami field survey of Chock et al. (2013) indicated that lateral pressure, uplift pressure and surge flow from the

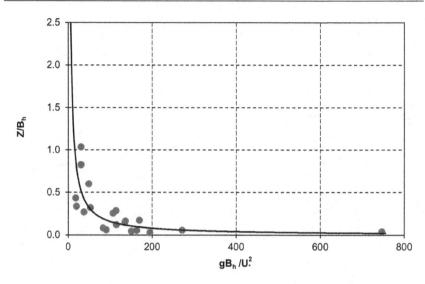

Figure 6.3.6 Relationship between relative scour depth and shear velocity $(=\sqrt{(CV)^2})$ from field datasets (Nicholas et al., 2016).

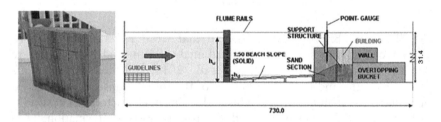

Figure 6.3.7 Tested rectangular model building (left), scour experimental set-up at UEL (right) (dimensions in cm) (Jayaratne et al., 2018).

tsunami-induced hydrodynamic forces were responsible for the destruction of many structures along the Tohoku coastline. The associated hydrodynamic parameters of a tsunami, such as pressures and velocities, also influence the tsunami-induced scour around buildings. In order to further explore the influence of various structure geometries (e.g., rectangular and circular) through small-scale laboratory experiments, Jayaratne et al. (2018) conducted a series of comprehensive small-scale experiments in order to reproduce a damaged rectangular building in Minamisanriku, Miyagi Prefecture, following the *2011 Tohoku Tsunami*. The experimental set-up was similar to Jayaratne et al. (2016), as shown in Figure 6.3.7. Table 6.3.1 illustrates hydraulic conditions (upstream and downstream water depths before overflowing the model dike) used in scour experiments at UEL.

Table 6.3.1 Three hydraulic conditions used in experiments at UEL (h_u and h_d are defined in Figure 6.3.2)

Hydraulic condition	h_u (m)	h_d (m)	$\alpha_i \left(= \dfrac{h_d}{h_u} \right)$
1	0.3	0.015	0.050
2	0.3	0.030	0.100
3	0.3	0.050	0.167

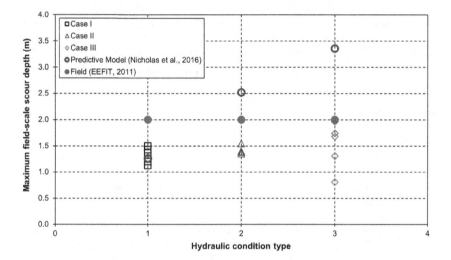

Figure 6.3.8 Comparison results of local scour depths of a building foundation under various hydraulic conditions used at UEL's hydraulic laboratory (Mugnaini, 2018).

For the case of hydraulic condition Type 1, the laboratory-measured maximum local scour depth and the value obtained from the scour depth predictive model of Nicholas et al. (2016) (8.8 mm and 8.4 mm, respectively) were comparable with the field-scale (1:150) value of 1.3 m, as illustrated in Figure 6.3.8. EEFIT (2011) measured the scour depth around a Minamisanriku building apartment and reported the value to approximately 2 m. For the case of hydraulic conditions Types 2 and 3, the predictive scour model over-predicts when compared against the results of experimental scour depths. This discrepancy was attributed to the fact that the experimental set-up was two-dimensional, which limited to some degree the scour obtained, though they are of the same order of magnitude as the field measurements of EEFIT (2011).

It is recommended to repeat the same hydraulic conditions at a larger geometrical scale in a 3D wave basin in order to reproduce the scour-induced tsunami damage to the building at Minamisanriku. Such experiments could result in a higher accuracy of measurements of the scour depths at the seaward corners of the building, which is expected to be closely comparable to the predictive model values of Nicholas et al. (2016).

6.3.5 Conclusions

Based on post-tsunami field surveys in Japan following the *2011 Tohoku Tsunami* and small-scale laboratory experiments at the University of East London (UEL), two new scour predictive models (overtopping scour depth by Jayaratne et al., 2016 and local scour depth by Nicholas et al., 2016) were developed, which are applicable as simple practical models for scour depth estimation. The interesting point in both models is that they were composed of measurable quantities in the field. Therefore, this work can be used by practising engineers. It is expected that the present work can be further improved with similar field measurements from future tsunami events.

REFERENCES

Aránguiz, R. et al., 2020. The 2018 Sulawesi tsunami in Palu city as a result of several landslides and coseismic tsunamis. *Coastal Engineering Journal*, 62(4): 445–459.

Arrighi, C., Alcèrreca-Huerta, J., Oumeraci, H., and Castelli, F., 2015. Drag and lift contribution to the incipient motion of partly submerged flooded vehicles. *Journal of Fluids and Structures*, 57: 170–184.

ASCE, 2016. *ASCE/SEI 7-16: Minimum design loads and associated criteria for buildings and other structures*. American Society of Civil Engineers, Reston, VA.

Bricker, J.D., Francis, M., and Nakayama, A., 2012. Scour depth near coastal structures due to the 2011 Tohoku tsunami. *Journal of Hydraulic Research*, 50(6): 637–641.

Bricker, J.D., Gibson, S., Takagi, H., and Imamura, F., 2015. On the need for larger manning's roughness coefficients in depth-integrated tsunami inundation models. *Coastal Engineering Journal*, 57(02): 1550005.

Ca, V.T., Yamamoto, Y., and Charusrojthanadech, N., 2010. Improvement of prediction methods of coastal scour and erosion due to tsunami back-flow. In *20th International Offshore and Polar Engineering Conference*, Beijing, pp. 1053–1060.

Chock, G., Robertson, I., Kriebel, D., Francis, M., and Nistor, I., 2013. *Tohoku Japan Tsunami of March 11, 2011 Performance of Structures*. ASCE Publications, Reston, VA.

Dehghani, A.A., Bashiri, H., and Meshkati, S.M.E., 2010. Local scouring due to flow jet at downstream of rectangular sharp-crested weirs. In E.D. Schmitter, N. Mastorakis (eds.). *Water and Geoscience 2010, 5th IASME WSEAS*, Cambridge, UK, pp. 127–131. WSEAS Press, Athens, Greece.

Derschum, C., Nistor, I., Stolle, J. and Goseberg, N., 2018. Debris impact under extreme hydrodynamic conditions part 1: Hydrodynamics and impact geometry. *Coastal Engineering*, 141: 24–35.

Earthquake Engineering Field Investigation Team (EEFIT), 2011. The Mw 9.0 Tohoku Earthquake and Tsunami of 11th March 2011, Field Report, Institution of Structural Engineers, UK, p. 204.

Ghobarah, A., Saatcioglu, M., and Nistor, I., 2006. The impact of the 26 December 2004 earthquake and tsunami on structures and infrastructure. *Engineering Structures*, 28(2): 312–326.

Goseberg, N., Stahlmann, A., Schimmels, S., and Schlurmann, T., 2009. Highly-resolved numerical modeling of tsunami run-up and inundation scenario in the city of Padang, West Sumatra. In *Proceeding of the 31st Int. Conference on Coastal Engineering*.

Haehnel, R.B., and Daly, S.F., 2004. Maximum impact force of woody debris on floodplain structures. *Journal of Hydraulic Engineering*, 130(2): 112–120.

Hom-Ma, M., 1940. Discharge coefficients of low overflow weirs (part 1). *Journal of the Civil Engineering Society*, 26(6): 635–645 (in Japanese).

Hui, G. et al., 2018. Linkage between reactivation of the Sinistral strike-slip faults and 28 September 2018 MW7. 5 Palu earthquake, Indonesia. *Science Bulletin*, 63(24): 1635–1640.

Ikeno, M., Kihara, N., and Takabatake, D., 2013. Simple and practical estimation of movement possibility and collision force of debris due to tsunami. *J. JSCE, B2 (Coastal Eng.)*, 69(2): 861–865.

Jayaratne, R., Mikami, T., Esteban, M., and Shibayama, T., 2013. Investigation of coastal structure failure due to the 2011 Great Eastern Japan Earthquake Tsunami. In *Coasts, Marine Structures and Breakwaters Conference, ICE*, Edinburgh.

Jayaratne, R., Premaratne, B., Abimbola, A., Mikami, T., Matsuba, S., Shibayama, T., Esteban, M. and Nistor, I., 2016. Failure mechanisms and local scour at coastal structures induced by tsunamis. *Coastal Engineering Journal*, 58(4): 1640017.

Jayaratne, R., Nicholas, M., Ghodoosipour, B., Mugnaini, S., Nistor, I., and Shibayama, T., 2018. Tsunami-induced hydrodynamics and scour around structures. In *36th International Conference on Coastal Engineering*. ASCE, Maryland.

Jordan, C., Jayaratne, R., Mugnaini, S., and Nicholas, M., 2019. Tsunami induced scour on structures. In *9th Coastal Sediments Conference*, Tampa, pp. 2521–2532.

Kato, F., Suwa, Y., Watanabe, K., and Hatogai, S., 2012. Mechanics of coastal dike failure induced by the Great East Japan Earthquake tsunami. In *33rd International Conference on Coastal Engineering*. ASCE, Santander.

Khowitar, E., Riggs, H.R., and Kobayashi, M.H., 2014. Beam response to longitudinal impact by a pole. *Journal of Engineering Mechanics*, 140(7): 04014045.

Larsen, B.E., Arbøll, L.K., Kristoffersen, S.F., Carstensen, S., and Fuhrman, D.R., 2018. Experimental study of tsunami-induced scour around a monopile foundation. *Coastal Engineering*, 138: 9–21.

Larsen, B.E., Fuhrman, D.R., Baykal, C., and Sumer, B.M., 2017. Tsunami-induced scour around monopile foundations. *Coastal Engineering*, 129: 36–49.

Lida, T., Kure, S., Udo, K., Mano, A., and Tanaka, H., 2016. Prediction of the 2011 Tohoku tsunami scouring near structures. *Journal of Coastal Research* 75: 872–876.

Link, O., Castillo, C., Pizarro, A., Rojas, A., Ettmer, B., Escauriaza, C., and Manfreda, S., 2016. A model of bridge pier scour during flood waves. *Journal of Hydraulic Research*, 55: 310–323.

Matsuba, S., Mikami, T., and Shibayama, T., 2015. Effect of coastal forest in reducing overflowing tsunami force around coastal dikes. *Journal of Japan Society of Civil Engineers, Ser. B2 (Coastal Engineering)*, 71(2): I_871–I_876 (in Japanese with English abstract).

Matsutomi, H., Okamoto, K., and Harada, K., 2010. Inundation flow velocity of tsunami on land and its practical use. In *32nd International Conference on Coastal
Engineering*, ASCE. Shanghai.

Mikami, T., Matsuba, S., and Shibayama, T., 2013. Fluid motion around coastal
dyke due to overflowing tsunami. *Journal of Japan Society of Civil Engineers, Ser.
B2 (Coastal Engineering)*, 69(2): I_991–I_995 (in Japanese with English abstract).

Ministry of Land, Infrastructure, Transport and Tourism, 2011. White paper on
land, infrastructure, transport and tourism in Japan, 2010. https://www.mlit.go.jp/
english/white-paper/2010.pdf

Mizutani, S., and Imamura, F., 2002. Design of coastal structure including the
impact and overflow on tsunamis. *Proceedings of Coastal Engineering, JSCE*,
49: 731–735.

Mugnaini, S., 2018. Laboratory modelling of tsunami-induced scour on tsunami-
induced scour on building structures, MSc thesis, University of East London, 99.

Nadal, N.C., Zapata, R.E., Pagan, I., Lopez, R., and Agudelo, J., 2010. Building
damage due to riverine and coastal floods. *Journal of Water Resources Planning
and Management, ASCE*, 136(3): 327–336.

Nicholas, M., Jayaratne, R., Suzuki, T., and Shibayama, T., 2016. Predictive model
for scour depth of coastal building failures due to tsunamis. In *35th International
Conference on Coastal Engineering*, ASCE, Antalya.

Nistor, I. et al., 2017a. Experimental investigations of debris dynamics over a horizontal plane. *Journal of Waterway, Port, Coastal, and Ocean Engineering*, 143(3):
04016022.

Nistor, I., Goseberg, N., and Stolle, J., 2017b. Tsunami-driven debris motion and
loads: A critical review. *Frontiers in Built Environment*, 3: 2.

Noguchi, K., Sato, S., and Tanaka, S., 1997. Large-scale model experiment of wave
overtopping and frontal surface scouring of revetment due to tsunami run-up.
Annual Journal of Coastal Engineering, 44: 296–300 (in Japanese).

Paczkowski, K., Riggs, H., Naito, C., and Lehmann, A., 2012. A one-dimensional
model for impact forces resulting from high mass, low velocity debris. *Structural
Engineering and Mechanics*. 42(6): 831–847.

Pagliara, S., and Carnacina, I., 2013. Bridge pier flow field in the presence of
debris accumulation. In *Proceedings of the Institution of Civil Engineers-Water
Management*, pp. 187–198.

Palermo, D., Nistor, I., Saatcioglu, M., and Ghobarah, A., 2013. Impact and damage
to structures during the 27 February 2010 Chile Tsunami. *Canadian Journal of
Civil Engineering*. 40(8): 750–758.

Parola, A.C., Apelt, C.J., and Jempson, M.A., 2000. *Debris forces on highway
bridges*. Transportation Research Board.

Rajaratnam, N., 1981. Erosion by plane turbulent jets. *Journal of Hydraulic
Research*, 19(4): 339–358.

Robertson, I.N., Riggs, H.R., Yim, S.C., and Young, Y.L., 2007. Lessons from hurricane katrina storm surge on bridges and buildings. *Journal of Waterway, Port,
Coastal, and Ocean Engineering*, 133(6): 463–483.

Schambach, L., Grilli, S.T., and Tappin, D.R., 2021. New high-resolution modeling of the 2018 palu tsunami, based on supershear earthquake mechanisms and
mapped coastal landslides, supports a dual source. *Frontiers in Earth Science*,
8: 627.

Schmocker, L., and Hager, W.H., 2013. Scale modeling of wooden debris accumulation at a debris rack. *Journal of Hydraulic Engineering*, 139(8): 827–836.

Shafiei, S., Melville, B.W., Shamseldin, A.Y., Beskhyroun, S., and Adams, K.N., 2016. Measurements of tsunami-borne debris impact on structures using an embedded accelerometer. *Journal of Hydraulic Research*, 54(4): 435–449.

Stein, O.R., Alonso, C.V., and Julien, P.Y., 1993. Mechanics of jet scour downstream of a headcut, *Journal of Hydraulic Research*, 31(6): 723–738.

Stolle, J. et al., 2018a. Experimental investigation of debris damming loads under transient supercritical flow conditions. *Coastal Engineering*, 139: 16–31.

Stolle, J. et al., 2019a. Debris transport over a sloped surface in tsunami-like flow conditions. *Coastal Engineering Journal*, 61(2): 241–255.

Stolle, J. et al., 2020a. Engineering lessons from the 28 September 2018 Indonesian tsunami: Debris loading. *Canadian Journal of Civil Engineering*, 47(1): 1–12.

Stolle, J., Derschum, C., Goseberg, N., Nistor, I. and Petriu, E., 2018b. Debris impact under extreme hydrodynamic conditions part 2: Impact force responses for non-rigid debris collisions. *Coastal Engineering*, 141: 107–118.

Stolle, J., Goseberg, N., Nistor, I., and Petriu, E., 2019b. Debris impact forces on flexible structures in extreme hydrodynamic conditions. *Journal of Fluids and Structures*, 84: 391–407.

Stolle, J., Nistor, I., Goseberg, N., and Petriu, E., 2020b. Multiple debris impact loads in extreme hydrodynamic conditions. *Journal of Waterway, Port, Coastal, and Ocean Engineering*, 146(2): 04019038.

Sumer, B.M., Christiansen, N., Fredsoe, J., 1992. Time scale of scour around a vertical pile. In *2nd International Offshore and Polar Engineering Conference*, San Francisco, pp. 308–315.

Suzuki, K., Seki, K., and Arikawa, T., 2018. Study on estimation of scouring behind the breakwater. In *36th International Conference on Coastal Engineering*, ASCE, Maryland.

The 2011 Tohoku Earthquake Tsunami Joint Survey Group, 2012. Survey dataset (release 20121229). http://www.coastal.jp/ttjt/

Timoshenko, S., 1914. Zur frage nach der wirkung eines stoßes auf einen balken. *Zeitschrift fur Mathematik und Physik*, 62: 1315–1318.

Tonkin, S., Yeh, H., Kato, F., and Sato, S., 2003. Tsunami scour around a cylinder. *Journal of Fluid Mechanics*, 496: 165–192.

Yamamoto, Y., Charusrojthanadech, N., and Nariyoshi, K., 2011. Proposal of rational evaluation methods of structure damage by tsunami. *Annual Journal of Coastal Engineering*, 67(1): 72–91 (in Japanese).

Yeh, H., Sato, S., and Tajima, Y., 2013. The 11 March 2011 east japan earthquake and tsunami: Tsunami effects on coastal infrastructure and buildings. *Pure and Applied Geophysics*, 170(6–8): 1019–1031.

Chapter 7

Numerical simulations and future predictions

CONTENTS

DOI: 10.1201/9781003156161-7

7.1 TSUNAMI SIMULATION

Takahito Mikami

Tokyo City University, Tokyo, Japan

7.1.1 Introduction

As the capacity of computers continues to increase, large-scale numerical simulations have become a powerful tool to analyse complex physical phenomena. Tsunami simulations are currently widely used for a variety of purposes, including tsunami hazard mapping, estimating damage patterns or evaluating forces on coastal structures. In many cases a two-dimensional tsunami simulation is employed, which generally consists of estimating the initial conditions around the source of the wave and then calculating its propagation and inundation based on the governing equations and bathymetry/topography data. Although there are several ways to conduct such simulations, the present chapter will describe the basic manner in which they can be done.

7.1.1.1 Governing equations

The governing equations of fluid motion consist of the equation of continuity (derived from the mass conservation law) and the equations of motion (derived from Newton's laws of motion). Two-dimensional tsunami motion is usually represented by the non-linear shallow water equations (or non-linear long wave equations), which can be obtained by integrating the governing equations of three-dimensional incompressible viscous fluid motion in the vertical direction. The non-linear shallow water equations are as follows:

$$\frac{\partial \eta}{\partial t} + \frac{\partial M}{\partial x} + \frac{\partial N}{\partial y} = 0 \tag{7.1.1}$$

$$\frac{\partial M}{\partial t} + \frac{\partial}{\partial x}\left(\frac{M^2}{D}\right) + \frac{\partial}{\partial y}\left(\frac{MN}{D}\right) + gD\frac{\partial \eta}{\partial x} + \frac{gn^2}{D^{7/3}}M\sqrt{M^2+N^2} = 0 \tag{7.1.2}$$

$$\frac{\partial N}{\partial t} + \frac{\partial}{\partial x}\left(\frac{MN}{D}\right) + \frac{\partial}{\partial y}\left(\frac{N^2}{D}\right) + gD\frac{\partial \eta}{\partial y} + \frac{gn^2}{D^{7/3}}N\sqrt{M^2+N^2} = 0 \tag{7.1.3}$$

where η is the vertical displacement of the water surface above the still water level, h is the still water depth, D is the total water depth (equals to $h + \eta$), t is time, x and y are horizontal coordinates, M and N are the x and y components of the discharge per unit width, g is the acceleration of gravity, and n is Manning's roughness coefficient. Eq. (7.1.1) is the equation of continuity and Eqs. (7.1.2) and (7.1.3) are the equations of motion for the x and y directions, respectively. For a deep ocean area the non-linear and bottom friction terms can be neglected, and thus the linear shallow water equations are sometimes used. The linear shallow water equations consist of Eq. (7.1.1) and the following:

$$\frac{\partial M}{\partial t} + gh\frac{\partial \eta}{\partial x} = 0 \tag{7.1.4}$$

$$\frac{\partial N}{\partial t} + gh\frac{\partial \eta}{\partial y} = 0 \tag{7.1.5}$$

In some cases additional terms, such as the Coriolis force terms, dispersion terms, and horizontal mixing terms, are included in the governing equations. When simulating a tsunami travelling a long distance (for example from Chile to Japan), it is necessary to rewrite the governing equations so that the spherical coordinate system is used instead of the Cartesian one.

7.1.1.2 Numerical method

To conduct a simulation using the governing equations mentioned above a numerical method is needed to solve them. The leap-frog method (Goto & Ogawa, 1982) is usually used to conduct two-dimensional tsunami simulations. In this method, η, M, and N are discretized with a staggered grid in both time and space, as shown in Figure 7.1.1. The method is an

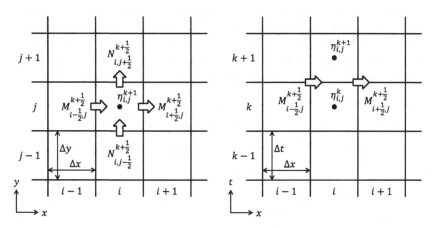

Figure 7.1.1 Discretized values with a staggered grid (Goto & Ogawa, 1982).

explicit method, and thus requires the following condition for obtaining stable results:

$$\frac{\Delta s}{\Delta t} \geq C \qquad\qquad (7.1.6)$$

where Δs is the spatial grid size, Δt is the time step, and C is the wave celerity. This is known as the CFL (Courant-Friedrichs-Lewy) condition.

7.1.1.3 Initial conditions

Various types of geological activities (such as submarine earthquakes, submarine landslides and volcanic eruptions) can generate tsunamis, though the most common events are those caused by earthquakes (see Section 5.1). When a submarine earthquake occurs it can cause a vertical deformation of the seabed, which creates an identical displacement of the sea surface above it. Then, this displacement of the sea surface becomes the source of the tsunami waves.

The deformation of the seabed caused by a submarine earthquake can be calculated by using the methods presented by Mansinha & Smylie (1971) or Okada (1985), which provide the initial conditions of the sea surface (η at $t = 0$). This calculation requires assumptions regarding the fault parameters of the earthquake, including its length, width, dislocation, dip angle, slip angle, and strike angle (see Figure 7.1.2), as well as its location (latitude/longitude and depth). If an earthquake is modelled using multiple faults, the initial condition can be obtained by superimposing the deformation caused by each fault segment. The initial conditions of the discharge (M and N at $t = 0$) are usually set to zero.

7.1.1.4 Boundary conditions

As some of the waves in a simulation may reach the offshore boundaries of the computational domain, free transmission conditions are usually used as the boundary conditions. If a simulation only focuses on propagation, then the boundaries between land and sea are considered as vertical walls, which completely reflect incoming waves. However, if a simulation includes

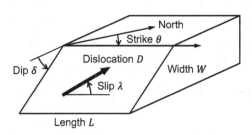

Figure 7.1.2 Fault parameters.

inundation over land then moving boundary conditions are applied to the boundaries between land and sea. In this case, the Manning's roughness coefficient of each grid of the land should be set according to the land use pattern at that point.

7.1.1.5 Bathymetry/topography data

Tsunami simulations require bathymetry/topography data that covers the entire computational domain, with one of the most widely used datasets being the gridded data of the General Bathymetric Chart of the Oceans (GEBCO). The latest one is the GEBCO_2021 Grid (GEBCO Compilation Group, 2021), which covers the entire surface of the Earth by a 15 arc-second interval grid. For the case of Japan, a finer gridded dataset is provided by the Global tsunami Terrain Model (GtTM) (Chikasada, 2020), which details the seas around Japan (E117-155, N20-49) by a 2 arc-second interval grid.

7.1.2 Application examples of two-dimensional tsunami simulation

There are many examples of research that has applied the use of two-dimensional tsunami simulations. Mikami et al. (2014) conducted a numerical simulation of the *2010 Mentawai Islands Tsunami* (see also Section 2.5) using linear shallow water equations to estimate the tsunami arrival time at the affected coastlines. As shown in Figure 7.1.3, on the western coastlines

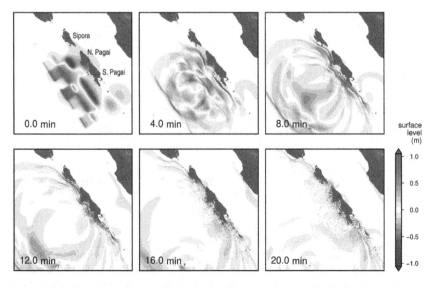

Figure 7.1.3 Results of a numerical simulation of the *2010 Mentawai Islands Tsunami*, see Mikami et al. (2014).

of the islands the sea level lowered and then the first tsunami wave arrived 10–20 min after the earthquake occurred. The results indicate that disaster education and evacuation drills are of great importance to people in these islands, given that it is very difficult to effectively disseminate a tsunami warning throughout every village before the arrival of the first tsunami wave.

Shibayama & Ohira (2010) and Takabatake & Shibayama (2012) conducted numerical simulations of various tsunami scenarios with non-linear shallow water equations to identify the most vulnerable areas along the coastline of Tokyo Bay. The results show that a tsunami of over 1 m in height could reach even the innermost part of the bay and some of the ports and river mouths in Tokyo could suffer inundation, highlighting the importance of preparation against tsunamis.

7.1.3 Three-dimensional fluid dynamics simulation

The general propagation of a tsunami from the source to coastal areas can be analysed by using two-dimensional tsunami simulations. However, when it is necessary to understand more detailed tsunami behaviour, such as wave breaking over shallow areas or the flow around coastal structures, three-dimensional simulations can be employed. Three-dimensional incompressible viscous fluid motion is governed by the following equations, namely the equation of continuity and Navier–Stokes equations:

$$\frac{\partial u}{\partial x} + \frac{\partial v}{\partial y} + \frac{\partial w}{\partial z} = 0 \qquad (7.1.7)$$

$$\frac{\partial u}{\partial t} + u\frac{\partial u}{\partial x} + v\frac{\partial u}{\partial y} + w\frac{\partial u}{\partial z} = X - \frac{1}{\rho}\frac{\partial p}{\partial x} + v\left(\frac{\partial^2 u}{\partial x^2} + \frac{\partial^2 u}{\partial y^2} + \frac{\partial^2 u}{\partial z^2}\right) \qquad (7.1.8)$$

$$\frac{\partial v}{\partial t} + u\frac{\partial v}{\partial x} + v\frac{\partial v}{\partial y} + w\frac{\partial v}{\partial z} = Y - \frac{1}{\rho}\frac{\partial p}{\partial y} + v\left(\frac{\partial^2 v}{\partial x^2} + \frac{\partial^2 v}{\partial y^2} + \frac{\partial^2 v}{\partial z^2}\right) \qquad (7.1.9)$$

$$\frac{\partial w}{\partial t} + u\frac{\partial w}{\partial x} + v\frac{\partial w}{\partial y} + w\frac{\partial w}{\partial z} = Z - \frac{1}{\rho}\frac{\partial p}{\partial z} + v\left(\frac{\partial^2 w}{\partial x^2} + \frac{\partial^2 w}{\partial y^2} + \frac{\partial^2 w}{\partial z^2}\right) \qquad (7.1.10)$$

where (u, v, w) are the components of the velocity, p is the pressure, ρ is the density, v is the kinematic viscosity, and (X, Y, Z) are the components of the external force. Although these equations can be solved directly, this requires a very small grid size to resolve the turbulent nature of flow. Therefore, turbulence models, which represent the effects of turbulence in the flow, are usually employed in the governing equations of three-dimensional simulations. RANS (Reynolds-Averaged Navier–Stokes equations) and LES (Large Eddy Simulation) are the two most widely used turbulence models.

Mikami & Shibayama (2012) applied LES with the CIP (Constrained Interpolation Profile) method to tsunami wave-breaking over a wide shallow reef flat. During the *2009 Samoan Islands Tsunami* (see also Section 2.3), local residents saw the wave breaking over the reef flat behind the edge of the reef, and this made them aware of an approaching danger. As shown in Figure 7.1.4 at some point along the reef flat the height of a wave starts decreasing, indicating that wave breaking occurs. The results highlight that tsunami wave breaking should be considered in coastal areas which have a shallow reef flat in front of them.

Recently, open source software which can be used for three-dimensional fluid dynamic simulations has become available. OpenFOAM is a widely used model which has been used by a number of studies applied to simulate complex tsunami behaviour. In OpenFOAM the governing equations are solved with the finite volume method, and the volume of fluid (VOF) method is used for capturing the free surface of the flow. Iimura et al. (2021) analysed the tsunami behaviour around two upright sea dikes with different heights using OpenFOAM. Ishii et al. (2021) investigated tsunami flows around different building layouts also using OpenFOAM. These studies

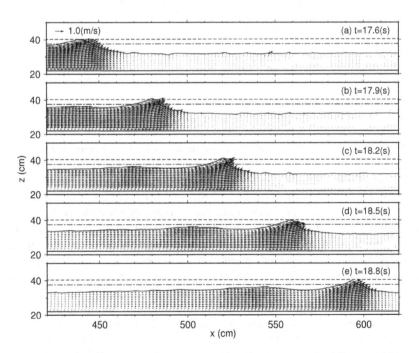

Figure 7.1.4 Velocity vectors of a wave propagating over a reef flat, see Mikami & Shibayama (2012).

illustrate how this type of software can complement the findings from three-dimensional laboratory experiments and contribute to provide a better understanding of the detailed characteristics of tsunamis as they propagate over land.

7.1.4 Conclusions

In recent times computers allow researchers to conduct large-scale tsunami numerical simulations. To analyse the tsunami propagation and inundation from the source area to coastal areas two-dimensional simulations based on the non-linear or linear shallow water equations are often used. The results of this type of simulations can help to estimate tsunami arrival time and wave heights along the coastline according to a variety of scenarios. To analyse more detailed tsunami behaviour along shallow coastal areas or land, three-dimensional simulations can be applied, which usually employ turbulence models such as RANS and LES. The results of such three-dimensional simulations can provide a better understanding of tsunami flows and forces, which can assist in formulating appropriate preparation and mitigation strategies against future tsunamis.

7.2 STORM SURGE SIMULATION AND GLOBAL WARMING

Ryota Nakamura

Niigata University, Niigata, Japan

7.2.1 INTRODUCTION

Similar to tsunamis, storm surges are categorized as long waves. Therefore, non-linear long wave equations that are solved by using a leap-frog scheme (Goto et al. 1997), usually used for the simulation of tsunamis, can also be applied to simulate storm surges. However, storm surges also exhibit the characteristics of ocean currents, as the increase in water level is partially driven by strong currents that are accelerated by strong surface winds. Thus ocean circulation models can be more adequate for the accurate simulation of storm surges. Indeed, significant improvements in storm surge simulations have been made after ocean and coastal circulation models such as ADCIRC (ADvanced CIRCulation; Luettich and Westerink, 1991) and FVCOM (Finite Volume Community Ocean Model; Chen et al. 2003) were developed. Also, the use of unstructured grid models such as ADCIRC and FVCOM can contribute to the accuracy of the simulation of such oceanic disturbances around complex coastal geometries, due to the flexibility they offer in simulating them.

7.2.1.1 Storm surge simulation schemes

The accurate calculation of storm surges can be achieved by using a combination of several numerical models, each simulating one of its four different components (wind-induced setup, pressure surge, wave set-up and astronomical tide, see Figure 7.2.1). First of all, a numerical model for simulating the ocean circulation is needed in order to reproduce the wind-induced setup and atmospheric pressure fields (including the calculation of the low pressure at the centre of the storm). Then, the astronomical tide should be simulated at the open boundaries of the coastal and ocean circulation model. Finally, wave setup can be included by using radiation stress calculated from sea surface wave models coupled with ocean circulation models.

7.2.1.1.1 Atmospheric numerical simulation model

Atmospheric simulations of weather systems can be performed using meteorological numerical models such as ARW-WRF (Advanced Research Weather Research and Forecasting; Skamarock et al., 2008). Surface meteorological variables, such as wind velocity and sea-level pressure, simulated from WRF, can be employed for forcing the boundary of the ocean circulation (storm surge) model. However, the simulation of extra-tropical and tropical cyclones can be difficult when attempting to obtain precise values of the surface wind velocity and low pressure at the centre of the storms, due to uncertainty in simulating the track and intensity of cyclones. Thus, finding the appropriate simulation settings is of critical importance, and in ARW-WRF the choice of microphysics and planetary boundary layer schemes can have a big influence on the intensity and trajectories of cyclones. Nakamura and Shibayama (2016a,b) indicated how the different combinations of cloud microphysics schemes (calculating the phase shift among water, ice and vapour contained in the cloud) and planetary boundary layers (calculating

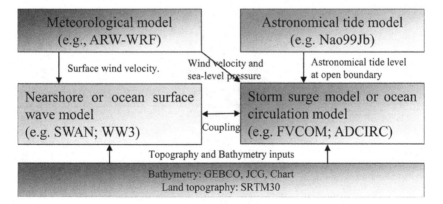

Figure 7.2.1 Typical arrangement of simulation methods for calculating storm surges.

the meteorological variables at the surface of the earth) can yield different results regarding the intensity and trajectories of the simulated storm. This highlights the importance of choosing appropriate physical schemes for the simulation of tropical cyclones.

7.2.1.1.2 Ocean circulation models (storm surge models)

Simulating storm surges with the use of unstructured grid model is of critical importance in order to consider a complex coastal geometry and to take into account the conservation of mass, as the water column is typically piled onto coastal areas by the surge. Unstructured numerical models such as ADCIRC and FVCOM have become popular for the simulation of storm surges, with the results obtained by using either of these models being comparatively similar to each other.

These models are typically run at the mesoscale level, over an area of between 100 km to 1000 km and at a computational grid resolution of 100 m to 10 km. From the beginning of the 21st century such models have often been used to conduct street-scale inundation simulations, in order to understand the characteristics of flows between buildings and the vulnerability of the residents of coastal settlements (e.g., Nakamura et al. 2019).

7.2.1.1.3 Mesoscale simulation

Meso-scale simulations can be carried out using atmospheric simulations. Figure 7.2.2 shows the unstructured computational grids employed to estimate the storm surge at Nemuro (Nakamura et al. 2019, see also Section 3.6), covering a part of the Japanese Archipelago, the northwestern part of the Pacific Ocean and the Sea of Okhotsk. Smaller grids were employed around areas where the slope of the bathymetry is steep, while

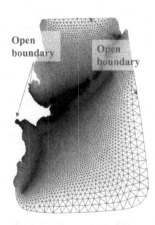

Figure 7.2.2 Unstructured grids for the storm surge simulation at Nemuro, Japan.

wide grids can be set in areas where precise outputs are not needed in order to economize computational resources.

Furthermore, as described in Section 5.2, the selection of an appropriate air sea drag coefficient is of great importance to calculate the wind-induced setup in Nemuro Bay. Many formulas have been proposed to estimate the momentum transfer between the air and sea surface, including those by Wu (1982), Honda and Mitsuyasu (1980) and Powell (2006). When calculating the air-sea drag coefficient, the density of the surface air has some influence on the wind stress (as the low air pressure at the centre of a cyclone results in lower air density, possibly leading the lower wind stress, Nakamura et al. 2020).

7.2.1.1.4 Street scale simulation

Street-scale simulations can be carried out by using the result of meso-scale simulations. The location, height and shape of structures located in the coastal area can be obtained from local authorities or OpenStreetMap. Coastal buildings imported from OpenStreetMap can be reproduced at different computational resolutions, with their locations being similar to those shown by aerial photos from Google Earth (Figure 7.2.3). A high resolution

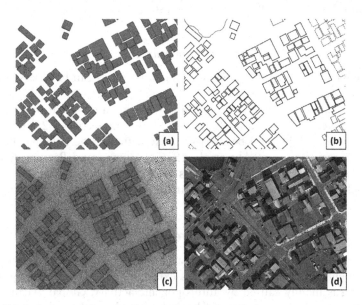

Figure 7.2.3 Process of compiling structural buildings on unstructured grids (Nakamura et al., 2019). First, (a) shows the original outline of local buildings extracted from the OpenStreetMap. Second, (b) shows how the outline is converted into an unstructured grid system. Third, (c) shows that the unstructured grids encompass the inside and outside of buildings. Finally, (d) shows the Google Earth image that was compared with the buildings in the unstructured grid system (credit of Google Earth is Data SIO, NOAA, US Navy, NGA, GEBCO).

is required for inundation simulations at the street level, as the results are highly sensitive to the quality and resolution of computational grids.

7.2.1.1.5 Coupling with astronomical tide and the sea surface wave model

The simulation of the astronomical tide can be carried out using tide models and applied as open boundary conditions to the storm surge simulation models, as shown in Fig 7.2.2. In terms of coupling with the wave model, two major coupling schemes have been proposed: the inclusion of radiation stress (Longuet-Higgins and Stewart, 1962) and vortex force formalism (McWilliams et al., 2004). Considering radiation stress is more popular than vortex force formalism. The influence of radiation stress from surface waves on ocean currents can be calculated by using four components: significant wave height, wave period, direction and length. In addition, sea surface waves are also influenced by the sea status and, for instance, if the direction of high-speed currents is the same as that of wave trains, wave trains can be accelerated by the currents. Also, wave height, especially that which is limited by wave breaking due to water depth, can be influenced by sea surface vertical fluctuations. These interactives are not negligible when attempting to accurately simulate storm surges.

7.2.1.1.6 Analysis of the impacts of storm surges

The impact of storm surges on coastal communities can be assessed based on numerical models, which should consider street-scale effects such as the topography and layout of buildings, which can have an influence on the characteristics of coastal floods. As shown in Section 5.2, an analysis of human instability can be carried out to estimate the risk that flooding poses to residents (e.g., Abt et al., 1989; Jonkman and Penning-Rowsell, 2008). In addition, momentum fluxes calculated from an ocean model can be employed to estimate the risk of damage or collapse of coastal structures.

7.2.2 Summary of global warming

According to the Intergovernmental Panel on Climate Change Fifth Assessment Report (IPCC AR5), it is extremely likely that ongoing global warming is mainly caused by anthropogenic activity. Global warming can have a severe impact on natural hazards and physical processes, some of which can negatively influence the sustainability of human societies around the planet. Particularly, in the case of coastal zones one of the main impacts is sea-level rise, which can result in higher vulnerability to coastal hazards (e.g., storm surge and tsunami) and influence coastal processes (e.g., coastal erosion). In addition, increases in sea surface temperature (SST) could have the potential to increase the intensity of tropical cyclones.

7.2.2.1 Global warming scenarios in the IPCC

The IPCC AR5 proposed a number of RCP (representative concentration pathway) scenarios (Assessment report 5: IPCC, 2013), based on the amount of radiative forcing (the ratio of energy fluxes absorbed by the atmosphere of the earth from the incoming solar radiation). An increase in radiative forcing means that, as a consequence of the increase in the concentration of greenhouse gases, the Earth's atmosphere absorbs and retains more of the energy it receives from the sun. In the IPCC AR5 four future scenarios are proposed: RCP2.6, 4.5, 6.0 and 8.5 scenarios, corresponding to a radiative forcing of 2.6, 4.5, 6.0 and 8.5 w/m². The RCP2.6 scenario represents a very stringent pathway of future greenhouse gas emissions and results in the lowest warming among four scenarios. RCP 4.5 and 6.0 are intermediate scenarios with greenhouse gas emissions peaking in 2040 and 2060, respectively. Finally, the RCP 8.5 scenario is the worst-case scenario, without humanity making significant efforts to reduce emissions until the end of 21st century.

7.2.2.2 Representative impacts on atmospheric and ocean extremes

Atmospheric circulation could be changed due to global warming, though there is still low confidence regarding the long-term trend of tropical cyclones under a changing climate (IPCC AR5 2013). However, from a thermodynamic physical point of the view, it is to be expected that warmer SST could lead to the intensification of tropical cyclones, as these conditions could enhance the sea-surface heat flux exchanges that feed tropical cyclones (Knutson et al., 2010). Indeed, it has been postulated that the genesis of super-typhoon Haiyan in 2013 (see Section 3.5) was a signal of global warming, as its rapid development and intensification was caused by warm SST (Schiermeier, 2013).

Sea-level rise is caused by both snow melting in polar regions and glaciers and the expansion of the volume of the sea (with the second effect being the most significant until recently). Sea-level rise can influence the vulnerability of coastal communities to tsunamis and storm surges, increasing the potential area that can be inundated by such events.

7.2.2.2.1 Tropical cyclones under pseudo global warming fields

The pseudo global warming (PGW) method considers the difference in meteorological variables between the past and expected future to perform hindcasting simulations (Schär et al., 1996), allowing to understand the effect that different atmospheric parameters and extremes can have on tropical cyclones (TCs). The difference in meteorological variables between future and present can be estimated by using the output of global circulation models (GCMs). Essentially, the difference between present and future meteorological variables is added to the initial and boundary conditions of meteorological simulations in ARW-WRF. The environmental parameters that should be considered include SST, AT (Atmospheric temperature), RH (relative humidity), geopotential height and wind velocity. More details on how the PGW method can be constructed is given in Nakamura et al. (2016, 2020) and Mäll et al. (2017, 2020). Mäll et al. (2020) mentioned the importance of carefully selecting GCMs for the construction of PGW fields, given that future projected extra-tropical cyclones exhibit a different genesis location and intensity according to various GCMs.

Indeed, there are variabilities in the atmospheric and oceanic environmental variables simulated from GCMs, with Figure 7.2.4 showing the variability of SST, and the vertical profile of AT and RH according to RCP 8.5 scenario around the Japanese archipelago. For instance, the SST by 2090 could be on average 3.16 °C higher (as high as 4.8 °C or as low as 1.8 3.16 °C), with uncertainty growing the further into the future that projections are made. It should be noted that these atmospheric and oceanic environmental variables, especially RH, could be highly dependent on the regions and seasonality. For example, around the Philippines, the relative humidity could decrease by 2–3% in October (Nakamura et al., 2016).

The result of TCs simulation under PGW fields is shown in Figure 7.2.5, indicating how they are likely to intensify in the future as a consequence of SST and the effect that this has on latent heat release and surface diabatic heating (which is supported by much research, for example, Emanuel 1987; Knutson et al., 2010). Indeed, the surface wind velocity of TCs affected by an increase in SST can be much greater than that at the present case (with the MSLP also likely to decrease more rapidly).

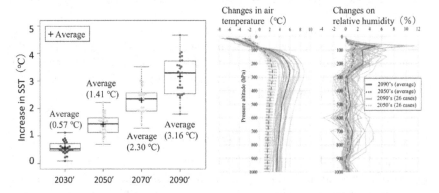

Figure 7.2.4 The ensemble-area-average difference of SST, AT and RH between future (2090' and 2050' time horizons) and present around Japan, around the peak typhoon season (September to November) (Nakamura and Shibayama, 2016).

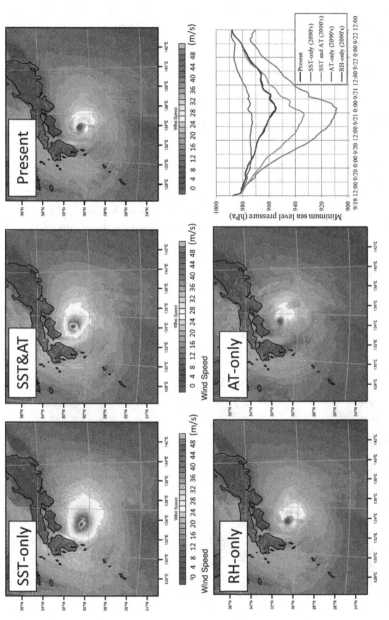

Figure 7.2.5 Changes in the intensity of wind velocity and MSLP considering changes in four different parameters (SST-only, SST&AT, AT-only and RH-only). Typhoon ROKE (2011) was employed as the target typhoon, simulated under the PGW method (Nakamura and Shibayama, 2016).

When considering future air temperature and SST, the increase in the intensity of TCs can be weaker than when only higher SST are considered. Though TCs are likely to still be stronger than at present, this dampening effect can be attributed to the decrease of convective circulation inside the cyclone area under high air temperature (see Wang et al., 2014, and Nakamura et al., 2016). Also, Figure 7.2.5 shows that the effect of the change on RH on the intensity of TC is very weak compared to the much stronger influence of changes in SST and AT.

7.2.2.2.2 Simulation of storm surges under PGW fields

Increases in the intensity of TCs are also expected to result in the generation of higher storm surges. Figure 7.2.6 shows that the future expected storm surges in Tokyo Bay are likely to increase in the future, especially when considering only an increase in SST projected by the RCP8.5 scenario (which could result in the height of the storm surge increasing by nearly 40% compared with present-day conditions). Even when also taking into account changes in RH the storm surge would likely be larger than at present, which in this case appears to be more linked to changes in the track of the storm than its intensity.

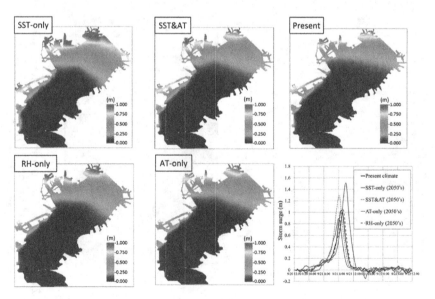

Figure 7.2.6 Changes in the storm surge height in Tokyo Bay associated with the typhoon ROKE (2011) under PGW fields (Nakamura and Shibayama, 2016).

7.2.3 CONCLUSIONS

Although there are still large uncertainties regarding how tropical cyclones and storm surges will respond to a future climate, it appears likely that the risk to shoreline communities will increase. This is likely to result in the rate of coastal erosion increasing and natural disasters becoming more common, unless significant adaptation countermeasures are implemented. Especially, future tropical cyclones and the storm surges that they generate could become more intense. Thus, coastal residents as well as coastal planners should start to increase their resilience against coastal hazards and processes, and more research should be conducted about how such events could intensify in a changing climate.

7.3 WIND WAVE SIMULATION AND GLOBAL WARMING

Tomoya Shibayama
Waseda University, Tokyo, Japan

Shinsaku Nishizaki
Waseda University, Tokyo, Japan

7.3.1 Introduction

Wind waves can be predicted using a variety of numerical simulations, including the coupling of the mesoscale weather prediction model WRF (Skamarock et al., 2008) and third-generation wave estimation models such as SWAN (Booij et al., 1999, see also Section 7.2). In order to obtain accurate results, it is crucial that a high wind resolution data is inputted from WRF into SWAN, and this chapter will explain how such simulations can be carried out. However, it is important to note how future climate change will likely affect wave patterns around the world, given for example the likely intensification of tropical cyclones due to rising sea surface temperature (SSTs) (see also Section 7.2). The accurate prediction and forecasting of wave patterns is crucial for the adequate design of coastal structures, which should take into account changing conditions in order not to be under-designed against future events.

7.3.2 Simulation of future wind waves under global warming conditions

In order to understand how future wave patterns might diverge from present-day conditions it is necessary to estimate the monthly average of expected significant wave heights. However, calculating the wind field associated with meteorological disturbances such as typhoons requires the analysis of huge amounts of data over prolonged durations of time. Typhoons typically occur between summer to autumn, affecting the coastlines of Japan that are facing the Pacific Ocean. However, the behaviour of these weather events is notoriously difficult to simulate, given how sensitive they are to the various atmospheric conditions around them, making monthly long-term wave estimation difficult.

The prediction of the waves that could affect the Japanese coastline can be estimated using the procedure shown in Figure 7.3.1., which can also be applied to future conditions using the pseudo-warming method (Nishizaki et al., 2018). In order to obtain monthly averages, the hindcasting of past climate was simulated taking into account the passage of typhoons, with the computations being interrupted if these weather events deviated significantly from recorded values (the moment at which the simulation was made to converge with the observed weather fields using actual historical data). This essentially resulted in the simulation being divided into irregular periods, whereas the hindcasting was fairly stable during most of the year but required shorter calculation periods during the typhoon season. Overall, the results of the wave hindcasting simulations were in good agreement with the observed data, which resulted in an improvement in how these parameters could be predicted.

The future climate was calculated based on the RCP8.5 scenario presented in the Intergovernmental Panel on Climate Change (IPCC) Fifth Assessment Report. The future climate field was constructed using geophysical quantities

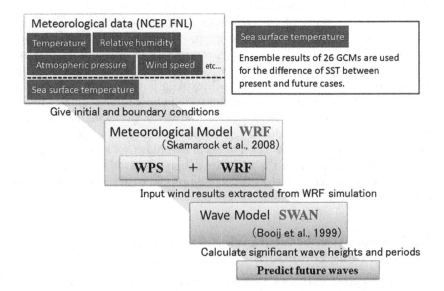

Figure 7.3.1 Numerical model structure for the future projection of wind waves under global warming conditions.

generated by the ensemble averaging of the output of 26 GCMs (Global Climate Model) from CMIP5 (the fifth phase of the Coupled Model Intercomparison Project). The difference between the temperature, SST, and relative humidity in the period 2081–2100, compared with the period 2006–2015, fed greater energy to typhoons in the areas around the Japanese seas, in turn leading to higher waves. Figure 7.3.2 provides the spatial distribution of seasonal mean significant wave heights and their changes over the Northwest Pacific Ocean for present and future conditions, with Figure 7.3.3 showing a time history comparison between two specific points, Shizuoka-Omaezaki on the Pacific side and Kanazawa on the Japan Sea side.

As can be seen from Figure 7.3.2., typhoon intensity is likely to increase in the future, with the mean significant wave heights of summer waves increasing over the entire Northwest Pacific Ocean (Figure 7.3.2(c-3)). On the other hand, the mean significant wave height is likely to decrease in the mid-latitude region around Japan during the rest of the year, though it could increase in the low-latitude region around the Philippines (Figure 7.3.2(c-1)

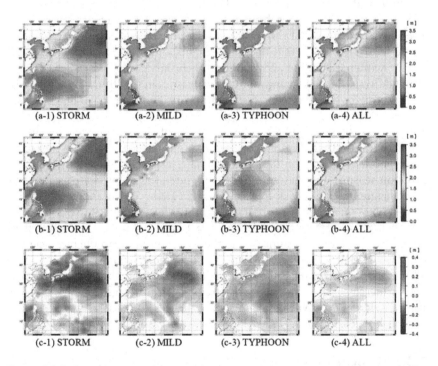

Figure 7.3.2 Distribution of seasonal mean significant wave heights over the Northwest Pacific Ocean for present and future conditions. (a) mean significant wave height in the present climate, (b) mean significant wave height in the future climate, and (c) change between future and current conditions. Three different seasons are considered: Storm: Nov to Feb; Mild: March to June; Typhoon: July to October; All: Average results for the year (Nishizaki et al., 2018).

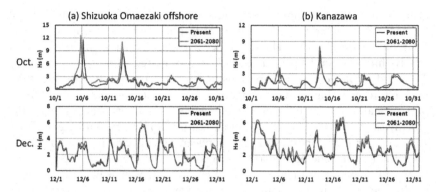

Figure 7.3.3 Comparison of the time history of significant wave height between present and future scenarios (Nishizaki et al., 2018).

and Figure 7.3.2(c-2)). In the northwestern Pacific, the change in mean wave height varied from –0.20 m to 0.13 m from March to June, from –0.03 m to 0.23 m from July to October, from –0.45 m to 0.27 m from November to February, and from –0.15 m to 0.20 m over the entire season year. Overall, in the future it is likely that the wave heights and periods will increase during the winter season (i.e., November to February).

In addition, the change in wave characteristics in the future climate is also likely to affect the frequency distribution of significant wave height along the Japanese coastal area. For example, off the coast of Shizuoka-Omaezaki, which faces the Pacific Ocean, the frequency of significant wave heights of 1.5 m or less decreased, and those of 4 m or more increased due to the stronger typhoons expected in the future climate (especially during the summer typhoon season). On the other hand, in winter, the frequency of significant wave heights above 2 m decreased, and that below 1.5 m increased significantly. The same trend was observed at other sites, such as Kanazawa in the Sea of Japan and Naha in the Okinawa Islands, although the amount of change was different. In summary, in the seas around Japan the mean significant wave height is projected to increase in summer and decrease in winter, indicating that the range of variations in wave heights throughout the year would increase. The increase in the frequency and height of waves in the future is likely to have significant consequences for coastal infrastructure such as breakwaters.

7.3.3 Conclusions

It is likely that the wave climate around the world will change in the future due to global warming, particularly as a result of the potential intensification of typhoons due to increasing SSTs. In order to predict such changes it is possible to use numerical simulation models, which were proven to be able to realistically hindcast past wave conditions. The increase in SSTs is likely to result in more intense typhoons during the typhoon season (July to October) and could lead to increases in monthly average expected significant wave heights on the Pacific Ocean. During the winter season the increase in SSTs does not greatly affect the development of extratropical cyclones, which appears to imply that future conditions might not significantly deviate from that at present. These potential changes in wave climate have important implications for the design and maintenance of coastal infrastructure, and could result in significant damage taking place unless adaptation countermeasures are implemented.

REFERENCES

Abt, S. R., Wittler, R. J., Taylor, A., and Love, D. J. 1989. Human stability in a high flood hazard zone. *Journal of the American Water Resources Association* 25(4): 881–890.

Booij, N., Ris R. C., and Holthuijsen, L. H. 1999. A third-generation wave model for coastal regions: 1. Model description and validation. *Journal of Geophysical Research: Oceans* 104: 7649.

Chen, C., Liu, H., and Beardsley, R. C. 2003. An unstructured grid, finite-volume, three-dimensional, primitive equations ocean model: application to coastal ocean and estuaries. *Journal of Atmospheric and Oceanic Technology* 20(1): 159–186.

Chikasada, N. 2020. *Global tsunami Terrain Model.* doi:10.17598/NIED.0021

GEBCO Compilation Group 2021. *GEBCO 2021 Grid.* doi:10.5285/c6612cbe-50b3-0cff-e053-6c86abc09f8f.

Goto, C., and Ogawa, Y. 1982. *Numerical Method of Tsunami Simulation with the Leap-Frog Scheme.* Department of Civil Engineering, Tohoku University, p. 52.

Goto, C., Ogawa, Y., Shuto, N., and Imamura, F. 1997. IUGG/IOC time project, numerical method of tsunami simulation with the leap-frog scheme. *IOC Manuals and Guides*, vol. 35, p. 130. UNESCO, Paris.

Honda, C., and Mitsuyasu, K. 1980. Laboratory study on wind effect to ocean surface. *Proceedings of the Japanese Conference on Coastal Engineering* 27: 90–93. 10.2208/proce1970.27.90 (in Japanese).

Iimura, K., Shibayama, T., Takabatake, T., and Esteban, M. 2021. Experimental and numerical investigation of tsunami behavior around two upright sea dikes with different heights. *Coastal Engineering Journal* 63(1): 1–16.

IPCC 2013. Summary for policymakers. In Stocker, T. F. et al. (eds.), *Climate Change 2013: The Physical Science Basis. Contribution of Working Group I to the Fifth Assessment Report of the Intergovernmental Panel on Climate Change.* Cambridge University Press, Cambridge, United Kingdom and New York, NY, USA.

Ishii, H., Takabatake, T., Esteban, M., Stolle, J., and Shibayama, T. (2021). Experimental and numerical investigation on tsunami run-up flow around coastal buildings. *Coastal Engineering Journal*, 63(4): 485–503.

Jonkman, S. N., and Penning-Rowsell, E. 2008. Human instability in flood flows. *Journal of the American Water Resources Association* 44: 5.

Knutson, T. R., McBride, J. L., Chan, J., Emanuel, K., Holland, G., Landsea, C., Held, I., Kossin, J. P., Srivastava, A. K., and Sugi, M. 2010. Tropical cyclones and climate change. *Nature Geoscience* 3: 157–163.

Longuet-Higgins, M. S., and Stewart, R. 1962. Radiation stress and mass transport in gravity waves, with application to surf-beats. *Journal of Fluid Mechanics* 13: 481–504.

Luettich, R. A., and Westerink, J. J. 1991. A solution for the vertical variation of stress, rather than velocity, in a three-dimensional circulation model. *International Journal for Numerical Methods in Fluids* 12: 911–928.

Mäll, M., Nakamura, R., Suursaar, Ü., and Shibayama, T. 2020. Pseudo-climate modelling study on projected changes in extreme extratropical cyclones, storm waves and surges under CMIP5 multi-model ensemble: Baltic Sea perspective. *Natural Hazards* 102(1): 67–99. doi:10.1007/s11069-020-03911-2.

Mäll, M., Suursaar, Ü., Nakamura, R., and Shibayama, T. 2017. Modelling a storm surge under future climate scenarios: case study of extratropical cyclone Gudrun (2005). *Natural Hazards* 89: 1119–1144.

Mansinha, L., and Smylie, D. E. 1971. The displacement fields of inclined faults. *Bulletin of the Seismological Society of America* 61(5): 1433–1440.

McWilliams, J. C., Restrepo, J. M., and Lane, E. M. 2004. An asymptotic theory for the interaction of waves and currents in coastal waters. *Journal of Fluid Mechanics* 511: 135–178.

Mikami, T., and Shibayama, T. 2012. Numerical analysis of tsunami propagating on wide reef platform using turbulence model. *Journal of Japan Society of Civil Engineers, Ser. B2 (Coastal Engineering)* 68(2): I_76–I_80 (in Japanese with English abstract).

Mikami, T., Shibayama, T., Esteban, M., Ohira, K., Sasaki, J., Suzuki, T., Achiari, H., and Widodo, T. (2014). Tsunami vulnerability evaluation in the Mentawai islands based on the field survey of the 2010 tsunami. *Natural Hazards* 71(1): 851–870.

Nakamura, R., and Shibayama, T. 2016a. Ensemble forecast of extreme storm surge: A case study of 2013 Typhoon Haiyan. *Coastal Engineering Proceedings* 35: 1–10.

Nakamura, R., and Shibayama, T. 2016b. Evaluation of atmospheric and oceanic environments governing intensity of typhoon and storm surge. *Journal of Japan Society of Civil Engineers, Ser. B2 (Coastal Engineering)* 72(2): I_1495–I_1500.

Nakamura, R., Mäll, M., and Shibayama, T. 2019. Street-scale storm surge load impact assessment using fine-resolution numerical modelling: a case study from Nemuro, Japan, *Natural Hazards* 99: 391–422.

Nakamura, R., Shibayama, T., Esteban, M., and Iwamoto, T. 2016. Future typhoon and storm surges under different global warming scenarios: Case study of Typhoon Haiyan (2013). *Natural Hazards* 82(3): 1645–1681. doi:10.1007/s11069-016-2259-3.

Nishizaki, S., Nakamura, R., Shibayama, T., and Stolle, J. (2018). Future wave projection during the typhoon and winter storm season. *Proceedings of the Coastal Engineering Conference*, 36: 17.

Okada, Y. 1985. Surface deformation due to shear and tensile faults in a half-space. *Bulletin of the Seismological Society of America* 75(4): 1135–1154.

Powell, M. D., Vickery, P. J., and Reinhold, T. A. 2006. Reduced drag coefficient for high wind speeds in tropical cyclones. *Nature* 422: 279–283.

Schär, C., Frei, C., Lüthi, D., and Davies, H. C. 1996. Surrogate climate-change scenarios for regional climate models. *Geophysical Research Letter* 23: 669–672.

Shibayama, T., and Ohira, K. 2010. A study on tsunami behavior in Tokyo Bay and estimation of damage under earthquake of Tokyo metropolitan. *Proceedings of Civil Engineering in the Ocean, JSCE* 26: 219–223 (in Japanese with English abstract).

Skamarock, W. C., Klemp, J. B., Dudhia, J., Gill, D. O., Barker, D., Duda, M. G., Xiang-Yu, H., Wang, W., and Powers, J. G. 2008. *A Description of the Advanced Research WRF Version 3*. No. NCAR/TN-475+STR, University Corporation for Atmospheric Research. doi:10.5065/D68S4MVH.

Takabatake, T., and Shibayama, T. 2012. Predicting the risk of storm surge and tsunami in Tokyo Port. *Journal of Japan Society of Civil Engineers, Ser. B3 (Ocean Engineering)* 68(2): I_894–I_899 (in Japanese with English abstract).

Wang, S., Camargo, S. J., Sobel, A. H., and Polvani, L. M. 2014. Impact of the tropopause temperature on the intensity of tropical cyclones-an idealized study using a mesoscale model. *Journal of Atmospheric Science* 71(11): 4333–4348. doi:10.1175/JAS-D-14-0029.1.

Wu, J. 1982. Wind-stress coefficients over sea surface from breeze to hurricane. *Journal of Geophysical Research: Oceans* 87(C12): 9704–9706. doi:10.1029/JC087iC12p09704.

Chapter 8

Disaster mitigation and the protection of residents

CONTENTS

DOI: 10.1201/9781003156161-8

8.1 EVACUATION SIMULATION AND PLANNING

Tomoyuki Takabatake

Kindai University, Osaka, Japan

8.1.1 Introduction

A significant tsunami can cause devastating damage to coastal communities situated along the shoreline. In the case of the *2004 Indian Ocean Tsunami*, a total of more than 230,000 people died in countries around the Indian Ocean (Shibayama 2015, see also Section 2.1). The *2011 Tohoku Tsunami* (see Section 2.6) devastated coastal communities in the Tohoku region, Japan and caused over 16,000 dead or missing (National Police Agency of Japan 2020). The *2018 Palu Tsunami* (see Section 2.7) and the *2018 Sunda Strait Tsunami* (see Section 2.8) also ravaged coastal areas in Indonesia (Mikami et al. 2019; Krautwald et al. 2021; Stolle et al. 2020; Harnantyari et al. 2020; Takabatake et al. 2019a).

In Japan, following the *2011 Tohoku Tsunami*, a new conceptual classification of tsunami events, based on their relative return period, was proposed (Shibayama et al. 2013a). Essentially, tsunamis whose return period ranges from every few decades to 100 years are defined as Level 1 (L1) events. The design of hard measures (e.g., coastal dykes, seawalls) is supposed to protect human lives and property against L1 events. Tsunamis whose return period is more than several hundreds to a thousand years are defined as a Level 2 (L2) event. As it is difficult to completely protect coastal communities against L2 events using hard measures (given that the height and energy of such waves can be considerable), in order to minimize casualties a combination of hard and soft measures (e.g., land use regulations, evacuation plans) should be employed.

This classification has been utilized when formulating the reconstruction plans for the tsunami-affected areas in the Tohoku region. Generally speaking, coastal risk managers first perform tsunami numerical simulations considering all earthquake scenarios that could potentially occur in the future, establishing return periods for certain wave heights and determining the design height of coastal defences against L1 tsunamis. Then, by considering the presence of coastal defences (designed against L1 tsunamis), the expected extent of inundation from L2 tsunami events can be estimated. The simulated maximum extent of the L2 tsunami inundation is used to determine land use in the settlements at risk. Essentially, many tsunami-affected municipalities have restricted the construction of new residential houses and other buildings in the areas where simulated inundation depths can be expected to exceed 2 m in L2 events (when the simulated inundation depth is lower than 2 m, some municipalities adopted the concept of an "on-site reconstruction", which allows residents to rebuild their houses where they had been situated prior to the 2011 events). The reasons behind this "2 m" limit resides in the fact that when inundation depth exceeded 2 m, the percentages of houses washed away substantially increased during the *2011 Tohoku Tsunami*. Thus, it has been reasoned that, if people who live in an area where inundation depth is lower than 2 m were to stay on the second floor or above the buildings that are strongly built, they could survive a tsunami. However, as tsunami simulations are typically performed using a number of assumptions (Takabatake et al. 2019b), the inundation depth during an actual L2 event could be higher than 2 m. Furthermore, tsunami simulation models generally do not consider the effects of debris displaced by a tsunami, such as containers, trees and vehicles (Takabatake et al. 2021a). In fact, debris is typically pushed inland by tsunami waves (see Section 6.2), causing significant damage to structures in past events (e.g., Ghobarah et al. 2006; Stolle et al. 2020; Takabatake et al. 2019a). As it is known that forces acting on structures would increase if debris are included in a tsunami flow (Stolle et al. 2017, 2018; Hamano et al. 2020) residential houses might be exposed to larger forces than anticipated in design. Thus, an adequate and swift evacuation of the population at risk is still crucial for saving lives in the event of L2 tsunamis (Shibayama et al. 2013).

Coastal communities at risk of being affected by a tsunami are required to establish effective evacuation plans. To do so, it is important to clarify potential complications and difficulties regarding evacuation, such as the total required evacuation time, locations of congested roads, and the required number of refuge areas/shelters available in a given coastal area. While one effective way to clarify such issues is to conduct evacuation drills, it is nearly impossible for all local residents and visitors to participate in such drills. Thus, tsunami evacuation simulations can be used to help coastal communities formulate effective evacuation plans and strengthen their resilience. Although several methodologies (e.g., GIS approach [Sugimoto et al. 2003; Wood & Schmidtlein 2013]) have been proposed to simulate the behaviour of evacuees during a tsunami event, an agent-based modelling approach arguably represents the best approach, as it is able to consider a variety of complex behaviour of evacuees and their interactions (Na & Banerjee 2019). This chapter introduces recent progress made in the development of agent-based evacuation simulation models, and also details progress in one such agent-based tsunami evacuation model developed by the author.

8.1.2 Recent progress in the development of agent-based evacuation simulation models

An agent-based modelling approach is able to simulate the complex behaviour and interactions of a large number of individual agents. In such models, each individual agent is modelled as if it were a human being, which can decide its own movement considering the environment surrounding it. As the agent-based model can also consider the interactions between the various individual agents, it is possible to obtain a response of a group of individual agents to a given phenomenon. In fact, a number of researchers have utilized agent-based modelling approaches to simulate evacuation procedures under various emergency situations. Especially, crowd evacuation from a building due to a fire or an earthquake has frequently been simulated using agent-based modelling approaches (Owen et al. 1996; Pelechano & Badler 2006; Pan et al. 2007; Liu et al. 2018). For instance, using this approach, Pan et al. (2007) simulated the human behaviour of a crowd (e.g., competitive, queuing, and herding), which would be observed during evacuation from a building. The agent-based modelling approach has also been applied to crowd evacuation in urban networks in the event of river flooding or storm surges (Liu et al. 2008; Chen and Zhan 2008; Dawson et al. 2011). As the window time for evacuation is relatively longer for these types of natural hazards, evacuation by vehicle generally needs to be appropriately modelled. For instance, Chen and Zhan (2008) developed an agent-based evacuation simulation model that can consider traffic flows and investigated the effectiveness of staged evacuation strategies. The influence of flooding inundation on evacuation behaviour was also simulated with the agent-based modelling approach by Dawson et al. (2011).

Notably, the number of applications of agent-based modelling approaches to tsunami evacuation have significantly increased after the *2011 Tohoku Tsunami*, including that of pedestrians (Mas et al. 2012; Nishihata et al. 2012; Kumagai 2014; Makinoshima et al. 2016), and also vehicles (Nishihata et al. 2012; Makinoshima et al. 2016), as the importance of proper evacuation has been widely acknowledged. The agent-based modelling approach is also effective to estimate the potential number of casualties that can result from such events, and evaluate the effectiveness of countermeasures in terms of the reduction in loss of lives. For instance, Johnstone and Lence (2009, 2012) developed an agent-based tsunami evacuation simulation model and investigated the effect of an early evacuation to decrease mortality rates in Vancouver Island, Canada. Uno et al. (2015) and Koyanagi and Arikawa (2016) analysed the decrease in mortality rates by implementing several types of countermeasures (e.g., elevating a coastal dyke, early evacuation). Mostafizi et al. (2017, 2019) utilized their own tsunami evacuation simulation model to assess the vulnerability of a network or to optimize the location of refugee shelters. Considering that the damage or collapse of buildings and structures (due to severe ground shaking) in a coastal area would lead to road blockage for the case of a near-field tsunami, the effects of road blockage have also been considered by some agent-based tsunami evacuation simulation models (see for instance Ito et al., 2020 or Wang and Jia, 2021).

The author of this chapter has also developed an agent-based tsunami evacuation simulation model and applied it to various tsunami-prone coastal areas, which will be briefly explained in the next section.

8.1.3 Development and application of an agent-based tsunami evacuation simulation model

Some coastal settlements that are at risk of suffering from tsunamis are famous sightseeing towns or have beaches that are popular with tourists. In such coastal areas, if a tsunami approaches during the daytime in a holiday period not only local residents but also many visitors would be forced to evacuate. As visitors generally do not know the area well, their evacuation behaviour would be different to that of local residents. However, many existing agent-based tsunami evacuation simulation models have ignored the presence and evacuation behaviour of visitors. Thus, Takabatake et al. (2017) developed an agent-based tsunami evacuation simulation model which considers the evacuation behaviour of both local residents and visitors. As the model is able to incorporate the simulated results of a tsunami propagation and inundation simulation model (Shibayama et al. 2013; Nagai et al. 2020), which is based on the 2D shallow water equations, it is possible to estimate the expected number of casualties during a tsunami event (as shown in Figure 8.1.1).

In Takabatake et al. (2017), the model was applied to Kamakura City, which is one of the most popular sightseeing towns situated along the shorelines of Japan. The simulated results indicated that the presence and behaviour of visitors significantly influenced the evacuation process, including total evacuation time, locations of bottlenecks, and the number of evacuees accommodated in each refugee area/building, and expected mortality rates. To simulate the evacuation behaviour of local residents and visitors more realistically, Takabatake et al. (2018a) incorporated various degrees of tsunami awareness and possible evacuation behaviour by utilizing the results from previous questionnaire surveys conducted among local residents (Enomoto et al. 2015), visitors (Carlos-Arce et al. 2017) and beach users (Yasuda et al. 2016). Using the model, the expected number of casualties in Kamakura City was estimated for different evacuation scenarios. As a significant number of casualties were estimated in Takabatake et al. (2018a), the effectiveness of various countermeasures was explored in the same study area by Takabatake et al. (2020a). Considering that the number and types of evacuees would greatly differ according to the time when a tsunami arrives at the study area, the impacts of different numbers and types of evacuees on the effectiveness of each countermeasure were investigated. The simulated results clarified that while the effectiveness of hard countermeasures (i.e., elevating the road embankment) did not greatly change according to the number and types of evacuees, that of soft countermeasures (i.e., minimizing the evacuation start time, optimizing the evacuation route, usage of cars for evacuation, and a combination of these measures) greatly influenced survival rates. Takabatake et al. (2020a) also found that prompting evacuation and guiding all evacuees to the closest evacuation place could even worsen mortality rates, especially during holiday periods when many visitors could

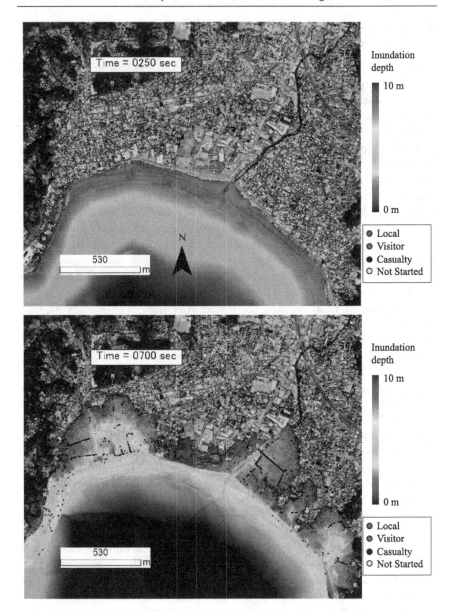

Figure 8.1.1 Example of the tsunami evacuation simulation results before (upper) and after (lower) tsunami arrival.

Source of satellite imagery: Esri, Maxar, GeoEye, Earthstar Geographics, CNES/Airbus DS, USDA, USGS, AeroGRID, IGN, and the GIS User Community.

be present in Kamakura (as a number of refugee buildings would reach their capacity and heavy congestions could take place in some roads). Such studies reached the conclusion that the presence and behaviour of visitors should be appropriately taken into account when formulating effective tsunami countermeasures.

The usage of vehicles for tsunami evacuation is generally not recommended, as it could lead to severe congestion on roads and hinder prompt evacuation. However, evacuation by vehicle could be helpful for those who are not able to walk fast (e.g., the infants, elderly, or disabled). Thus, recently the usage of vehicles for evacuation has started to be considered at locations where there are many such vulnerable residents. Considering the importance of simulating the effectiveness of car evacuation appropriately, Takabatake et al. (2020b) improved their previous model so that it can consider the behaviour of both pedestrian and car evacuees, and also the interactions between them. To validate the model, Takabatake et al. (2020b) reproduced the actual traffic jams observed along roads in Tagajyo City, Japan during the *2011 Tohoku Tsunami*. Simulated results demonstrated that the model was able to reproduce the location of car congestion along the roads relatively well. Later, the effectiveness of soft tsunami countermeasures (e.g., early evacuation, restriction of car usage) in Tagajyo City was investigated by Takabatake et al. (2020d). The model was also applied to another coastal area, Shinguu City, Japan to investigate the effectiveness of car usage for evacuation (Takabatake et al. 2020b), showing that the expected number of casualties increased as the car usage ratio increased when all the evacuees headed to the closest refuge areas/shelters. However, when evacuees chose to go to refuge areas outside of the inundation zone, which have a large capacity to accommodate evacuees, the expected number of casualties did not significantly increase even if 50% of evacuees used vehicles for evacuation. The results suggested that considering appropriate evacuation routes and refuge areas is important to achieve a successful evacuation that considers the use of vehicles for residents with poor mobility.

This agent-based tsunami evacuation simulation model developed by the author has also been applied to a coastal area outside of Japan. The western coastline of Vancouver Island, Canada, has a considerable risk of being affected by a significant tsunami, which would be generated by an earthquake in the Cascadia Subduction Zone (Takabatake et al. 2019b, 2020c). The simulated results indicated that mortality rates would be higher when an earthquake occurs at night, and that the locations and capacity of refuge areas/shelters should be reconsidered to decrease the expected number of casualties from a major tsunami in the future. Recently, the model was further improved by Takabatake et al. (2021b) to incorporate the effects of road blockage due to collapse of buildings by an earthquake. This new model was applied to three coastal cities in Japan (Kamakura, Zushi, and Fujisawa City), with the results showing that the expected number of casualties from a tsunami would significantly increase in all the cities when considering the effects of road blockage.

8.1.4 Conclusions

Appropriate evacuation procedures are important to protect human lives from a significant tsunami, and agent-based tsunami evacuation simulation models can be helpful to formulate effective evacuation plans. In this chapter, recent progress made to develop agent-based modelling approaches was reviewed, and the evolution and application of an agent-based tsunami evacuation model developed by the author of this chapter were introduced. As a consequence of the various tsunamis that have taken place around the world following the *2004 Indian Ocean Tsunami* (as detailed in other chapters of this book), the awareness and preparedness of coastal residents in many countries have improved as shown in Harnantyari et al. (2020), Takabatake et al. (2018b) and Sections 2.7 and 2.8. Thus, it would not be surprising if the number of people that evacuate early if a strong earthquake occurs in the future is high. However, previous agent-based tsunami evacuation simulations conducted by the author demonstrated that, even if evacuees start moving relatively quickly after an earthquake, residents and tourists alike still have chances of being caught by a tsunami, especially if they do not know the appropriate evacuation routes and refuge area/shelters (including information about their capacity). This finding highlights the need to utilize agent-based tsunami evacuation simulation models to establish effective evacuation plans, which ensures that all evacuees can successfully reach refugee areas/shelters during a significant tsunami event.

8.2 RECOVERY PROCESS AND DISASTER MITIGATION EFFORTS

Ryo Matsumaru

Toyo University, Tokyo, Japan

8.2.1 Introduction

Measures to mitigate disaster damage should be proactive. While disaster mitigation countermeasures, especially the construction of infrastructure that can reduce the natural forces that cause disasters, require funding and time, they are typically not prioritized by policy-makers as they do not generate immediate economic benefits. This is particularly the case in developing countries, where financial constraints are high and there are many competing projects vying for public funding. As a result of this lack of countermeasures to mitigate the large forces that cause disasters, extensive damage can take place.

In the aftermath of such regrettable events, the minds of policy-makers and the population of the country get fixated by the disaster and abundant funding is typically allocated for reconstruction. Therefore, it can be said that recovery from large-scale disasters represents an opportunity for developing countries to invest in disaster mitigation measures and enhance the resilience of at-risk communities. This concept is also emphasized in the 2015 Sendai Framework for Disaster Reduction, which has adopted the *Build Back Better* (BBB) concept to improve resilience in the aftermath of such events.

This book mainly deals with coastal disasters. Coastal areas are one of the most convenient places for people to live, though many stretches of the shoreline have been severely damaged in the past by events such as storm surges or tsunamis. In this chapter, the recovery process from two large-scale coastal disasters in developing countries (the *2004 Indian Ocean Tsunami*, see Section 2.1 and 2013 Typhoon Haiyan disaster, see Section 3.5) will be reviewed and compared from the point of view of disaster reconstruction and mitigation planning.

8.2.2 Case of Banda Aceh, Indonesia - Recovery from Indian Ocean Tsunami

8.2.2.1 Tsunami and damages

The *2004 Indian Ocean Tsunami* was one of the most catastrophic tsunami events ever recorded (see also Section 2.1). In Indonesia, the tsunami travelled inland, sometimes several kilometers, washing away thousands of buildings and severely damaging urban infrastructure. According to a BRR report (Badan Rehabilitasi dan Rekonstruksi, or the Agency for the Rehabilitation and Reconstruction in Aceh and Nias, BRR NAD-Nias, 2008), more than 166,000 people died or went missing in Indonesia, and an estimated 1,500 km of coastline was affected (Matsumaru et al., 2012). Banda Aceh was the city most severely damaged, as the waves that hit it were 6–12 m in height, resulting in approximately 15,000 casualties (nearly 10% of the population at the time). Figure 8.2.1 shows how residential areas along the coastline of Banda Aceh were completely levelled by the tsunami.

Figure 8.2.1 Coastline of Banda Aceh one month after the *2004 Indian Ocean Tsunami*.

8.2.2.2 Urban planning

To create a safe city against future tsunamis, the Indonesian Government tried to make fundamental changes to the urban structure by making modifications the spatial planning of the city. As a result, at the end of February 2005 the Indonesian Government prepared the first draft of a new spatial plan, which was referred to as a "Blueprint" (see Figure 8.2.2).

As shown in Figure 8.2.2, the city was divided into multiple zones, and the land use in the tsunami inundation area (Figure 8.2.3, left) was designated as Zona Perikanan/Tambak (Fishery/Pond Zone) and Permukiman Terbatas (Limited settlement), where economic and residential activities

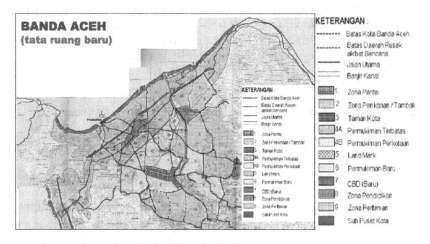

Figure 8.2.2 The first draft of spatial plan (referred to as the "Blueprint"), as prepared by BAPPENAS (2005).

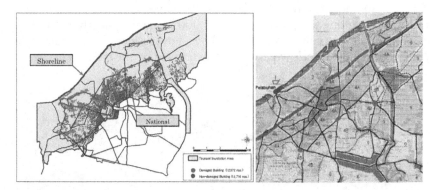

Figure 8.2.3 Inundation Area and proposed land use according to the Blueprint.

Source: JICA (2005), BAPPENAS (2005).

other than the fishery sector were strongly restricted. The outer part of the inundation area was proposed as Permukiman Perkotaan (Urban Settlements), which would form the main residential area (Figure 8.2.3 right).

The local people who were briefed about the proposal strongly opposed it, so the Indonesian Government abandoned the zoning plan proposed in the Blueprint and allowed residents and communities in the tsunami-affected areas to return to their original places of residence on their own responsibility, with this community-led resettlement plan later being referred to as the "village plan".

From April to December 2005, the Government of Japan, through the Japan International Cooperation Agency (JICA), assisted the city of Banda Aceh in formulating the urban reconstruction plan, which basically followed the Blueprint concept and zoning plan. However, the area designated as "limited settlement" in the Blueprint was also designated as a residential area, to be consistent with the village plan (Figure 8.2.4).

The original "Build Back Better" concept of regulating activities through zoning had to be changed. Due to this change, several countermeasures such as the construction of a seawall, and placement of evacuation routes and facilities were necessary, with these measures being included in the Master Plan for Reconstruction (Figures 8.2.5 and 8.2.6).

8.2.2.3 Issues with disaster mitigation strategies

As mentioned earlier, the Master Plan proposed defensive structures along the coast to protect against tsunamis and other coastal hazards. However, the construction of such structures has not made any significant progress, while residents resettled back along the coastline and economic activities returned to levels similar to that prior to the tsunami. As a result, the settlements in the area have returned to having same levels of vulnerability they had before the tsunami (see Figure 8.2.7). As a consequence of this, the establishment of effective evacuation procedures is crucial to minimize human losses, and in that sense evacuation routes and tsunami shelters (which also serve as community centres) have been constructed (see Figure 8.2.8). However, even then a number of issues have been identified with such strategies, such as an insufficient number of tsunami shelters being constructed and the ineffective network of escape routes (Matsumaru et al., 2012).

When focusing on the community level, in some villages land ownership was restored to what it had been prior to the tsunami. For example, Blang Oi village, located about 1 km from the coast, has a narrow (approx. 5 m) and complex road network and this type of resettlement recreates vulnerable conditions against future disasters (Matsumaru et al., 2012).

In Banda Aceh city the reconstruction of infrastructure and the restoration of livelihoods has continued to progress, and there were few traces left

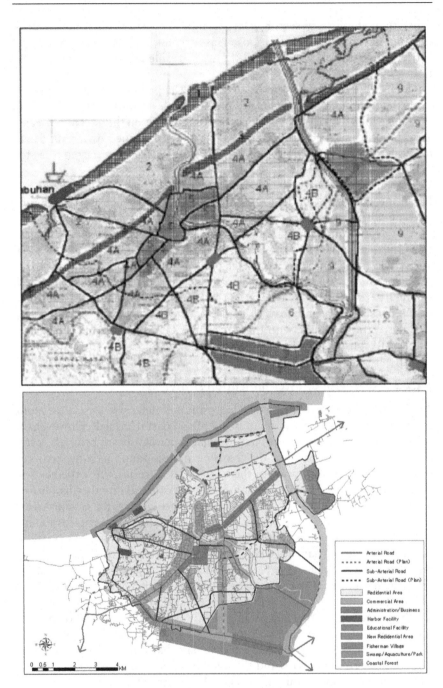

Figure 8.2.4 Comparison of Blueprint (top) and Land Use Plan by JICA (bottom), where it can be seen that area 4A was changed to being a residential area.

Source: JICA (2005), BAPPENAS (2005).

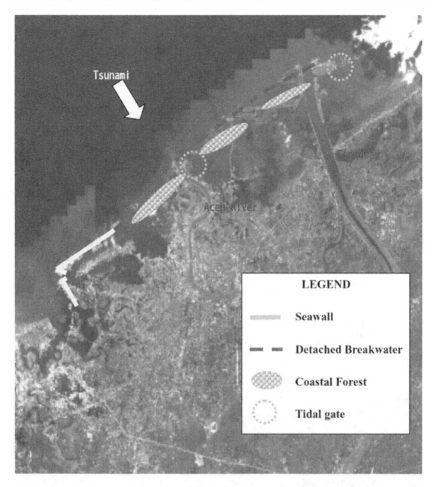

Figure 8.2.5 Proposed structural countermeasures at the shoreline (JICA, 2005).

of the tsunami disaster in 2020. However, the reconstruction of houses (supported by the Government and international organizations) did not involve the elevation of the ground under them or the construction of any disaster prevention facilities, even in the coastal areas that were worst affected by the tsunami. Essentially, coastal residents who wanted to rebuild their livelihoods as soon as possible forced the Government to approve a plan that allowed them to settle back to where they had originally lived, without waiting for the land use plan elaborated as part of the reconstruction plan. As a result, it is clear that the community remains as vulnerable as it was originally, highlighting the dilemma that exists in post-disaster reconstruction between the time required for planning and the wishes of residents to rebuild their livelihoods as quickly as possible.

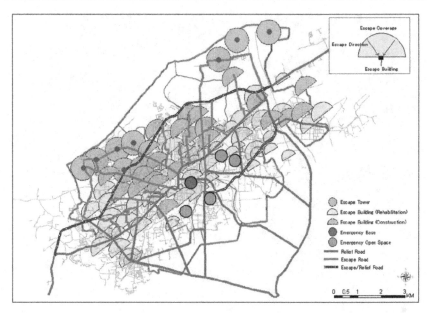

Figure 8.2.6 Proposed emergency road network and emergency public facilities
(JICA, 2005).

Figure 8.2.7 Housing rebuilt along the coast, as of 2008.

Figure 8.2.8 Tsunami Evacuation Centres. Left: Constructed with Japanese assistance. Right: Constructed by the Indonesian government.

8.2.3 Recovery from the storm surge of Typhoon Haiyan

8.2.3.1 Storm surge and damage

On the 8th of November 2013, category five typhoon Haiyan (Philippine's name: Yolanda) made landfall in Leyte Island (see also Section 3.5). Tacloban city, the capital and economic hub of Eastern Visayas, suffered severe damage to its infrastructure, houses and industries due to the combination of heavy rain, strong winds and storm surge. At the national level 7,000 people lost their lives, and the total economic loss was estimated to be 89.5 billion PHP (JICA, 2015), with over 2,600 people killed by the typhoon in Tacloban.

The storm surge generated by the typhoon was over 5 m in some areas of the centre of Tacloban City (Mikami et al., 2016), and the settlements along the coast of the city were heavily damaged. Approximately 29,000 houses were swept away by the storm surge, and it was said that about 90 % of them were informal settlements (Iuchi 2018), as this coastal area was the receiving area for the population flowing into Tacloban City as part of rural-urban migration patterns.

8.2.3.2 Recovery planning for a safer city

On December 16, 2013, the Philippines Government published "Reconstruction Assistance on Yolanda (RAY): Build-Back-Better", a strategic plan to guide the recovery and rebuilding of the economy, lives, and livelihoods in the affected areas. This contained a rapid action plan to address critical, immediate needs, as well as aid in the development and implementation of a full set of recovery and reconstruction interventions over the medium-term (NEDA, 2013).

In the RAY, the concept of *Build Back Better* was clearly stated as a principle that should guide reconstruction. Under this concept, with the aim to improve the resiliency of coastal settlements, the Government applied both land use regulations that included the relocation of vulnerable communities along the coast and the construction of seawalls to protect the city from storm surges and high waves.

8.2.3.3 Land use regulation and relocation

In 2014, the City of Tacloban enacted an ordinance to establish a No Build Zone up to 40 m from the coastline. Since people living in the No Build Zone, many of whom were illegal residents, were forced to relocate, the city government developed housing complexes in the northern area of the city to accommodate them (Figure 8.2.9). As a result, people who had been living illegally along the coastal areas of the centre of Tacloban City were able to obtain their own houses in newly developed settlements in the north (Ong

Figure 8.2.9 Left. Location of new settlements, created based on Open Street Map data and Tacloban City CLUO 2013 (Iuchi, 2018). Right. New housing complex for relocated settlers.

et al., 2016), and did not have to resort to rebuilding their houses in their original locations.

This relocation policy reduced the vulnerability of the coastal areas, and residents who relocated to the north were no longer exposed to coastal hazards. However, many of the residents were forced to change their lives and livelihoods due to this relocation, creating new challenges and issues, such as reduced opportunities for income earning, increased expenses for daily life, and difficulties in the schooling of children (Ong et al., 2016). Furthermore, residents from each of the communities were relocated to a number of different housing complexes, disrupting the original social bonds between them. Due to such difficulties many people wanted to stay in the coastal areas where they originally lived, highlighting how comprehensive relocation plans need to be formulated (taking into account the rebuilding of lives and livelihoods) for a successful relocation to take place.

8.2.3.4 Construction of the seawall

A total of about 27 km of seawall construction have been planned for the areas affected by the storm surge, including the entire coastline of Tacloban City, as shown in Figure 8.2.10 (JICA, 2015). The required height of the crest of the seawall was determined by an analysis of the 50-year return probability of storm surge height. As a result, the elevation of the crest varies from 2.8 m to 4.5 m depending on the location (see Figure 8.2.10, left). The construction of the seawall is the responsibility of the national government. Due to changes in the development plan of Tacloban, there are uncertainties regarding the execution of the seawall in the centre of the city, but construction was in progress in sections 4 and 5 in 2018 (Figure 8.2.10), significantly increasing the resilience to storm surges of the communities behind them.

Figure 8.2.10 Left. Location and sections of the proposed seawall (JICA, 2015). Right: Progress of seawall construction, as of 2018.

8.2.3.5 Discussion

The Philippine government adopted the concept of the *Build Back Better* immediately after the storm surge disaster. Technical assistance for formulating the Recovery Plan and the restoration of some damaged facilities was carried out with Japanese funding and other international assistance to help with the recovery of the affected area. However, the government of the Philippines led the resettlement of people away from the coastal areas to the north and the construction of the seawall, which are at the core of the *Build Back Better* for the area. Although there have been some issues in implementing the resettlement and the construction of the seawall, generally speaking the process of improving the resilience of the communities affected has been moving forward. Based on that, it is clear that the government has shown some degree of leadership and will to *Build Back Better*.

8.2.4 Conclusion

This chapter discussed two different case studies of the formulation of disaster countermeasures in the recovery process from large-scale disasters in Indonesia and the Philippines. In both cases, the governments aimed to *Build Back Better* in the immediate aftermath of a disaster, though with strikingly different mid-term results.

Although it is not possible to make a simple comparison between the two cases due to the differences in the type, scale, and social characteristics of the communities that suffered each disaster, it is clear that the intention of the government and their willingness to stick to their original plans differed markedly. The pressure by disaster-affected residents to return to their normal daily lives and rebuild in the place they had originally lived in is very strong. Unless a given government is willing and able to withstand such pressures, and offer meaningful and quick alternatives, it is difficult to rebuild settlements that are more resilient to future disasters.

8.3 GRADUAL CHANGE OF LAND USE FOR RISK REDUCTION

Tomoya Shibayama and Naoki Hoshiyama

Waseda University, Tokyo, Japan

8.3.1 Introduction

The Japanese islands are also known as the "disaster archipelago", given the range and frequency with which natural hazards affect human settlements within them. The oldest recorded tsunami in Japan was the *Hakuho Tsunami* of 797 (see Section 9.1), and since this time many other tsunami and storm surge events have been recorded in the country (see for example Section 3.6). Particularly, during the period of industrialization and modernization following the Meiji Restoration in 1867 and the rapid economic growth after World War II, coastal areas became more extensively used, increasing the vulnerability and damage suffered during major disasters.

However, given the phenomenon of population ageing and decline, there is an ongoing general decrease in the population density of many coastal areas in Japan which, together with improved risk management strategies, means the country is moving away from the label of being a "disaster archipelago". In that sense, encouraging the movement of population to areas with a smaller chance of suffering from a given hazard is an important risk management strategy (Figure 8.3.1, see also Section 8.1). In that sense, the author of this chapter has been advising his students for the last 30 years that "When you graduate from university, enter the workforce and get married, moment at which you will have to choose a new house. When you do, please choose a place not only because of the low rent and land price

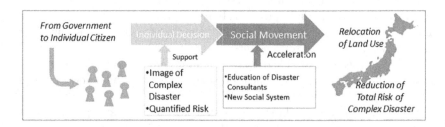

Figure 8.3.1 Gradual change in disaster risk management in Japan.

(economic efficiency) and the short commute to work (convenience), but also due to it being safe from hazards (vulnerability to disasters)". If such a mindset on how to choose a house gradually expands it will be possible to slowly build a society that is more resilient to disasters.

However, in order to provide adequate information so that individuals can make good choices it is necessary to perform a risk assessment of the various residential areas to a range of hazards. To do so it is necessary to understand the return probabilities of various hazards, estimate the potential losses, multiply these by the probability of occurrence, convert it into a monetary value, and calculate it on the same scale as economic efficiency and convenience. Nowadays, this calculation can be done with some degree of accuracy, and such hazard maps and computations can be referred to by prospective house-buyers when choosing a new home.

8.3.2 Calculation of disaster risk in Japan

For the case of Japan, it is possible to calculate the expected value of annual disaster losses by using databases prepared by government agencies, such as those by the Cabinet Office, Disaster Prevention (2003), Ministry of Land, Infrastructure, Transport and Tourism (n.d), Government of Japan, Earthquake Research Promotion Headquarters (2019), Bureau of Construction, Tokyo Metropolitan Government (2013, 2018) or Tokyo Metropolitan Government Disaster Prevention (2012). Waseda University developed a new index for the quantitative evaluation of natural disaster risk using simulation data and numerical geographical information, which provides the expected annual damage to houses due to natural disasters. To do so, first the annual probability of each of the seven major types of hazards affecting Japan (namely earthquake-induced building collapse, liquefaction, fires, tsunamis, river flooding and sediment-related disasters caused by rainfall, and storm surges) was quantified and determined, and then the expected damage to housing stock for a given return period could be calculated. These values were multiplied for each of the seven disasters and finally added together to obtain an overall picture of disaster risk, which can be mapped for all regions of Japan (though it should be noted that not all types of data are available in each prefecture).

Next, specific data to determine the annual probability of each of the seven types of disasters was collected. First, the probability of occurrence of earthquakes is based on the distribution of the probability of experiencing a seismic intensity greater than a specified value in the next 50 years, as shown in the probabilistic seismic motion prediction map by the Earthquake Headquarters of the Cabinet Office. For liquefaction and fire spread, the probability of these happening is calculated based on the chance of earthquakes taking place, according to the estimation by the Headquarters of Earthquake Research Promotion of the Japanese Government (2019) for each prefecture. Figure 8.3.2 shows an example of liquefaction estimation

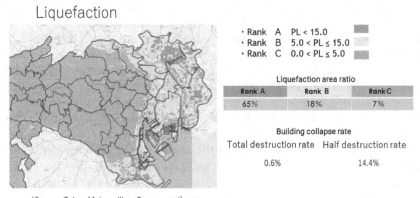

Liquefaction

- Rank A   PL < 15.0
- Rank B   5.0 < PL ≤ 15.0
- Rank C   0.0 < PL ≤ 5.0

Liquefaction area ratio

Rank A	Rank B	Rank C
65%	18%	7%

Building collapse rate
Total destruction rate   Half destruction rate

0.6%   14.4%

(Source: Tokyo Metropolitan Government)

Figure 8.3.2 Liquefaction calculation (modified from Tokyo Metropolitan Government, 2013). PL represents the "Liquefaction PL" parameter, which provides an overall evaluation of the possibility of liquefaction taking place.

based on the data of the Bureau of Construction, Tokyo Metropolitan Government (2013). The liquefaction ratio is classified into three different ranks. In Figure 8.3.2, PL represents the "Liquefaction PL" parameter, which provides an overall evaluation of the possibility of liquefaction taking place (Iwasaki et al., 1980). The percentage (%) (Liquefaction area ratio) indicates the ratio of land surface over which liquefaction would take place.

The flood damage caused by heavy rainfall was calculated using each river's flood inundation area map. Two digital maps were used, one for rainfall with an expected return period of 50 to 200 years and the other for rainfall that occurs every 1,000 years, each of which is considered as an independent event. The storm surge is assumed to be caused by a typhoon of the largest possible size striking through the worst possible path it can take, with a return probability of once in 3,000 years (see also Sections 5.2 and 7.3). The probability of a landslide was estimated according to the hazard map of landslide occurrence, based on topographical and geological factors by the National Institute of Land and Infrastructure Management. Finally, the return period of tsunamis was assumed to be once in 2000 years, using the inundation hazard map assuming the largest class of tsunami (Level 2 tsunami, see also Section 5.1).

The damage to houses was classified into collapsed damage and flooding damage. In the case of collapsed houses, total damage was set at 34.63 million yen, with half damage being 17.32 million yen (which is the standard way of evaluating such damage in Japan according to the methodology of the Ministry of Land, Infrastructure and Transport, 2020). A two-storey wooden house with a total floor area of 100 m² was assumed for inundation damage, with the monetary losses being calculated for a given depth of inundation employing the methodology of the Flood Control Economy Manual by the Ministry of Land, Infrastructure and Transport, 2020).

It was possible to obtain sufficient data to perform the damage simulations and quantify all seven types of disasters in major urban areas such as Tokyo, Osaka, and Kanagawa. However, many local governments in Hokkaido, Tohoku, Hokuriku or other regions have insufficient data that could allow a detailed damage simulation to be performed. This highlights the need to gather further data to improve the accuracy of such exercises and allow better risk management practices to be implemented.

8.3.3 Introduction of disaster zoning

In Japan the Building Standards Law now has a zoning system that regulates the construction of buildings in areas with a high probability of disaster occurrence. The designation of tsunami disaster warning areas has been implemented to encourage residents to withdraw from such zones, and local governments can use these designations to prohibit the building of new dwellings. However, this process takes a long time and may have some negative social side effects, such as some residents (including many elderly) being left behind in their original houses while others relocate. This can aggravate certain social problems and lead to isolation and exclusion, which can be particularly problematic in an ageing society such as that in Japan. This highlights the importance of taking a step-by-step approach when relocating residents from disaster-prone areas by using strong regulations and a system that induces gradual changes, as described above, while preventing residents from being left behind in disaster-prone areas.

8.3.4 Conclusions

It is necessary that disaster risk managers and academics prepare sufficient scientific information that allows citizens to carefully choose the area in which they want to live, taking into account the potential monetary losses due to natural disasters. The gradual change in the location and density of residential areas, gradually moving more of the population to safer areas, can improve the resilience of Japanese settlements to natural hazards. Nevertheless, this process can take time, maybe even a full generation, and can lead to temporary social problems with some older residents being left in disaster prone-areas while their children relocate to safer areas. It is thus clear that a holistic approach to disaster risk mitigation is required, one which not only considers engineering and disaster risk management aspects but involves sociologists, in order to ensure the sustainability of coastal settlements in the future.

REFERENCES

Carlos-Arce, R.S., Onuki, M., Esteban, M., & Shibayama, T. (2017). "Risk Awareness and Intended Tsunami Evacuation Behavior of International Tourists in Kamakura City, Japan." *Int. J. Disaster Risk Reduct.*, 23: 178–192. https://doi.org/10.1016/j.ijdrr.2017.04.005

Badan Perencanaan Pembangunan Nasional (BAPPENAS) (2005). *A Draft of "Blueprint" Prepared by BAPPENAS for Local Explanation*, BAPPENAS.

Bureau of Construction, Tokyo Metropolitan Government (2013). Forecast Map of Liquefaction in Tokyo (2012 Revised Version). http://doboku.metro.tokyo.jp/start/03-jyouhou/ekijyouka/pdf/00_zenbun__.pdf

Bureau of Construction, Tokyo Metropolitan Government (2018). Tokyo Storm Surge Inundation Forecast Area Map. https://www.kouwan.metro.tokyo.lg.jp/yakuwari/files/WFD.jpg

Cabinet Office, Disaster Prevention (2003). Damage Assumption Methodology for the Tonankai and Nankai Earthquakes. http://www.bousai.go.jp/jishin/tonankai_nankai/pdf/sankou_siryou.pdf

Chen, X., & Zhan, F. B. (2008). Agent-Based Modelling and Simulation of Urban Evacuation: Relative Effectiveness of Simultaneous and Staged Evacuation Strategies. *J. Oper. Res. Soc.*, 59(1): 25–33. https://doi.org/10.1057/palgrave.jors.2602321

Dawson, R.J., Peppe, R., & Wang, M. (2011). "An Agent-Based Model for Risk-Based Flood Incident Management." *Nat. Hazards*, 59: 167–189. https://doi.org/10.1007/s11069-011-9745-4

Enomoto, T., Takanashi, N., & Ochiai, T. (2015). "Research for Enforcement of Tsunami Evacuation Measures in the Historical Tourist City, Kamakura, Based on Findings by Surveys." *J. Soc. Safety Sci*, 27: 1–10. https://doi.org/10.11314/jisss.27.75

Ghobarah, A., Saatcioglu, M., & Nistor, I. (2006). "The Impact of the 26 December 2004 Earthquake and Tsunami on Structures and Infrastructure." *Eng. Struct.*, 28(2): 312–326. https://doi.org/10.1016/j.engstruct.2005.09.028

Government of Japan, Earthquake Research Promotion Headquarters (2019). Probabilistic Seismicity Forecast Map (Revised January 2019). https://www.jishin.go.jp/main/chousa/18_yosokuchizu/yosokuchizu2018_chizu_2.pdf

Hamano, G., Ishii, H., Iimura, K., Takabatake, T., Stolle, J., et al. (2020). "Evaluation of Force Exerted by Tetrapods Displaced by Tsunami on Caisson Breakwater Return Wall." *Coast. Eng. J.*, 62(2): 170–181. https://doi.org/10.1080/21664250.2020.1723194

Harnantyari, A., Takabatake, T., Esteban, M., Valenzuela, P., Nishida, Y., et al. (2020). "Tsunami Awareness and Evacuation Behaviour during the 2018 Sulawesi Earthquake Tsunami." *Int. J. Disaster Risk Reduct.*, 40: 101389. https://doi.org/10.1016/j.ijdrr.2019.101389

Ito, E., Kawase, H., Matsushima, S., & Hatayama, M. (2020). "Tsunami Evacuation Simulation Considering Road Blockage by Collapsed Buildings Evaluated from Predicted Strong Ground Motion." *Nat. Hazards*, 101: 959–980. https://doi.org/10.1007/s11069-020-03903-2

Iuchi, K. (2018). *Rhetoric of Recovering Resilient Unveiling How Building Back Safer Transforms into Development for Prosperity: A Case of Post-Yolanda Rebuilding, Descriptive Case Study*. Lincoln Institute of Land Policy.

Iwasaki, T., Tatsuoka, F., Tokida, K., & Yasuda, S (1980). Estimation of Degree of Soil Liquefaction during Earthquakes. *Soil and Foundations*, 28(4): 23–29.

Japan International Cooperation Agency (JICA) (2005). *The Study on the Urgent Rehabilitation and Reconstruction Support Program for Aceh Province and Affected Areas in North Sumatra*, Urgent Rehabilitation and Reconstruction Plan for Banda Aceh City, JICA.

Japan International Cooperation Agency (JICA) (2015). *The Project on Rehabilitation and Recovery from Typhoon Yolanda*, JICA.

Johnstone, W. M., & Lence, B. J. (2009). Assessing the Value of Mitigation Strategies in Reducing the Impacts of Rapid-Onset, Catastrophic Floods. *J. Flood Risk Manag.*, 2(3): 209–221. https://doi.org/10.1111/j.1753-318X.2009.01035.x

Johnstone, W. M., & Lence, B. J. (2012). "Use of Flood, Loss, and Evacuation Models to Assess Exposure and Improve a Community Tsunami Response Plan: Vancouver Island." *Nat Hazards Rev.*, 13(2): 162–171. https://doi.org/10.1061/(ASCE)NH.1527-6996.0000056

Koyanagi, Y., & Arikawa, T. (2016). "Study of the Tsunami Evacuation Tower using the Tsunami Evacuation Simulation." *J. JSCE, B2 (Coastal Eng.)*, 72(2): 1567–1572. https://doi.org/10.2208/kaigan.72.I_1567

Krautwald, C., Stolle, J., Robertson, I., Achiari, H., Mikami, T., et al. (2021). "Engineering Lessons from September 28, 2018 Indonesian Tsunami: Scouring Mechanisms and Effects on Infrastructure." *J. Waterw. Port Coast. Ocean Eng.*, 147(2): 04020056. https://doi.org/10.1061/(ASCE)WW.1943-5460.0000620

Kumagai, K. (2014). "Validation of Tsunami Evacuation Simulation to Evacuation Activity from the 2011 off the Pacific Coast of Tohoku Earthquake Tsunami." *J. Japan Soc. Civil Eng. Ser. D*, 70(5): 187–196. https://doi.org/10.2208/jscejipm.70.I_187

Liu, H., Liu, B., Zhang, H., Li, L., Qin, X., et al. (2018). "Crowd Evacuation Simulation Approach Based on Navigation Knowledge and Two-layer Control Mechanism." *Inf. Sci.*, 436–437: 247–267. https://doi.org/10.1016/j.ins.2018.01.023

Liu, Y., Okada, N., & Takeuchi, Y. (2008). "Dynamic Route Decision Model-Based Multi-Agent Evacuation Simulation – Case Study of Nagata Ward, Kobe." *J. Nat. Disaster Sci.*, 28(2): 91–98. https://doi.org/10.2328/jnds.28.91

Makinoshima, F., Imamura, F., & Abe, Y. (2016). "Behavior from Tsunami Recorded in the Multimedia Sources at Kesennuma City in the 2011 Tohoku Tsunami and its Simulation by Using the Evacuation Model with Pedestrian – Car Interaction." *Coast. Eng. J.*, 58(4): 1640023. https://doi.org/10.1142/S0578563416400234

Mas, E., Suppasri, A., Imamura, F., & Koshimura, S. (2012). "Agent-Based Simulation of the 2011 Great East Japan Earthquake/Tsunami Evacuation: An Integrated Model of Tsunami Inundation and Evacuation." *J. Nat. Disaster Sci.*, 34(1): 41–57. https://doi.org/10.2328/jnds.34.41

Matsumaru, R., Nagami, K., and Takeya, K. (2012). Reconstruction of the Aceh Region following the 2004 Indian Ocean Tsunami Disaster: A Transportation Perspective. *IATSS Research*, 36(1): 11–19.

Mikami, T., Shibayama, T., Esteban, M., Takabatake, T., Nakamura, R., et al. (2019). "Field Survey of the 2018 Sulawesi Tsunami: Inundation and Run-Up Heights and Damage to Coastal Communities." *Pure Appl. Geophys.*, 176(8): 3291–3304. https://doi.org/10.1007/s00024-019-02258-5

Mikami, T., Shibayama, T., Takagi, H., Matsumaru, R., Esteban, M., Nguyen, D. T., De Leon, M., Valenzuela, V. P., Oyama, T., Nakamura, R., Kumagai, K., & Siyang, L. (2016). Storm Surge Heights and Damage Caused by the 2013 Typhoon Haiyan along the Leyte Gulf Coast. *Coastal Engineering Journal*, 58: 1640005. https://doi.org/10.1142/S0578563416400052

Mostafizi, A., Wang, H., Cox, D., & Dong, S. (2019). "An Agent-based Vertical Evacuation Model for a Near-field Tsunami: Choice Behavior, Logical Shelter Locations, and Life Safety." *Int. J. Disaster Risk Reduct.*, 34(3): 467–479. https://doi.org/10.1016/j.ijdrr.2018.12.018

Mostafizi, A., Wang, H., Cox, D., Cramer L. A., & Dong, S. (2017) "Agent-based Tsunami Evacuation Modeling of Unplanned Network Disruptions for Evidence-Driven Resource Allocation and Retrofitting Strategies." *Nat. Hazards*, 88(3): 1347–1372. https://doi.org/10.1007/s11069-017-2927-y

Na, H. S., and Banerjee, A. (2019). "Agent-based Discrete-event Simulation Model for No-notice Natural Disaster Evacuation Planning." *Comput. Ind. Eng.*, 129: 44–55. https://doi.org/10.1016/j.cie.2019.01.022

Nagai, R., Takabatake, T., Esteban, M., Ishii, H., & Shibayama, T. (2020). "Tsunami Risk Hazard in Tokyo Bay: The Challenge of Future Sea Level Rise." *Int. J. Disaster Risk Reduct.*, 45: 101321. https://doi.org/10.1016/j.ijdrr.2019.101321

National Police Agency of Japan (2020). "Damage Report of 2011 Tohoku Earthquake and Tsunami." Accessed on 29 March 2020. https://www.npa.go.jp/news/other/earthquake2011/pdf/higaijokyo.pdf

Nishihata, T., Moriya, Y., Anno, K., & Imamura, F. (2012). "Modeling of Car Evacuation from Tsunami Attacking and Evaluation for Accompanied Traffic Jam." *J. JSCE, B2 (Coastal Eng.)*, 68(2): 1316–1320. https://doi.org/10.2208/kaigan.68.I_1316

Ong, J. M., Jamero, M. L., Esteban, M., Honda, R., & Onuki, M. (2016). Challenges in Build-Back-Better Housings Reconstructions Programs for Coastal Disaster Management: Case of Tacloban City, Philippines. *Coastal Engineering Journal*, 58(1): 1640010.

Owen, M., Galea, E. R., & Lawrence, P. J. (1996). "The Exodus Evacuation Model Applied to Building Evacuation Scenarios." *J. Fire Prot. Eng.*, 8(2): 65–86. https://doi.org/10.1177/104239159600800202

Pan, X., Han, C., Dauber, K., & Law, K. (2007). "A Multi-Agent Based Framework for the Simulation of Human and Social Behavior during Emergency Evacuations." *AI & Soc.*, 22(2): 113–132. https://doi.org/10.1007/s00146-007-0126-1

Pelechano, N., & Badler, N. (2006) "Modeling Crowd and Trained Leader Behavior during Building Evacuation." *IEEE Computer Graphics and Applications*, 26(6): 80–86. https://doi.org/10.1109/MCG.2006.133

Shibayama, T. (2015). "2004 Indian Ocean Tsunami." In Esteban, M., Takagi, H., & Shibayama, T. (eds.). *Handbook of Coastal Disaster Mitigation for Engineers and Planners*. Butterworth-Heinemann (Elsevier), Oxford, pp. 3–19.

Shibayama, T., Esteban, M., Nistor, I., Takagi, H., Nguyen, D. T., et al. (2013a). "Classification of Tsunami and Evacuation Areas." *Nat. Hazards*, 67(2): 365–386. https://doi.org/10.1007/s11069-013-0567-4

Shibayama, T., Ohira, K., & Takabatake, T. (2013b). Present and Future Tsunami and Storm Surge Protections in Tokyo and Sagami Bays. In *Proceedings of the 7th International Conference on Asian and Pacific Coasts (APAC 2013)*, pp. 764–766.

Stolle, J., Krautwald, C., Robertson, I., Achiari, H., Mikami, T., et al. (2020). "Engineering Lessons from the 28 September 2018 Indonesian Tsunami: Debris Loading." *Can. J. Civ. Eng.*, 47(1): 1–12. https://doi.org/10.1139/cjce-2019-0049

Stolle, J., Takabatake, T., Mikami, T., Shibayama, T., Goseberg, N., et al. (2017). "Experimental Investigation of Debris-induced Loading in Tsunami-like Flood Events." *Geosciences*, 7: 74. https://doi.org/10.3390/geosciences7030074

Stolle, J., Takabatake, T., Nistor, I., Mikami, T., Nishizaki, S., et al. (2018). "Experimental Investigation of Debris Damming Loads under Transient Supercritical Flow Conditions." *Coast. Eng.*, 139: 16–31. https://doi.org/10.1016/j.coastaleng.2018.04.026

Sugimoto, T., Murakami, H., Kozuki, Y., & Nishikawa, K. (2003) "A Human Damage Prediction Method for Tsunami Disasters Incorporating Evacuation Activities." *Nat. Hazards*, 29(3): 587–602. http://doi.org/10.1023/A:1024779724065

Takabatake, T., Chenxi, D. H., Esteban, M., & Shibayama, T. 2021b. "Influence of Road Blockage on Tsunami Evacuation: A Comparative Study of Three Different Coastal Cities in Japan." *Int. J. Disaster Risk Reduct.*, 102684. https://doi.org/10.1016/j.ijdrr.2021.102684

Takabatake, T., Esteban, M., Nistor, I., Shibayama, T., & Nishizaki, S. (2020a). Effectiveness of Hard and Soft Tsunami Countermeasures on Loss of Life under Different Population Scenarios. *Int. J. Disaster Risk Reduct.*, 45: 101491. https://doi.org/10.1016/j.ijdrr.2020.101491

Takabatake, T., Fujisawa, K., Esteban, M., & Shibayama, T. 2020b. "Simulated Effectiveness of a Car Evacuation from a Tsunami." *Int. J. Disaster Risk Reduct.*, 47: 101532. https://doi.org/10.1016/j.ijdrr.2020.101532

Takabatake, T., Mäll, M., Esteban, M., Nakamura, R., Kyaw, T. O., et al. (2018b). "Field Survey of 2018 Typhoon Jebi in Japan: Lessons for Disaster Risk Management." *Geosciences*, 8(11): 412. https://doi.org/10.3390/geosciences8110412

Takabatake, T., Nistor, I., & St-Germain, P. (2020c). "Tsunami Evacuation Simulation for the District of Tofino, Vancouver Island, Canada." *Int. J. Disaster Risk Reduct.*, 48: 101573. https://doi.org/10.1016/j.ijdrr.2020.101573

Takabatake, T., Shibayama, T., Esteban, M., & Ishii, H. (2018a) "Advanced Casualty Estimation Based on Tsunami Evacuation Intended Behavior: Case Study at Yuigahama Beach, Kamakura, Japan." *Nat. Hazards*, 92(3): 1763–1788. https://doi.org/10.1007/s11069-018-3277-0

Takabatake, T., Shibayama, T., Esteban, M., Achiari, H., Nurisman, N., et al. (2019a). "Field Survey and Evacuation Behaviour during the 2018 Sunda Strait Tsunami." *Coast. Eng. J.*, 61(4): 423–443. https://doi.org/10.1080/21664250.2019.1647963

Takabatake, T., Shibayama, T., Esteban, M., Ishii, H., & Hamano, G. (2017). "Simulated Tsunami Evacuation Behavior of Local Residents and Visitors in Kamakura, Japan." *Int. J. Disaster Risk Reduct.*, 23: 1–14. https://doi.org/10.1016/j.ijdrr.2017.04.003

Takabatake, T., St-Germain, P., Nistor, I., & Shibayama, T. (2019b). "Numerical Modelling of Coastal Inundation from Cascadia Subduction Zone Tsunamis and Implications for Coastal Communities on Western Vancouver Island, Canada." *Nat. Hazards*, 98: 267–291. https://doi.org/10.1007/s11069-019-03614-3

Takabatake, T., Stolle, J., Hiraishi, K., Kihara, N., Nojima, K., et al. (2021a). "Inter-Model Comparison for Tsunami Debris Simulation." *J. Disaster Res.*, 16:1030–1044. https://doi.org/10.20965/jdr.2021.p1030

The National Economic and Development Authority (2013). Reconstruction Assistance on Yolanda (RAY): Build-Back-Better.

Tokyo Metropolitan Government Disaster Prevention (2012). Estimated Damage in Tokyo due to an Earthquake Directly Beneath the Tokyo Metropolitan Area. https://www.bousai.metro.tokyo.lg.jp/_res/projects/default_project/_page_/001/000/401/assumption.part1-2-4.pdf

Uno, Y., Shigihara, Y., & Okayasu, A. (2015). "Development of Crowd Evacuation Simulation for Risk Evaluation of Human Damage by Tsunami Inundation." *J. JSCE, B2 (Coastal Eng.)*, 71(2): 1615–1620. https://doi.org/10.2208/kaigan.71.I_1615

Wang, Z., and Jia, G. (2021). "A Novel Agent-Based Model for Tsunami Evacuation Simulation and Risk Assessment." *Nat. Hazards*, 105:2045–2071. https://doi.org/10.1007/s11069-020-04389-8

Wood, N. J., & Schmidtlein, M. C. (2013). "Community Variations in Population Exposure to Near-field Tsunami Hazards as a Function of Pedestrian Travel Time to Safety." *Nat. Hazards*, 65(3): 1603–1628. https://doi.org/10.1007/s11069-012-0434-8

Yasuda, T., Hatayama, M., & Shimada, H. (2016). "National Questionnaire Survey of Surfers Awareness about Tsunami Evacuation." *Safety Science Review*, 6: 61–80.

Chapter 9

Case reports of selected countries

CONTENTS

9.1 EAST AND SOUTHEAST ASIA

9.1.1 Coastal erosion problems in Vietnam: Present status, causes and proposed solutions

Nguyen The Duy and Nguyen Danh Thao
Ho Chi Minh City University of Technology (HCMUT), Ho Chi Minh City, Vietnam

Nguyen Ngoc An
New CC Construction Consultants Co., Ltd., Ho Chi Minh City, Vietnam

Le Van Cong
Vietnam Administration of Seas and Islands (VASI), Hanoi, Vietnam

9.1.1.1 Introduction

Vietnam has a long coastline that is over 3,260 km in length. Coastal areas with rich natural resources currently provide livelihoods for about half of the population of the country, though these are often heavily affected by natural disasters, causing significant human and economic losses (Thao et al., 2014).

Coastal erosion is a common phenomenon in the coastal areas of all three regions of Vietnam: the Northern, Central and Southern coastlines. Coastal erosion can have very serious consequences, including the direct loss of life, property, infrastructure and land (as discussed elsewhere in this book, see for example Section 9.4). It can also contribute to exacerbate other coastal hazards, and degrade the environment through the process of inundation, salinization and pollution of lands, resulting in unsustainable socio-economic development, low efficiency of investments and ultimately the migration of those affected. From the end of the 20th century to the present, the level of coastal erosion in Vietnam has increased in scale and intensity, probably related to both the rapid economic and social changes that have taken place since the promotion of the Doi Moi (reform) policy, which started in 1986. The average erosion rate is about 5–10 m per year, but can reach 50–100 m per year or more. About 12 million people in coastal provinces are at risk of being affected by flooding and more than 35% of settlements have shorelines that are being eroded. The coastal tourism industry mainly relies on beaches and primary ecosystems to attract visitors, but up to 42% of resorts and hotels are built in coastal areas that are located near beaches that are being eroded.

Generally speaking, the causes of coastal erosion in Vietnam can be divided into two groups:

- **Natural factors:** including the impacts of wind, waves, tides, nearshore currents, geological structure of the coastal area, and position of the shoreline.
- **Human impacts:** besides natural factors, negative impacts from humans are also increasing the process of coastal erosion, such as:
  - Destruction of watershed protection forests and mangroves. Many coastal communities are not aware that mangrove forests form an ecosystem that helps to reduce the intensity of wind, waves and currents as well as creating favourable conditions for sediment accumulation that prevent the erosion of the coast. As a result, these are often cut in order to provide space for aquaculture, build tourist resorts or construct other projects close to the coast (e.g., the mangrove area of the Mekong Delta decreased from 250,000 hectares in 1950 to 46,000 hectares in 2001).
  - The decrease of sand and sediment supply from rivers to estuaries due to the construction of dams for hydropower generation, flood protection measures or the exploitation of sand as a construction material.
  - The uncontrollable growth of marine tourism with the building of many beach resorts has also increased the coastal erosion process. The reason for this is that most resorts have constructed groynes to trap sand and enlarge their beaches, which causes other areas further along the longshore drift to be eroded.

9.1.1.2 Erosion status of the Vietnamese coastline

The erosion patterns vary greatly depending on the topography of each area, though generally speaking the country can be divided into three zones (see Figure 9.1.1.1): North (from Quang Ninh to Ninh Binh), Central (from Thanh Hoa to Binh Thuan) and South (from Vung Tau to Kien Giang). According to published survey data (Van, 2013), nationwide there are 2,229 points which are affected by riverbank and coastal erosion, over a total

Figure 9.1.1.1 The three areas into which the Vietnam coastline can be divided (Cong et al., 2014).

length of more than 2,837 km, of which the provinces in the Northern area and Thanh Hoa represent 439 points (394 km), the Central coastal provinces from Nghe An to Binh Thuan 815 points (1,200 km), the Southeast provinces 117 points (160 km) and the Mekong Delta provinces 470 points (689 km). In the subsections below examples of some of the worst affected areas in each of these stretches will be highlighted.

9.1.1.2.1 Northern coastal area (from Quang Ninh to Ninh Binh)

From Mong Cai to Nam Dinh, Cat Hai (Hai Phong) and Hai Hau (Nam Dinh) are the most severely eroded areas. Since 1955, when a hydro-electric dam was built on a tributary of the Red River, and then in 1987 when the Hoa Binh hydroelectric dam was brought into operation, the Hai Hau coast has suffered intensive erosion, with the shore advancing inland by more than 20 m per year. This area is composed of a fine sand, and some experts believe the impact of sea waves and tides is partially to blame for this erosion. Hai Hau is affected by 5–6 big storms every year which, if they coincide with high tides, can result in waves up to 6–7 m high. An increase in the number of storms in the region has been recorded, together with the perpendicular direction of the coastline to the North-East monsoon, are all factors that can increase the impact of sea waves, damaging structures and changing the topography of the shore-line (see also Section 7.3).

9.1.1.2.2 Central coastal area (from Thanh Hoa to Binh Thuan)

Waves and wind are the two main driving forces behind coastal erosion in the Central coast, in which 93% of the coastal soil is composed of sand, gravel, mud, or muddy sand. In the Central coastal area rivers are typically short, have a steep bed slope and supply small amounts of sediments to the sea, providing fewer materials that can attenuate the incoming wave energy. In addition, the Central region suffers many storms every year, with the number of such storms increasing gradually: 117 storms from 1901 to 1930, 134 from 1931 to 1960, 171 from 1961 to 1990, and from 1991 to 2020 there were 183 storms. In addition, the position of the shoreline relative to the wave and monsoon directions is another important factor causing coastal erosion to the Central area of Vietnam.

The storm surges resulting from such systems can increase the water level and allow waves of up to 4–5 m high to hit the coastline (see Section 5.2). For example, from the end of September to November of 2020, severe and widespread coastal erosion took place due to the consecutive impact of eight storms from the East Sea, six of which made landfall in the Central region, generating large waves which resulted in severe beach erosion at many coastal locations.

The coast from Thanh Hoa to Nha Trang has suffered intense erosion, covering an area of nearly 9,000 hectares, of which 94% were sandy banks (Figure 9.1.1.2, left, shows the erosion of a sandy beach in Hoi An city, Quang Nam province). In places with rough terrain, inside of bays, and at sections with coastal cliffs the rate of erosion is slower than along the sandy shores and capes that are facing the wind and waves, where erosion is much faster. Going further to the south of Nha Trang, the coastlines of Phu Yen and Binh Thuan have also suffered severe erosion. Figure 9.1.1.2, right, shows an image of collapsed houses due to beach erosion in Phan Thiet city, Binh Thuan province.

In general, the scale and speed of beach erosion in the Central coastal area have been increasing. The mean erosion speed in some areas reaches 10 m per year, such as at Tuy Phong, Ham Tien (Binh Thuan), Cua Dai (Hoi An),

Figure 9.1.1.2 Left: Erosion of a sandy beach (Hoi An City, Quang Nam Province). Right: Collapsed houses due to beach erosion (Phan Thiet City, Binh Thuan Province).

Sa Huynh (Quang Ngai), and up to 25 m per year at some places such as Phu Thuan, Phu Vang (Thua Thien-Hue), Cua Dai (Hoi An).

9.1.1.2.3 Southern coastal area (from Vung Tau to Kien Giang)

In the South, the coast from Vung Tau to Kien Giang is composed of loose alluvial sediments from the Holocene period, and are thus relatively young geomorphological structures. The main structural component is brown clay mud, which is very easily broken even by moderate wave energy, and is easily moved by the nearshore currents. Before 1940 this coastline was completely free of erosion, though from 1940 to 1950 the estuaries started to slowly be eroded, and from 1960 until the present date the intensity of this has increased.

In the Can Gio area (Ho Chi Minh City) the most severe erosion has been observed at the coasts of Can Thanh and Thanh An. Tidal river flows and nearshore currents are the main driving forces of coastal erosion. Since the Tri An hydroelectric dam was built the amount of sediment transported to the estuaries decreased, with this area suffering 10–20 m of erosion per year.

The coast of Go Cong Dong has been partially eroded due to the impact of waves and nearshore currents due to the northeast monsoon season, as well as the digging of irrigation canals, the excavation of flood discharge channels towards the Gulf of Thailand, the cutting of mangrove forest to build ecotourism projects, and the development aquaculture production, resulting in an erosion rate of 10–30 m per year. In this area the protective forest layer in front of the sea dike is gradually narrowing, and in 2000 a 3-km-long stretch of forest in the Tan Thanh area disappeared, with the sea eroding to the base of the dyke.

Cape Ca Mau is being eroded due to overexploitation of the underground water system (in spite of local regulations) and the increased energy of coastal waves. Essentially, the amount of river sediment being transported to the coast is gradually decreased, partly due to the impact of hydroelectric dams built in the upstream Mekong, and partly due to the excavation of irrigation canals in the delta. The west coast of Ca Mau is eroded due to the cutting of mangrove forests.

The coast of Kien Giang is influenced by the hydrological regime of the West Sea. Due to the fact that there are fewer estuaries in this area than in the East Sea, the Kien Giang coast is quite stable and accretion dominates. However, in recent years the area around Ranh cape on the south bank of the Cai Lon river has been encroached by the sea by about 200 m and around Vam Ray (Hon Dat district) by more than 200 m.

9.1.1.3 Countermeasures to prevent erosion

Countermeasures for erosion control and coastal protection can be divided into two groups: structural solutions (or hard solutions) and non-structural solutions (or soft solutions). The types of countermeasures implemented to prevent coastal erosion should have two functions: reducing wave and wind energy and controlling nearshore currents to minimize the transport of sediment along the shore.

9.1.1.3.1 Structural solutions (hard solutions)

These include the construction of seawalls/dikes, groynes, submerged breakwaters, and groynes combined with submerged breakwaters or artificial headlands. Figure 9.1.1.3 shows a sea dike designed to protect the shoreline from storm surges in Central Vietnam (Hoi An City, Quang Nam Province).

Figure 9.1.1.3 Sea dike designed to protect the shoreline from storm surges in Central Vietnam (Hoi An City, Quang Nam Province).

Some experts advise that hard solutions should be prioritized in areas where very severe coastal erosion is taking place, such as Cat Hai (Hai Phong), Hai Hau (Nam Dinh), Hai Duong, Hoa Duan (Thua Thien), Mui Ne (Binh Thuan), Go Cong Dong (Tien Giang) or Ganh Hao (Ca Mau). However, if a hard solution is used in these areas, it is necessary to ensure that scouring does not take place under the toe of the structure (as this can otherwise lead to its collapse, see also Section 6.3) and the ecosystem of the coastal area behind it is not destroyed. These undesirable impacts have been already observed when building shore protection structures along the coast of Doi Duong (Phan Thiet), Ca Na beach and Cape La Gan (Binh Thuan).

Up to now, due to limitations in construction budgets, seawall/dike systems in Vietnam have been only designed to sustain the attack of typhoons up to level 9 on the Beaufort scale (wind speeds = 75–88 km/h) and under the condition of normal tides. However, the Vietnam coastline may be exposed to the attack of typhoons with intensities of up to level 12 on the Beaufort scale (wind speed = 118–133 km/h). Under the simultaneous attacks of high tides and storm waves with greater intensity than the design wave and tide of a structure, seawalls/dikes can suffer severe damage (Figure 9.1.1.4 shows the damage to a seawall after a storm in Hoi An city, Quang Nam province). The damages or failures of these structures can bring about further erosion of the shoreline, flooding of coastal areas, and heavy losses of human life and property.

So far, two-thirds of the total length of seawalls/dikes in Vietnam (about 2,659 km) have been designed and constructed under the above-mentioned

Figure 9.1.1.4 Damage to a seawall after a storm (Hoi An City, Quang Nam Province).

limitation (i.e., only designed for storms up to level 9 in the Beaufort scale). Therefore, the risk of shoreline erosion due to waves and storm surges is still very large (Duy, 2005).

9.1.1.3.2 Non-structural solutions (soft solutions)

These include beach nourishment, planting mangrove forests, casuarina forests or sand dunes, etc. Figure 9.1.1.5 shows a mangrove forest planted along the shoreline at the Can Gio coast, in Ho Chi Minh city.

Soft solutions are less expensive but require some time to become effective (for instance, young mangrove trees can be damaged and uprooted by moderate waves until they are fully developed). Casuarina forest and palm trees can be planted along the sandy Central coast. Mangrove forests that include cork trees, mangroves and melaleuca trees can be planted in the coastal areas of the Red River and Mekong deltas.

Mangrove forests help to protect the coast from erosion and can help to strengthen dike systems. However, planting mangroves can also be difficult, as which species to plant depends on the climatic conditions of each region, the hydrological regime and the physical and chemical properties of the land. For example, a mangrove planting project in the Vam Ray area (Kien Giang province) achieved encouraging results, but the same type of project failed when applied to Long Phu (Soc Trang province) and Go Cong districts (Tien Giang province). This shows that soft solutions can be challenging to introduce, and that each solution must be carefully designed for its target area.

Figure 9.1.1.5 Mangrove forest along the Can Gio coast, Ho Chi Minh City.

9.1.1.3.3 Combined hard and soft solutions

Due to the challenges identified above, some experts have proposed that the optimal solution to protect the coast of Vietnam is to combine both hard and soft solutions (COBSEA, 2013), which carefully considering the characteristics of each region so that the project can contribute to:

- Reducing the incident wave energy;
- Providing natural windscreens for the shoreline (including the planting of mangroves, for example);
- Creating conditions similar to the natural pattern of shoreline establishment and development;
- Ensuring that erosion in one area is not transferred to nearby places.

Also, it is necessary to map out coastal areas at high risk of suffering from erosion in order to implement coastal no build zones (areas adjacent to the shoreline where new houses and buildings cannot be built) (COBSEA, 2013). In the near future, a firm sea dike system must be built in areas at high risk to ensure safety against nearshore waves and currents. The construction of coastal projects, in particular beach resorts, must be carefully planned so as not to increase coastal erosion. At present, a number of provinces and cities along the coast of Vietnam have planned pilot areas to trial solutions to coastal erosion such as Hai Phong, Nam Dinh, Thanh Hoa, Thua Thien – Hue, Binh Thuan, Tra Vinh, Tien Giang, and Ca Mau, in which different measures have been applied depending on the local conditions.

Regarding the experiences of developed countries in solving coastal environment problems, the Japanese experienced big changes to their coastal environment due to the rapid economic growth since 1945, which can provide useful lessons to solve the coastal environment problems that Vietnam is facing now, including coastal erosion. By analyzing the interlink between economic development and coastal environment problems in Japan, and assuming that the development of each country goes through a common process, it could be possible to improve the sequence of events and obtain better results by using different developmental policies based on the Japanese experience (Shibayama, 2008).

9.1.1.4 Conclusions

Coastal erosion represents a serious problem for the coasts of Vietnam, a country with a shoreline that is over 3,000 km in length. In order to limit the damage caused by such issues, countermeasures for erosion control and coastal protection are currently being studied and implemented nationwide, with their success or failure being crucial to the sustainable development of the country in the future.

9.1.2 Numerical modelling of boundary conditions for Typhoon Haiyan (2013) driving forces on a public school building

Justin Valdez

Waseda University, Tokyo, Japan

9.1.2.1 Introduction

The Philippines is a tropical country frequently affected by tropical cyclones generated in the Northwest Pacific Basin, with an average of 20 typhoons entering the Philippine Area of Responsibility annually and 9 making landfall. In 2009, the Metropolitan Manila region was devastated by Typhoon Ketsana (local name: Ondoy), which caused massive flooding due to torrential rainfall and led the government to issue a state of calamity. The strongest typhoon that hit Mindanao, in the southern part of the Philippines where tropical cyclones rarely make landfall, was Bopha (local name: Pablo), which originated at a very low latitude and rapidly intensified before making landfall as a Category 5 typhoon at Davao Oriental in 2012. At the end of 2021, Typhoon Rai (local name: Odette) passed through the Visayas region and heavily affected the provinces of Cebu and Bohol. Given the COVID-19 pandemic, government agencies, NGOs and academics found it was difficult to conduct relief operations and field surveys to assess the actual damage that took place, especially that to remote areas.

Perhaps one of the most well-known destructive typhoons – in terms of casualties and economic cost – that hit the Philippines was Haiyan (local name: Yolanda) in 2013. With a record low central pressure and high maximum sustained winds, Haiyan generated storm surges of up to 6 m high at San Pedro Bay (Mikami et al. 2016, see also Section 3.5). Widespread damage to structures was observed in Tacloban city and nearby towns, including public school buildings that are usually used as evacuation centres during disasters. As a response, in order to improve structures and make them more resilient against such weather events, the National Structural Code of the Philippines (NSCP) was revised in 2015 to increase the design wind speed throughout the country (ASEP 2016). Accordingly, the Department of Public Works and Highways (DPWH) and the Department of Education (DepEd) of the Philippines released modified standard public school building design plans using the updated provisions of the NSCP. However, these school buildings were designed using dead, live, wind, and seismic load provisions and have not considered other driving forces such as flood loads from surges. Obviously, considering the driving forces that can be experienced by a structure during a storm surge event is important for building back better, particularly in the case of essential facilities such as school buildings.

9.1.2.2 Numerical modelling

A three-part model (see Figure 9.1.2.1) consisting of meteorological, storm surge, and structural numerical programs was used to determine the driving forces from a storm surge event similar to that generated by Typhoon Haiyan, together with how this would be experienced by the structure of a public school building (Valdez et al. 2022). Several studies have been conducted for simulating tropical cyclones and storm surges using the Weather Research and Forecasting or WRF model (Skamarock et al. 2008) and Finite Volume Coastal Ocean Model or FVCOM (Chen et al. 2003), respectively (Ohira et al. 2012; Nakamura et al. 2016; Nakamura et al. 2019). Nakamura et al. (2016) simulated Typhoon Haiyan under present climate and different global warming cases, and employed the Tropical Cyclone Bogussing (TC-Bogus) Scheme (Hsiao et al. 2010) to produce realistic outcomes.

Using the output data (inland wind speed, flood horizontal speed, and flood depth) from WRF and FVCOM, the horizontal wind, hydrostatic, and hydrodynamic forces can then be computed and analyzed for a school building model in STAAD. Pro (Bentley Systems 2019). Table 9.1.2.1 summarizes some of the numerical conditions used in the simulations. The location of Panalaron Central School in downtown Tacloban was selected for the study, with fine meshes of up to 5 m around the area (Figure 9.1.2.2).

9.1.2.3 Analysis and discussion

The hindcasting of Typhoon Haiyan in WRF, with the improvements derived from using the TC-Bogussing scheme, provided good results in terms of minimum sea-level pressure, maximum wind speed, and track (Figures 9.1.2.3 and 9.1.2.4). The wind and pressure data from WRF were then used as boundary conditions for the FVCOM storm surge simulation. Measured

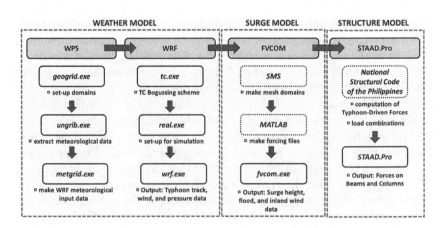

Figure 9.1.2.1 Weather-Surge-Structure Models (Valdez et al., 2022).

Table 9.1.2.1 Domain set-up (Valdez et al., 2022)

WRF Set-up

	Domain 1 (Parent)	*Domain 2 (Nest)*
Simulation Start Time (UTC)	2013 November 04, 18:00	2013 November 07, 06:00
Simulation End Time (UTC)	2013 November 08, 18:00	2013 November 08, 18:00
Latitude and Longitude	2° S – 24° N, 113° E – 158° E	9.1° N – 12.1° N, 123.4° E – 127.1° E
Grid Size	9,000 m x 9,000 m	3,000 m x 3,000 m

FVCOM Set-up

	Domain (Propagation and Inundation)
Simulation Time (UTC)	2013 November 07, 06:00 – 2013 November 08, 18:00
Mesh Size	5 m – 1,000 m

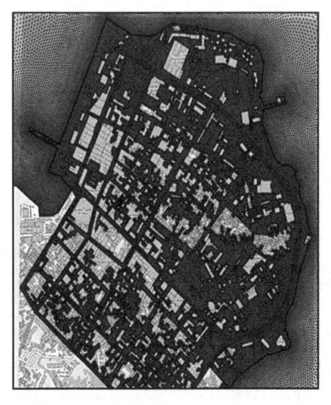

Figure 9.1.2.2 (a) Downtown Tacloban Inundation Mesh with buildings (Valdez et al., 2022).

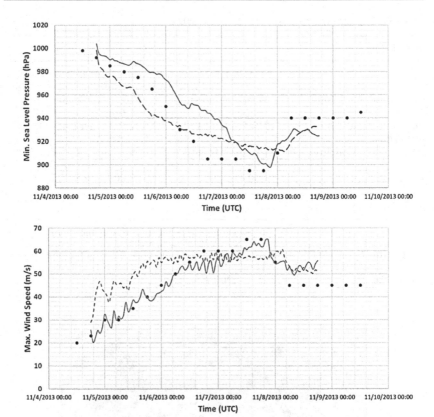

Figure 9.1.2.3 WRF Simulated minimum sea-level pressure and maximum wind
speed, indicating observed data by JMA and simulated results
(Valdez et al., 2022).

watermark levels from the post-Haiyan field survey data of Mikami et al.
(2016) were used to validate the storm surge simulation results (see also
Section 3.5). Afterwards, the time-history data of inland wind speed, flood
horizontal speed, and flood depth at the location of Panalaron Central
School were obtained (see Figure 9.1.2.5). These data provided the boundary conditions for wind and surge (hydrostatic, hydrodynamic, and breaking wave) forces that were applied to the public school building model.

A four-storey DPWH-DepEd public school building was modelled in
STAAD. Pro, with the driving forces due to the storm surge generated by
Typhoon Haiyan being considered (Figure 9.1.2.6), together with the design
load combinations provided in the NSCP. Using the boundary conditions
from the WRF and FVCOM weather and surge simulations, this methodology has the capability to check the strength of beams and columns of a

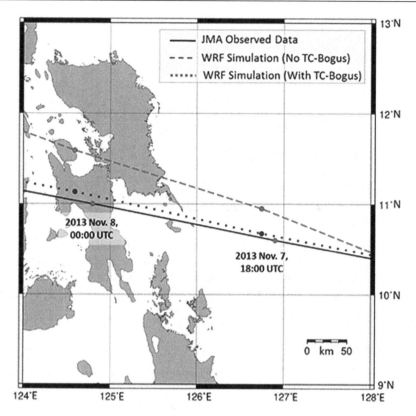

Figure 9.1.2.4 WRF simulated typhoon track.

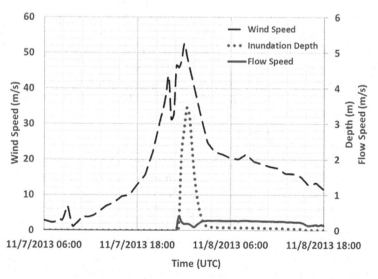

Figure 9.1.2.5 FVCOM simulated inland wind speed, inundation depth, and flow
        speed (Valdez et al., 2022).

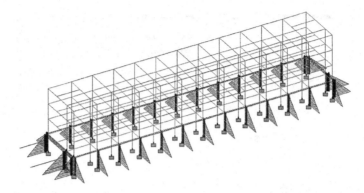

Figure 9.1.2.6 STAAD.Pro public school building model.

structure against a storm surge event similar to that which took place during Typhoon Haiyan. This could be considered as the worst-case scenario, and considering it would mean that essential facilities such as public school buildings would be resilient enough to withstand even the most extreme typhoon events.

9.1.2.4 Conclusions

Given the flexibility of the numerical models used (in terms of set-up and boundary conditions), such methodologies have the potential to simulate other typhoons and storm surges, including future climate scenarios. They represent a useful tool for building back better, by allowing disaster risk managers and practising engineers to make changes to structural codes in order to design more resilient structures, which is particularly important for the case of public schools that are used as evacuation centres. However, further simulations should be carried out to check the validity of the models in the case of different typhoon events and locations. Also, numerical modelling can help in determining whether the existing design code provisions are sufficient against another significant storm surge event like Typhoon Haiyan.

Given that the design code provisions were updated based on the intensity of Haiyan, it is of concern that recurring significant damage to structures takes place when other significant storms hit the country, such as in the case of Typhoon Rai in 2021, which also heavily affected the Visayas region eight years after Haiyan. Thus, it is clear that other factors should also be considered for increasing the resilience of structures, such as how to ensure the proper implementation of the design code, the quality of materials used in houses, and the lack of coastal protection measures, among others. Further coordination among engineers, local officials, and other relevant personnel is clearly needed to discuss these lingering issues and build back better effectively.

9.1.3 Coastal hazards in Myanmar

Thit Oo Kyaw

Taisei Rotec's Research Institute, Saitama, Japan

9.1.3.1 Introduction

Myanmar is a country that is situated in Southeast Asia, bordering the North Indian Ocean to the west and south. The country possesses over 2,000 km of contiguous coastline from the Bay of Bengal to the Andaman Sea, with the Arakan Mountains situated along the northwestern coast, the Ayeyarwady and Yangon deltas in the middle part, and the high terrains of Bilauktaung (Taninthayi coast) to the south (see Figure 9.1.3.1). The total population is estimated to be 54 million, with approximately 9.2% of it living in low-lying coastal areas that are less than 5 m above mean sea level (MIMU, 2019). A tropical to subtropical monsoon climate is dominant in Myanmar, with three main seasons: i) the hot season from mid-February to mid-May, ii) the rainy season from mid-May to October and iii) the cold season from November to mid-February (CFE-DM, 2017).

In general, Myanmar is at risk of suffering the effects of tsunamis, tropical cyclones (TCs), and the storm surges associated with them. The country was affected by the *2004 Indian Ocean Tsunami*, where tsunami heights of

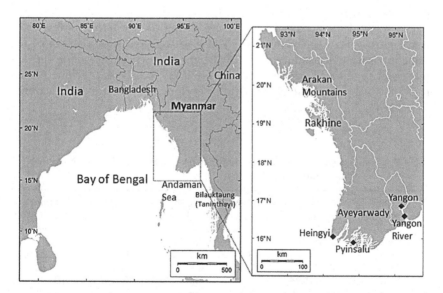

Figure 9.1.3.1 Left: Location of Myanmar in the Indian Ocean. Right: The Rakhine coast and Ayeyarwady-Yangon deltaic coast are often affected by tropical cyclones.

0.4–2.9 m took place along the Ayeyarwady-Yangon delta and Taninthayi coast of Myanmar (Satake et al., 2006). Myanmar is also affected by TCs that originate in the Bay of Bengal mainly during the pre-monsoon period (April or May) and the post-monsoon period (around October or November). Essentially, the arrival of the South Asian monsoon season at the Bay of Bengal during April and early May produces a combination of high sea surface temperatures, high mid-tropospheric humidity and low vertical wind shear that is highly conducive to TCs development (Vishnu et al., 2016). However, during the monsoon (from June to September), the enhanced vertical wind shear produced by the prevailing south-westerlies (in lower) and easterlies (in the upper troposphere) interferes with the development of tropical disturbances over the Bay of Bengal. Thus, between June to September, tropical disturbances over the Bay of Bengal are mostly limited to monsoon depressions that seldom develop into TCs (Yoon and Huang, 2012, Tasnim et al., 2015; Fosu and Wang, 2015).

Cyclone tracks in the Bay of Bengal basin generally do not veer towards Myanmar, and therefore the country is less exposed to these events than other countries in the region (such as Bangladesh, Tasnim et al., 2015, see also Section 8.2). According to Myanmar's Department of Meteorology and Hydrology (DMH), on average 10 TCs formed each year over the Bay of Bengal between 1887 and 2005, though only 80 out of the 1,248 TCs recorded events (i.e., only 6.4%) made landfall in Myanmar. Table 9.1.3.1 shows a list of the most severe historical cyclones that have affected Myanmar since 1950, as reported by JICA (2015). Among them, only two cyclones, Pathein (1975) and Nargis (2008), made landfall at the Ayeyarwady deltaic coast, with the rest hitting the western Rakhine coast (see Figure 9.1.3.1). Myanmar's DMH also reported that a storm of the same intensity will produce twice the height of storm surges along the deltaic coast than the Rakhine coast. As Table 9.1.3.1 shows, cyclone Nargis in 2008 was by far the most devastating cyclone in the history of Myanmar (see Section 3.3).

Table 9.1.3.1 List of historical cyclones that have severely affected Myanmar since 1950

No.	Name	Date	Peak Surge (m)	Landfall Point	Death toll	Damage (million USD)
1.	Sittwe	7/5/1968	4.25	Sittwe	1037	2.5
2.	Pathein	7/5/1975	3.00	Pathein	304	No data
3.	Gwa	4/5/1982	3.70	Gwa	31	No data
4.	Maungdaw	2/5/1994	3.66	Maungdaw	10	10
5.	Mala	29/4/2006	4.57	Gwa	37	6.7
6	Nargis	2-3/5/2008	7.01	Heingyi	138,000	4000
7.	Giri	22/10/2010	3.7	Kyaukphyu	2157	57

9.1.3.2 Tropical cyclone Nargis

Tropical cyclone Nargis developed over the Bay of Bengal during the pre-monsoon season in late April and early May 2008, and went on to make landfall on the Ayeyarwady deltaic coast on 2nd May 2008 as a category four cyclone on the Saffir–Simpson scale (Lin et al., 2009). The Regional Specialized Meteorological Center (RSMC) for the Indian Ocean under the India Meteorological Department (IMD) stated that it reached a minimum central pressure of 962 hPa and maximum sustained surface wind speed of 47 m/s on 2nd May 2008, just before making landfall. The minimum hourly atmospheric pressure value recorded at the inland weather stations (based on DMH data) in Yangon (Kaba Aye and Mingaladon) was approximately 974 hPa at around 23 UTC on 2nd May 2008. An interesting characteristic of Nargis was its unusual track (see Figure 9.1.3.2), as during the monsoon period cyclones forming over the Bay of Bengal usually make landfall at the northern and northeastern coast of the Bay of Bengal (Pattnaik and Rama Rao, 2008).

At the point of landfall, Nargis generated a storm surge greater than 5 m (see Section 3.3), with a maximum recorded water level of 7 m at Pyinsalu, as reported by Myanmar's DMH. Also, according to the field survey conducted by Shibayama et al. (2009, 2010), a maximum storm surge of 4.33

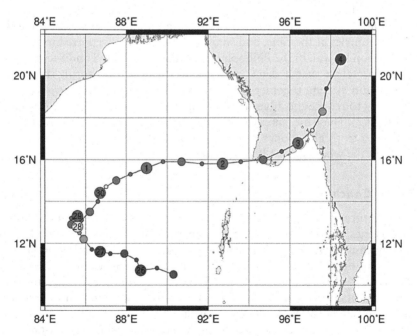

Figure 9.1.3.2 Observed track of cyclone Nargis from 26th April to 4th May 2008.

Source: Digital Typhoon.

m was reported near the Yangon river (see Section 3.3). The paddy fields in Yangon and Ayeywarwady deltas have a dense network of irrigation channels, and the salty water from the storm surge due to typhoon Nargis intruded inland through this network (see also Section 3.3). In the offshore region Level 2 Preprocessed (L2P) from the historical Geophysical Data Record of the GlobWave satellite database (Farquhar et al., 2013) recorded a significant wave height (*Hs*) of approximately 4.5 m at 4 UTC on 1st May 2008. Also, computer simulations indicate a maximum *Hs* of 7.3 m offshore of the deltaic coast of Myanmar (Ayeywarwady and Yangon) during the passage of Nargis (Kyaw et al., 2021). This is four to five times higher than the seasonal monsoon waves usual along this coastline, where low to moderate energy waves are the norm. Overall, the cyclone affected 50 townships along the deltaic coast, including Yangon, the largest city and former capital of Myanmar. Nargis caused over 85,000 deaths, with around 54,000 people still missing, and a 25% loss in the annual rice production of the entire country. The total damage to the infrastructure, social and productive sectors was estimated to be over 4 billion USD (Baker et al., 2008).

In the case of cyclone Nargis, the absence of public awareness about the risk of storm surges, the lack of cyclone shelters, the occurrence of destructive strong winds, together with the surge all contributed to the high casualty levels that were recorded (Kyaw et al., 2021). In this respect, it is worth noting how this lack of awareness was not an uncommon issue among the Asian countries at the time, and similar problems were reported in the case of Typhoon Haiyan in the Philippines (Esteban et al., 2015 and Section 3.5). Thus, it is clear that a dense network of weather monitoring and wave observation system, together with the establishment of cyclone shelters is necessary to reduce the adverse effects of coastal disasters (JICA, 2013 and Section 9.2). These measures can not only help mitigate the impacts of disaster due to tropical cyclones, but also enhance the weather monitoring networks in the whole Bay of Bengal area.

9.1.3.3 Conclusions

In the future, it is feared that as a consequence of global warming, tropical cyclones will increase in intensity and destructiveness. Cyclone Nargis already caused widespread destruction to the Bay of Bengal, especially in Myanmar, due to its abnormal track and rapid intensification. As cyclone Nargis is estimated to have had a return period of around 100 years, the consequences of cyclones with such levels of intensity becoming more frequent clearly represents a challenge to the sustainable development of Myanmar. Due to this, a standardized disaster preparedness scheme for coastal planning and emergency response should be organized and established at the region-wide scale in the Bay of Bengal, to increase the resilience of coastal communities against such events.

9.1.4 Indonesia

Hendra Achiari

Bandung Institute of Technology, Bandung, Indonesia

9.1.4.1 Introduction

Indonesia has suffered several natural disasters experiences in the recent past, including a number of tsunami events. The country is at the meeting point of three major continental plates – the Pacific, Eurasian, and Indian-Australian plates – and the much smaller Philippine plate, meaning that it is within what is known as the "Ring of Fire", a nearly continuous series of subduction zones lined by volcanoes, where earthquakes frequently occur. When these earthquakes occur in subduction zones under the water (such as, for example, in areas of the Indian Ocean to the west of Sumatra Island or to the south of Java Island), they can potentially generate tsunamis due to the vertical movement of the fault plate line (which displaces the water column on top of it), that can then propagate and destroy settlements in adjacent coastlines (See also Section 5.1). Indonesia has experienced an average of 290 significant natural disasters annually over the last 30 years (Reliefweb. Thomson Reuters Foundation 2018). This includes the *2004 Indian Ocean* Tsunami which killed approximately 220,000 people across four countries, 167,000 in Indonesia alone, and cost an estimated $10 billion in damages (see also Section 2.1).

9.1.4.2 Post-tsunami surveys and lessons learnt

Several post-tsunami field surveys, led by Prof. Shibayama (lead editor of the book) were carried out to assess the damage and record inundation heights and run-ups after various events in Indonesia. In this chapter four of these events will be described, including the *2006 Java (Pangandaran) Tsunami, 2010 Mentawai Tsunami, 2018 Sulawesi Tsunami* and *2018 Sunda Strait Tsunami.*

9.1.4.2.1 Java (Pangandaran) post-tsunami survey

The earthquake that generated the tsunami took place on July 17 at 15:19:27 local time, along the portion of the subduction zone of the Indian Ocean off the coast of west and central Java. The earthquake had a moment magnitude of 7.7 and a maximum perceived intensity of IV (Light) in Jakarta, the capital and largest city of Indonesia.

A survey was carried out between July 21–24, 2006, led by Prof. Shibayama and included members of Yokohama National University in Japan and Bandung Institute of Technology in Indonesia (see Section 2.2). The methodologies followed to survey the tsunami inundation and run-up

Figure 9.1.4.1   Activities conducted during the post-tsunami survey in Pangandaran, 2006.

were similar to those detailed in other chapters of this book (see Section 2.1.), and are detailed in Section 2.2 (so will not be repeated here). Figure 9.1.4.1 shows some of the activities that were carried out as part of this survey.

Amongst the lessons learnt from this event are the importance of considering the effect of landscape features and terrains on inundation patterns, such as the presence of small islands or tombolos (see Section 2.2). Also, the number of casualties can be highly influenced by population density and land use (as shown also in vulnerability studies of other areas such as Canada, see Section 9.4). Sand barriers can be effective in protecting residents against some of the power of the waves, and in tropical zones coral reefs can also reduce the incoming tsunami energy. Finally, this event showcased the importance of carefully considering how residents and tourists should evacuate, which is of paramount importance for saving lives (see Section 8.1).

9.1.4.2.2 Mentawai post-tsunami survey

The magnitude 7.8 *2010 Mentawai Earthquake* occurred on 25 October off the western coast of Sumatra at 21:42 local time. The earthquake took place along the same fault that produced the *2004 Indian Ocean Tsunami* (see Section 2.1), and was widely felt across the provinces of Bengkulu and West Sumatra. As a result, a substantial –yet localized – tsunami was generated, which struck the Mentawai Islands.

The survey was carried out on November 18–19, 2010, led by the lead editor of this book (Tomoya Shibayama), and included members from Waseda University, Yokohama National University and Bandung Institute of Technology. The survey visited the villages of Bosua, Gobik, Masokout, and Bere-Bereleu, as detailed in Section 2.5. Figure 9.1.4.2 shows the surveying activities that took place during the survey.

In terms of disaster risk management, it was clear that the remoteness of the islands made rescue and rehabilitation operations difficult. Interviews

Figure 9.1.4.2 Surveying of tsunami inundation height in Mentawai, 2010.

with local residents also revealed that there had been limited tsunami awareness dissemination activities, yet conducting disaster drills is of paramount
importance in order to reduce casualties from such events. Also, while some
of the forests between the sea and villages might have contributed to shielding them slightly from the waves, it is necessary to improve building materials and construct some vertical evacuation buildings to provide refuge in
areas where residents cannot reach any hills.

9.1.4.2.3 Sulawesi (Palu) post-tsunami survey

On 28 September 2018, a shallow, large earthquake struck the Minahasa
Peninsula, Indonesia, with its epicentre located in the mountainous Donggala
Regency, in Central Sulawesi. The magnitude 7.5 quake was located 70 km
(43 miles) away from the provincial capital Palu. This event was preceded by
a sequence of foreshocks, the largest of which was a magnitude 6.1 tremor
that occurred earlier that day. Following the main earthquake, a tsunami
alert was issued for the nearby Makassar Strait and a localized tsunami
struck Palu and other nearby towns, sweeping houses and other infrastructure situated along the shore.

The survey was carried out on October 27–31, 2018, led by the lead edi-
tor of this book (Tomoya Shibayama), and included members of Waseda
University, Tokyo City University, Toyohashi University of Technology,
Bandung Institute of Technology, State Institute for Islamic Studies (in Palu,
Indonesia), the University of Hawaii, the University of Ottawa and Technical
University of Braunschweig, Germany. Figure 9.1.4.3 shows the tsunami
inundation surveys that were carried out as part of this fieldwork.

The survey visited all areas situated along Palu Bay and its surrounding
towns, as summarized in Section 2.7. The patterns of tsunami damage,
arrival times and aerial images that were compiled indicated that this was a
rather unusual event (Mikami et al. 2019). A number of different factors
might have contributed to its generation, mostly related to the effect of a
number of landslides and co-seismic effects (Aranguiz et al., 2020), but

Figure 9.1.4.3 Post-tsunami survey in Palu, 2018.

possibly also including the influence of the fault dislocation and sloshing inside of Palu Bay (see also Section 5.3.)

9.1.4.2.4 Sunda strait post-tsunami survey

On 22 December 2018, a tsunami struck several coastal regions of the Sunda Straight of Banten in Java and Lampung in Sumatra, Indonesia. At least 426 people were killed and 14,059 were injured. The tsunami was caused by an undersea landslide that followed an eruption of Anak Krakatau, the "Child of Krakatoa", with much of this island collapsing into the sea. In the months leading up to the 2018 tsunami, Anak Krakatau had seen increased activity, with an eruption on 21 December 2018.

Two different surveys were carried out, as summarized in Section 2.8. The first on January 12–17, 2019, led by the lead editor of this book (Tomoya Shibayama), and included members from Waseda University, Tokyo City University, Toyohashi University of Technology, Bandung Institute of Technology, and Sumatera Institute of Technology-Indonesia. This field survey visited the Palu Lampung coast and Banten coast (Takabatake et al., 2019). The second field survey took place between August 15–18, 2019 and concentrated on the Krakatoa archipelago. Figure 9.1.4.4 showcases the surveying activities that took place during the fieldwork of the coastlines of Sunda Straight.

Figure 9.1.4.4 Post-tsunami survey along Sunda Strait, 2019.

The field surveys found evidence of the dramatic landscape alterations that took place following the eruption of Anak Krakatau, with the collapse of the island triggering a massive tsunami in its vicinity. The shielding effect that some of these islands provided helped to explain the tsunami inundation patterns that could be observed along the Kiluan coast and Legundi (Esteban et al., 2021).

9.1.4.3 Risk management coordination in Indonesia

In Indonesia, the National Disaster Management Agency (BNPB or Badan Nasional Penanggulangan Bencana) is the primary agency responsible for coordinating preparedness, response, prevention and mitigation, and rehabilitation and recovery. For the case of each individual province the Regional Disaster Management Agency (Badan Penanggulangan Bencana Daerah, or BPBD) is also established under the Regional Governor's Authority. Under the umbrella of the BNPB, BNPB and BPBD are responsible for the mobilization of equipment for disaster response that may be provided by international assistance. The BPBDs are currently using the Disaster Information Management System (DIMS) application, which manages damage and shelter information, provides digital maps, and can send messages to staff and other disaster management agencies. The application plays a vital role in the time and dissemination of information through its quick early warning system.

Indonesia uses the Incident Command System (ICS) in disaster response. The ICS facilitates inter-operability between disaster response personnel and other agencies in different jurisdictions. ICS is a standardized, on-scene, all-hazard, incident management concept. Traditionally an Incident Commander in Indonesia is a representative from the Indonesian Army (or TNI), who is appointed by the Head of the District/Municipality or the relevant BPBD as the On-Site Coordinator and reports to the head of local government. The Incident Commander has the authority to deploy all available resources and is responsible for overseeing the mobilization of human resources, equipment, logistics, and rescue operations. The Incident Commander also prepares a disaster plan to be used during response faces (GFDRR, 2018). When a natural disaster takes place in Indonesia, the BNPB and BPBD will be in charge of the coordination of this ICS system.

9.1.4.4 Conclusions

Indonesia is situated along the "Ring of Fire" in the Pacific, and is thus frequently affected by a number of natural hazards such as volcanic eruptions, earthquakes and tsunamis. These events have led to major disasters in recent decades, including the four major tsunamis that were described in this chapter. Post-tsunami surveys of the most recent ones to have taken place revealed an increase in overall awareness since the *2004 Indian Ocean*

Tsunami, though still many casualties were recorded. While the risk management coordination in Indonesia has been getting better, it is clear that major improvements are still needed in order to enhance the resilience of coastal communities against tsunamis so that they can be better protected against future events.

9.1.5 Coastal erosion in Thailand

Thamnoon Rasmeemasmuang
Burapha University, Saen Suk, Thailand

Wudhipong Kittitanasuan
Wishakorn Co., LTD., Bangkok, Thailand

Winyu Rattanapitikon
Thammasat University, Pathum Thani, Thailand

Sittichai Naksuksakul
PTT Research and Technology Institute, Bangkok, Thailand

9.1.5.1 Introduction

Thailand has a coastline of approximately 3,148 km, which runs across 23 coastal provinces, as shown in Figure 9.1.5.1. Within them, the shorelines of the Gulf of Thailand stretch for approximately 2,055 km, while the Andaman Coast is around 1,093 km long. It can be divided into 4 zones: (1) the East Coast covering 3 provinces: Trat, Chanthaburi and Rayong; (2) Upper Gulf of Thailand, 7 provinces: Chon Buri, Chachoengsao, Samut Prakan, Bangkok, Samut Sakhon, Samut Songkhram and Phetchaburi; (3) Southern Gulf of Thailand,7 provinces: Prachuap Khiri Khan, Chumphon, Surat Thani, Nakhon Si Thammarat, Songkhla, Pattani, and Narathiwat, and (4) Andaman Coast, 6 provinces: Ranong, Phang Nga, Krabi, Trang, Satun and Phuket.

While relatively rare, the country has been in the past affected by some severe coastal disasters, such as tsunamis and storm surges. For instance, the *2004 Indian Ocean Tsunami* (Shibayama, 2015; Choowong et al., 2008 see also Section 2.1), which was caused by a powerful earthquake with an epicentre off the west coast of northern Sumatra in Indonesia, hit six Thai provinces: Phuket, Phangnga, Ranong, Krabi, Trang and Satun, resulting in 5,400 dead, more than 8,000 injured and many missing. Typhoon Linda in 1997 (Aschariyaphotha et al., 2011), which had weakened to a tropical storm by the time it reached Thailand, generated storm surges that hit Chumphon, Prachuap Khiri Khan and Phetchaburi coasts and caused flooding that affected wide stretches of the coastline.

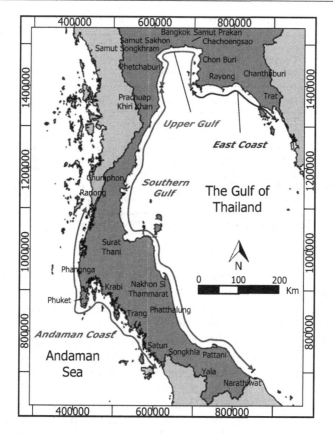

Figure 9.1.5.1 Geographical classifications of the coastlines of Thailand.

However, such destructive events remain statistically rare, and the more chronic and persistent disaster that affects the coastline of Thailand remains coastal erosion, which can be quite severe in several locations (as in other parts of the world such as Vietnam or the USA, see Sections 9.1 and 9.4).

9.1.5.2 Coastal erosion in Thailand

According to an analysis of aerial photos and satellite images between 1952 and 2008 of the East Coast and Upper Gulf Coast, and between 1967 and 1995 for the Southern Gulf Coast and the Andaman Coast (DMCR, 2011), it was found that over 830 km stretches of the coastline of the country suffered erosion, including 730 km of the Gulf of Thailand (with 228 km being classified as severe erosion – erosion rate of over 5 m per year – and 502 km as moderate erosion, with an erosion rate less than 5 m per year) and 100 km of the Andaman Coast (25 km of severe coastal erosion and 75 km of moderate erosion), as summarized in Table 9.1.5.1. The central, regional, and local government agencies have implemented projects to solve coastal

Table 9.1.5.1 Summary of coastal erosion in Thailand

Zone	Coast length (km)	Coastal Erosion					
		2008 and 1995[a]			2015		
		Moderate	Serious	Sum	Moderate	Serious	Sum
Gulf of Thailand	**2,055**	**502**	**228**	**730**	**90.05**	**69.47**	**159.52**
East	391	124	12	136	37.62	0.00	37.62
Upper	404	92	67	160	32.28	0.43	32.71
South	1,260	286	149	435	20.15	69.04	89.19
Andaman	**1,093**	**75**	**25**	**100**	**7.92**	**0.00**	**7.92**
Total	**3,148**	**577**	**253**	**830**	**97.97**	**69.47**	**167.44**

[a] 2008 for the East and the Upper Gulf of Thailand and 1995 for the southern Gulf and the Andaman Coast.

erosion problems by using various engineering structures such as seawalls, revetments, groins, and non-structural solutions such as mangrove reforestation or beach nourishment. Some of these have proven successful, though others have had a more moderate impact.

In 2015 (DMCR, 2017), an investigation of coastal changes was undertaken and a coastal erosion map of Thailand was created by processing data from the results of studies of coastal erosion in each area, along with site visits to monitor the status of the erosion and a comparative analysis of historical satellite images. It was found that 167.44 km of the shoreline had been facing erosion problems and no action had been taken to solve such issues. This consists of 160 km of the Gulf of Thailand (classified as approximately 70 km of severe erosion and 90 km of moderate erosion) and approximately 8 km of the Andaman coast (all moderate erosion). The erosion problems of 484 km of the Gulf of Thailand and 81 km of the Andaman Coast have been mitigated by preventive structures.

According to the physical mechanism behind them, the causes of coastal erosion can be classified into two main groups:

(1) Imbalance of coastal sediment. When the incoming sediment is less than that outgoing (due to either alongshore or cross-shore transport) for a given point coastal erosion can take place.
(2) Relative vertical movement between the level of land and the average level of the sea. If the sea level rises or the land subsides, the shoreline will retreat landward, leading to coastal erosion.

Regarding the coasts of Thailand, there are a variety of factors that can directly induce or indirectly promote these physical mechanisms, depending on the season and area, and which can be either natural or anthropogenic in nature, as outlined below.

9.1.5.2.1 Natural factors

- *Waves and currents*. These are the fundamental driving forces that dominate the incoming and outgoing sediment budgets. Breaking waves and nearshore currents can change in magnitude and direction based on the topography, water depth and season.
- *Monsoon*. The coasts of Thailand are affected by prevailing seasonal winds, including the northeast and southwest monsoons. The Southern Gulf of Thailand is affected by the northeast monsoon from mid-October to February. The Andaman Coast is affected by the southwest monsoon from mid-May to mid-October, triggering rain and strong wind waves that can cause coastal erosion.
- *Storm*. Most of the storms affecting the country take place in the Gulf of Thailand, as they are formed and move from the South China Sea and move westwards towards this coastline. Most of them are at the level of tropical depressions (with wind speeds near the centre less than 63 km/h) followed by tropical storms (63–118 km/h) and typhoon levels (118 km/h or more). In the last seven decades, there was only one storm that made landfall in Thailand at typhoon-level intensity (Typhoon Gay in 1989).
- *Sea level*. When the sea level rises coastlines become more geomorphologically active, with both the actions of waves and currents increasing, which can lead to higher levels of erosion and sediment transport. Sea-level rise may be transient and cyclical, such as that due to tides, or permanent, such as that due to warming ocean temperatures, which results in the relative movement between the coast and sea.
- *Coastal geological features*. Different geographic features cause unequal levels of coastal erosion and, for example, muddy and sandy beaches can retreat more quickly than rocky shores. Coastlines with very steep slopes are also more likely to retreat quickly than those with gentler inclinations.

9.1.5.2.2 Human factors

- *Coastal area development*. The construction of coastal structures (such as coastal roads, bridges or other structures protruding into the sea) can affect ecosystems and cause changes to the sediment budget balance, resulting in coastal erosion.
- *Mangrove forest encroachment*. Mangrove forests are usually a natural barrier against waves and promote sedimentation. Human encroachment into mangrove forests (to transform them into agricultural or aquaculture areas), can reduce the protection that these ecosystems offer against winds and waves. As a consequence, coastal erosion is likely to intensify.

- *Land subsidence.* Groundwater pumping can cause land subsidence, especially around coastal areas. If subsidence leads to the level of the land being lower than that of the sea then significant coastal erosion can take place, together with saltwater intrusion and other problems.
- *Construction of dams or weirs.* Normally rivers or canals deliver sediment to the coast, though when a dam or weir is built much of the sediment is trapped behind it, no longer reaching the coastline. This, together with other factors that reduce the availability of sediments (such as dredging for various purposes) can also lead to coastal erosion.
- *Construction of coastal protection projects.* Construction of poorly thought or designed coastal protective structures can result in changes in waves, currents and coastal sediment deposition and movement, all of which can contribute to an increase in coastal erosion.

9.1.5.3 Countermeasures to prevent coastal erosion

Due to limited knowledge, the formulation of guidelines for the prevention of coastal erosion in Thailand has not been very clear. The number of specialists in the field of coastal engineering, the knowledge and understanding of civil servants involved in coastal zone management, local residents and stakeholders remains limited, which together with constrained budgets to address such problems present significant problem (Kittitanasuan, 2020). Thus, both technical and trial-and-error countermeasures (i.e., those not actually being designed by coastal engineers) have been used. Measures to address coastal erosion in Thailand include both structural (hard) and nonstructural (soft) countermeasures (see Figure 9.1.5.2), such as the following:

(a) *Offshore breakwaters* are structures constructed offshore and parallel to the coastline. They are used to reduce the energy of waves arriving from the ocean, though they can be expensive and involve construction methods that require modern machinery. They also impact the aesthetics and view of the ocean, and surrounding environment.
(b) *Revetments* are land armouring structures that prevent soil erosion. Most of them are constructed by local governments or residents of areas where erosion occurs. Along muddy coasts they are built using medium-sized rocks to prevent quick subsidence, and can be efficient at protecting the land behind them. Their construction is not very difficult, though they are usually not able to withstand large wind waves.
(c) *Seawalls* are land protection structures that are similar to revetments. They can be vertical or inclined walls, constructed using concrete or rocks. When waves collide with them they are typically reflected back to the sea, though issues with scouring around their toe can lead to structural collapse if not designed correctly.

Figure 9.1.5.2 Common types of countermeasures to address coastal erosion in Thailand. a) Offshore breakwater at Rayong, b) revetment at Samut Prakan, c) seawall at Songkhla, d) groin at Rayong, e) geo-textile breakwaters at Samut Prakan, f) concrete-pole breakwaters at Bangkok, g) bamboo breakwaters at Samut Sakhon, h) mangrove reforestation at Chon Buri, (i) Beach fill at Chon Buri.

(d) *Groins* are structures that are perpendicular to the shoreline, con-structed using rocks or other materials in order to trap sediments behind them and build a beach area, reducing coastal erosion (see Section 9.4). They are commonly constructed in series along the coastline, with a certain spacing between them, though they can result in erosion taking place further down the longshore drift direction.

(e) *Geotextile breakwaters* are constructed using sandbags made from geo-textiles of high toughness and flexibility, typically placed 200–400 m offshore. They are easy to transport and install, given their light weight, and can reduce subsidence and wave intensity. However, they can block maritime traffic and the sandbags can easily break, leading to a leak of sand and their destruction.

(f) *Concrete-pole breakwaters* are made using concrete piles that are stuck into the soil layer to help dissipate wave energy. Sometimes used tires are attached to concrete poles, and these types of structures can often be seen in mud flats or estuaries in Thailand. Typically, they are built using several rows, with gaps between the poles allowing water and sediment to pass through. They can help to reduce subsidence, though if the poles are placed too far apart they might not work efficiently, failing to reduce wave intensity while impacting landscape aesthetics and presenting an obstacle to navigation.

(g) *Bamboo breakwaters* are similar to concrete-pole breakwaters, though they use bamboos to reduce the intensity of the waves. These are commonly used in mud flat areas in many countries because the cost to construct them is low. Local residents can install them by themselves using natural locally available materials, and such interventions have little impact on neighbouring areas. However, they can have short lifespans, as the bamboo is easily broken or toppled and regular maintenance is needed. Also, the decayed bamboo can become debris along the coast, and the performance of this type of structure in terms of wave reduction is not very good, especially when faced with large wind waves.

(h) *Mangrove reforestation* involves the restoration of the original mangrove forest, a non-structural countermeasure (see also Section 9.1, about Vietnam). This provides advantages to the environment, and such interventions can be implemented by local residents. However, planted mangrove trees can have low survival rates, as small trees cannot withstand strong winds or waves, requiring a sufficiently thick sediment for root adhesion. This measure is suitable for areas that do not face strong wind waves and should be used in conjunction with temporary measures to reduce wave action until the roots of the trees become strong.

(i) *Beach fill* involves the replenishment of sand from other sources that is similar to that of eroded areas. Particle size and sand properties must be taken into consideration when conducting such interventions, and care must be taken that they do not cause any changes in beach conditions and affect any living organisms in such areas. These kind of projects have been popular in Thailand, especially in popular tourist beaches.

Studies from lessons learned and experiences in coastal erosion solution in Thailand (e.g., Rasmee and Rasmeemasmuang, 2013) reveal that no measure can be effective in any location. Furthermore, it is important to remember that each measure has some degree of impact on the environment and human society around it. Measures that appear to be efficient at reducing a given problem can sometimes have a high financial cost and environmental impact, whereas eco-friendly measures can take a longer time to implement, might be less efficient (particularly in the short term) and require constant maintenance, though they seldom lead to other problems in nearby areas. Balancing the efficiency of coastal erosion solutions with the environmental impacts has become a challenge for the sustainable management of coastal erosion in Thailand.

9.1.5.4 Conclusions

Coastal erosion represents a continuous and protracted issue throughout the coasts of Thailand, though its intensity varies from place to place. This problem has been addressed through a variety of different countermeasures, with some of the solutions causing other side effects due to a lack of appropriate engineering assessment. At the present, the situation appears to be improving, though increased technical knowledge, experiences and experts to assist in solving the problem are required in order to sustainably manage the coastlines of the country.

9.1.6 Storm disasters in China

Shaowu Li

Tianjin University, Tianjin, China

9.1.6.1 Introduction

China has a continental coastline that is 18,000 km in length, with that of its islands adding another 14,000 km. The coastal zone in China is the most densely populated and developed area of the country and frequently suffers from storms (You, 2016). These storms can be caused either by cold air fronts, originating from Siberia from late autumn to spring or by tropical cyclones generated in the west Pacific from early summer to late autumn (see Sections 3.5 and 3.7). There is an average of 7.5 cold air fronts per year, among which 1.6 events reach the intensity of a "cold wave" (i.e., a strong cold air front, see Kang et al., 2010). Statistics regarding tropical cyclones that made landfall along the Chinese coast from 1949 to 2020 indicate that the average number of events declined at a rate of 0.31 per decade (see Figure 9.1.6.1). The classification of cyclones in the figure is according to the standard defined in the Chinese National standard GB/T

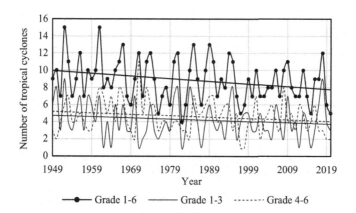

Figure 9.1.6.1 Number of tropical cyclone events making landfall along the coasts of China between 1949 and 2020 (Ying et al., 2014; Lu et al., 2021).

19201–2006, as shown in Table 9.1.6.1. This decline is due to drops in the number of cyclone events with higher intensity rated below grade 4 and beyond grade 4. Figure 9.1.6.2 shows the number of tropical cyclone events generated in the West Pacific, which also shows a declining tendency at both the lower and higher grades at a faster rate of 1.69 per decade. Statistics of the maximum wind speeds and the lowest central pressures of tropical cyclones making landfall on Chinese coasts are shown in Figure 9.1.6.3. Both the average and maximum wind speeds exhibit a declining tendency, despite the relatively stable tendency in the central pressure of the weather systems in this period.

Table 9.1.6.2 summarizes the occurrence frequency of tropical storms, maximum wind speed, name of the strongest typhoon for a given year, direct economic loss during the tropical cyclones, direct economic loss due

Table 9.1.6.1 Grade of tropical cyclones (Chinese National standard GB/T 19201–2006)

Grade	Designation and acronym	Typical wind speed (m/s)	Maximum wind speed (m/s)
0	Below tropical depression (TD) or intensity unknown		
1	Tropical depression (TD)	10.8	17.1
2	Tropical storm (TS)	17.2	24.4
3	Strong tropical storm (STS)	24.5	32.6
4	Typhoon (TY)	32.7	41.4
5	Strong typhoon (STY),	41.5	50.9
6	Super typhoon (Super TY)	≥51.0	

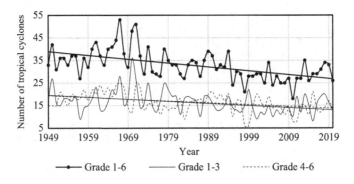

Figure 9.1.6.2 Annual number of tropical cyclones generated in the West Pacific Ocean between 1949 and 2020 (Ying et al., 2014; Lu et al., 2021).

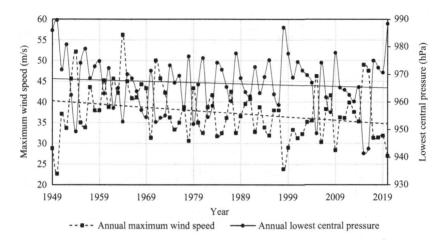

Figure 9.1.6.3 Highest wind speeds and lowest central pressures taken from any typhoons making landfall on China's coasts each year during 1949–2020 (Ying et al., 2014; Lu et al., 2021).

to storm surges, and casualties caused by waves and storm surges in the period from 1989 to 2020. The total economic loss caused by storms was lower level during the early 1990s (Figure 9.1.6.4) due to the lower level of economic development of the country, and increased in the period from the late 1990s to the first decade of the 21st century because of the rapid development of the economy and relatively slow implementation of protection measures. The direct economic loss is not strongly correlated with the wind speeds of the storms and has declined steadily since the late 2000s, possibly owing to the enhancement of anti-storm countermeasures. The casualties caused by coastal disasters, on the other hand, were quite high before 2000 (Figure 9.1.6.5), peaked at 1815 in 1994, and kept declining since then to become lower than 100 after 2014.

According to the statistics of the Bulletin of China Marine Disaster (as shown in Figure 9.1.6.6), tropical cyclones mainly cause destruction or damage to coastal buildings and aquacultures, flooding over residential areas. However, unlike tropical storms, most temperate cold fronts are less damaging and seldom directly destroy facilities, though they can instead deposit sediments along the navigation channels of seaports located on silty coasts that have no effective protection measures against storm waves. For example, Port Huanghua and Port Jingtang in Bohai Bay suffered a cold wave episode in October 2003, which lasted for more than 24 hours. During the storm, the deeply dredged navigation channels of the two ports became filled by sediment deposition almost to the natural sea bottom, leading to significant disruption to their activities.

Table 9.1.6.2 Statistics of storms and related economic loss and death toll during 1989–2020

Year	Occurrence frequency	Maximum wind speed (m/s)	Name of the strongest typhoon	Direct economic loss (100 million yuan)	Economic loss due to storm surges (100 million yuan)	Total deaths	Death toll caused by waves	Death toll caused by storm surges
1989	11	72	Orson	54.83	[a]	608	44	564
1990	9	77	Mike	47.33	[a]	933	28	905
1991	5	77	Yuri	23.63	[a]	333	27	306
1992	8	82	Gay	99.86	[a]	330	28	302
1993	7	72	Ed	84.08	[a]	174	59	115
1994	11	77	John	193.00	[a]	1819	4	1815
1995	10	79.7	Angela	87.02	[a]	292	75	217
1996	7	53	NO.9615 Sally	290.28	[a]	1699	269	1430
1997	4	60	NO.9711 Winnie	308.00	[a]	348	1	347
1998	3	24.5~28.4	NO.9806	14.50	[a]	71	65	6
1999	6	47.1	NO.9914	47.90	[a]	700[a]	[a]	259
2000	4	60	Saomai	120.80	115.40	79		15
2001	6	40	Chebi	100.00	87.00	401	63	136
2002	8	45	Sinlaku	66.00	63.10	124	265	30
2003	10	40	Dujuan	80.50	78.77	128	94	25
2004	10	50	Mindulle	54.00	52.15	140	103	49
2005	11	60	Longwang	332.40	329.80	371	91	137
2006	9	60	Saomai	218.45	217.11	492	234	327
2007	13	57	Sepat	88.37	87.15	161	165	18
2008	11	54.2	Fung-wong	206.05	192.24	152	143	56
2009	10	43	Morakot	100.23	84.97	95	96	57

2010	10	68	Megi	132.76	65.79	137	132	5
2011	9	60	Nanmadol	62.07	20.03	76	68	0
2012	13	50.9	Kaikui	155.25	126.29	68	59	9
2013	14	75	Haiyan	163.48	153.96	121	121	0
2014	5	60	Rammasun	136.14	134.69	24	18	6
2015	6	55	Chan-hom	72.74	72.62	30	23	0
2016	10	65	Meranti	50.00	45.94	60	60	0
2017	13	52	Hato	63.98	55.77	17	11	6
2018	12	65	Mangkhut	47.77	44.56	73	70	3
2019	9	62	Lekima	117.03	116.38	22	22	0
2020	10	38	Hagupit	8.32	8.10	6	6	0

Source: Data extracted from Bulletin of China Marine Disaster.

a Data is unavailable.

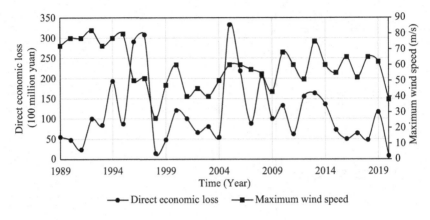

Figure 9.1.6.4 Direct economic loss caused by storms.

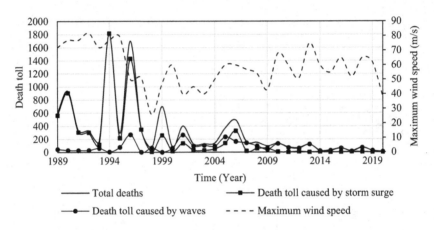

Figure 9.1.6.5 Deaths caused by coastal disaster.

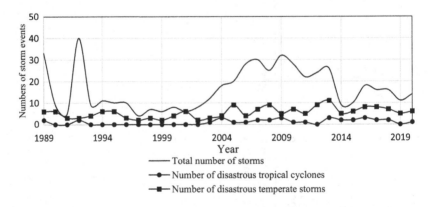

Figure 9.1.6.6 Total and disastrous numbers of storm events.

9.1.6.2 Damage due to typhoon events

Typhoon Hagupit was generated in the northwest Pacific Ocean to the east of the Philippines on the night of the 19th of September 2008, and landed on the coastal area of Chencun town, Dianbai County, Guangdong Province on the morning of the 24th of September (see Figure 9.1.6.7). The central pressure at landfall was 950 Pa, and the maximum wind force reached 15 on the Beaufort scale. According to the statistics of the Ministry of Civil Affairs on the 26th of September, typhoon Hagupit had affected nearly 115,354 people in Guangdong, Guangxi and Hainan provinces, 751,800 people were evacuated, 13 people were killed, 27,700 low-rise houses were damaged or collapsed, and the direct losses amounted to 13.456 billion RMB Yuan (Li et al., 2010). High water levels with a return period of 100-year, or even 200-year were observed at several tide observation stations, and a wave with a height of 13.7 m was observed at the Zhujiang River estuary (Fu et al., 2009). From Shantou to the east coast of Leizhou Peninsula, including the northern part of Hainan Island, the storm surge ranged between 0.55–2.70 m, and in most sections of Zhujiang river estuary it exceeded 2 m. The storm surge along the Beibu Gulf in the coast of the Guangxi Zhuang Autonomous Region was 0.70–1.0 m. Due to the coincidence of Typhoon Hagupit and a high astronomical tide, 10 tide stations in Guangdong, Hainan and Guangxi observed high tides beyond the local warning levels. One of the tide stations even observed a high tide 1.75 m above the warning level in the early morning of the 24th of September.

Typhoon Kalmaegi made landfall at Wengtian town, Wenchang County, Hainan Province at 9:40 a.m. on the 16th of September 2014. At this time

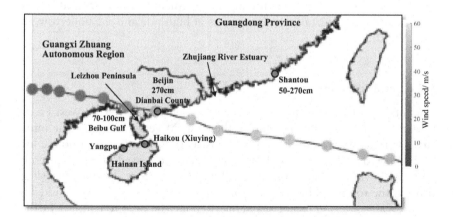

Figure 9.1.6.7 Path of typhoon Hagupit from 8:00 a.m. (UTC+8) on September 23 to 8:00 a.m. (UTC+8) on September 25 in 2008 (Fu et al., 2009), including measured storm surges at Guangdong, Guangxi, and north Hainan province of China.

Figure 9.1.6.8 Coastal erosion at the west shoreline near Haikou, Hainan Province of China, damaged by typhoon Kalmaegi in 2014 (Zou, 2020). (a) Sand bar before the storm, (b) Eroded profile in front of the revetment.

its central pressure was 960 Pa, and the maximum wind force reached 13 on the Beaufort scale. According to the statistics of the Hainan Provincial Department of Ocean and Fisheries, typhoon Kalmaegi caused serious damage to fishery facilities, resulting in a total direct economic loss of 926 million RMB Yuan, including the destruction of 1,567 fishing boats (Shi et al., 2015). This typhoon caused a record-breaking high tide setup of 1.47 m above the warning tide level at Xiuying Station (see Figure 9.1.6.7 for the location of this station). The powerful waves generated by the storm also resulted in serious beach erosion in Haikou Bay and Puqian Bay, where existing berms and revetment were seriously damaged due to the washing-out of sediment from the beach (Figure 9.1.6.8). A sand bar, which had formed in front of the revetment before the typhoon arrived, was complexly removed, exposing the toe of the revetment to the waves (likely due to the combined effect of standing waves at high tide and the undertow, see Zou, 2020).

9.1.6.3 Conclusions

The total number of tropical cyclone events generated over the West Pacific Ocean has declined slightly during the past 70 years, though their intensity has remained relatively stable intensity. Both the direct economic loss and the death toll caused by these storms have decreased in China recently, compared with the situation in the late 20th century. Despite the slight drop in the total number of typhoon events, storm surges have resulted in record water levels in recent years, highlighting the need to conduct further research into such events and improve disaster risk management throughout the country. Ports on silty coasts should be protected using anti-sediment dikes to avoid them being infilled by sediments due to wave action during cold waves.

9.1.7 Prevention of coastal erosion in Korean beaches

Hyuck-Min Kweon

Sungkyunkwan University, Seoul, Korea

Sun-Yong Lee, Yil-Seob Kim, and Jae-Hun Jeong

Sekwang Engineering Consultants Co., LTD Seoul, Korea

9.1.7.1 Introduction

The problem of coastal erosion in Korea, especially along the east coastline of the country, is an important social and economic issue that is highlighted by the media of the country every year. As a result of this, a number of specialized institutions in coastal maintenance are currently making great efforts to find potential ways to solve such problems.

Particularly, in recent times the construction of the third stage of the Donghae Port has drawn attention to the possibility that the white sandy beaches of Samcheok and Jeungsan could be eroded, and the necessity to formulate adequate erosion reduction measures. The present section will introduce a number of erosion reduction technologies that achieved good results in mitigating the potential adverse impacts of the construction of Donghae port (The Province of Gangwon, 2018).

9.1.7.2 Agreement with the residents and application techniques

9.1.7.2.1 Development project of Donghae Port

Donghae Port is a state-managed trade port located in Donghae-Si, Gangwon-Do, which opened on February 8, 1979. It is not a natural port, and relies on its breakwaters to provide shelter to ships that transport cement, bituminous coal, and limestone. The third stage of the East Sea Port development project has been implemented in stages since 2016 with the aim to expand port infrastructure in accordance with the 3rd National Trade Port Basic Plan (Ministry of Land, Transport and Maritime Affairs, 2011) in Korea to increase capacity and reduce delays. As a consequence of this, the port is periodically monitored to investigate sedimentation patterns and changes in the marine environment caused by the East Sea Port development project, including Samcheok Beach, which also records and analyses the effect of erosion reduction measures (see Figure 9.1.7.1).

9.1.7.2.2 Resident complaints and countermeasures

The third stage development project of Donghae Port (Donghae Regional Office of Oceans and Fisheries, 2018) was originally intended to create a hub port in the Hwandong Sea area. However, in the early stages of the

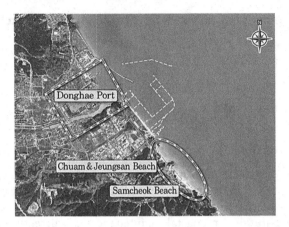

Figure 9.1.7.1 Location of Donghae Port, Jeungsan Beach and Samcheok Beach.

development plan, residents of the surrounding Samcheok, Chuam and Jeungsan coasts were concerned that this would lead to coastal erosion. Complaints were continuously filed so that the development project was disrupted and could not make meaningful progress. In order for development to continue a public-private agreement was reached regarding the installation of coastal protection countermeasures and the monitoring of the beach width during and after construction. The total amount of sand that should exist within the coastline system in this area was agreed upon between the Ministry of Oceans and Fisheries and residents in order to maintain the environment of Samcheok, Chuam, and Jeungsan beaches.

9.1.7.2.3 Oblique detached breakwater

In accordance with the agreement regarding the total amount of sand that should exist in the system, sedimentation monitoring for five years was conducted by the Ministry of Oceans and Fisheries in South Korea. Also, countermeasures were constructed to stabilize the coastline, including an oblique detached breakwater (see Figure 9.1.7.2) and groins (Losada et al., 1992,

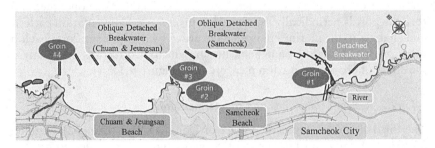

Figure 9.1.7.2 Layout of the oblique detached breakwaters that were constructed.

Figure 9.1.7.3 Beaches before and after the construction of the detached breakwaters.

Kweon et al., 2014, Lee et al., 2020). As a consequence of these interventions, the wave energy reaching the coastline could be reduced, effectively stabilizing it, with Figure 9.1.7.3 showing the coastline before construction of the project (in 2018) and afterwards.

9.1.7.2.4 Proposal for installation of the jetty moving to South

Another planned erosion reduction countermeasure was for a jetty to be installed in the future on the southern coast of Samcheok beach. However, residents in 2015 suggested moving the jetty about 75 m north of the river mouth in order to prevent floating waste flowing out of the river from moving to the front of the beach in the summer. However, there were concerns that the sand source from the river would be blocked if the northern jetty proposal requested by residents went ahead. In consultation with the Ministry of Oceans and Fisheries, experts, and representatives of the residents, a new plan was agreed to construct the south jetty. As a result, the amount of sand in Samcheok beach increased (see Figure 9.1.7.4).

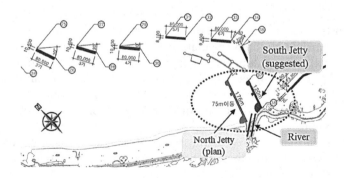

Figure 9.1.7.4 Change in construction plan for the jetty.

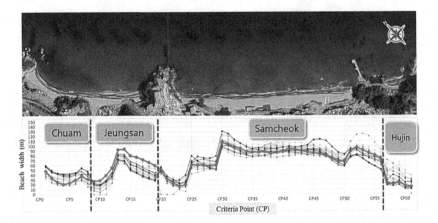

Figure 9.1.7.5 Change in beach widths in the project area (2017–2021).

9.1.7.2.5 Seasonal changes and total volume changes of sediment

The sedimentation and deposition monitoring at Samcheok, Chuam and Jeungsan beaches revealed that the northern beach was accreting and the southern beach eroding around June 2017. On the other hand, around June 2020, the beach was accreting due to the rubble mound breakwater constructed in front of Samcheok Beach.

Moreover, as a consequence of the countermeasures implemented, the beach width was maintained at 80–100 m from the reference coastline (the highest high water level), which was generally stable despite slight seasonal oscillations, with the overall amount of sand in it increasing slightly (see Figure 9.1.7.5).

9.1.7.3 Conclusions

The construction of Donghae Port was carried out after an agreement was signed regarding the implementation of measures to prevent beach erosion and enhance sedimentation in response to requests from residents of Samcheok. This agreement was reached following consultations with Samcheok city residents, which stipulated the total amount of sand to be supplied and the types of countermeasures to be constructed, which solved the erosion problem and provided a good example of a coastal maintenance project. In particular, based on the results of sedimentation and monitoring, it appears that the coastline can be maintained, and the amount of sand in the beach could be increased.

9.1.8 History of tsunami and storm surge in Japan

Naoto Inagaki and Tomoya Shibayama

Waseda University, Tokyo, Japan

9.1.8.1 Introduction

In Japan, disasters have taken place frequently since ancient times, with them being described in early historical archives dating back to as far as the 8th century. One of the oldest archives, "Nihonshoki (Chronicles of Japan)", compiled in 720 suggested that a great tsunami hit Japan at the end of the 7th century. Recent geological research has proven the existence of this tsunami, which makes it the oldest recorded event of this type in the country (e.g., Fujiwara et al. 2020).

There are many myths and folklore that are considered to be related to flood disasters, and vestiges of these myths can be found in existing local culture, region names, and town planning. In particular, significant facilities such as shrines and temples were built in places that were less likely to be flooded. In Japan, there are several shrines named "Hikawa Shrine" or "Yasaka Shrine", which worship a water deity known as "Susano-no-mikoto", with the location of the shrines thought to be related to flooding.

Given the influence that natural hazards have had on the ancient culture and literature of the country, it is useful to understand some of their characteristics in order to improve disaster prevention in Japan.

9.1.8.2 Tsunami

9.1.8.2.1 History

Most of the floods described in ancient literature relate to tsunamis, given the large damage they cause to society. As shown in Figure 9.1.8.1, the Japanese archipelago is located along a plate boundary, with powerful earthquakes occurring along the trenches adjacent to it (see Section 2.1). In particular, the Sanriku region and the Nankai Trough have repeatedly triggered tsunami-inducing earthquakes that have affected the country since ancient times.

Table 9.1.8.1 summarizes the tsunamis that have occurred in the Sanriku region, which has experienced such events 50–100 years since 1611. The *2011 Tohoku Tsunami* was the largest in the history of Japan (see Sections 2.1 and 2.6), with the country still recovering from it. Table 9.1.8.2 lists the tsunamis caused by earthquakes along the Nankai Trough. This region has recorded major tsunami events every 100–200 years, with the oldest recorded one being the Hakuho Tsunami. A tsunami named "Keicho" also occurred in the Sanriku region, which was considered to be caused by a series of plate movements. Essentially, a large-scale geological event in one region can trigger an earthquake in another region in a short period of time. The most recent tsunami event in this region was in 1707 and was

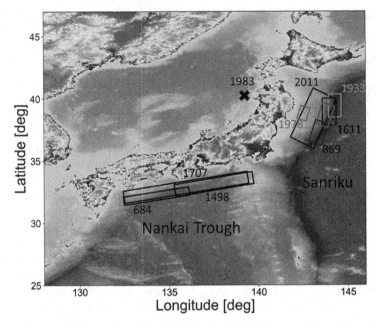

Figure 9.1.8.1 Bathymetry contours around the Japanese Archipelago, with the darker region representing the deeper sea (plate boundary). The rectangles and the cross-marks show epicentral regions that caused historical tsunamis from the Sanriku and Nankai Troughs.(with reference to Suppasri et al. (2013), Ishibashi (2004).)

Table 9.1.8.1 List of major tsunamis in the Sanriku region

Year	Known as	Magnitude (M_w)	Casualties
869	Jogan Tsunami	8.3	—
1611	Keicho Sanriku Tsunami	8.1	5,000
1677	Enpo Tsunami	7.5	500
1763	Horeki Tsunami	7.4	—
1854	Ansei Tsunami	7.5	—
1896	Meiji Sanriku Tsunami	8.5	26,360
1933	Showa Sanriku Tsunami	8.1	1,330
2011	Tohoku Tsunami	9.0	15,899 (2,526 unknown)

accompanied by the eruption of Mt. Fuji, the tallest mountain in Japan. In addition to the damage due to the earthquake and tsunami, volcanic ash was also problematic and resulted in a severe famine.

In Japan, tsunamis generated by the collapse of mountain sides due to volcanic eruptions have also been reported (see Table 9.1.8.3). The one in 1596 is the first in the series of the "Keicho" tsunamis. Although this type of landslide-generated tsunamis has not occurred in recent years in Japan, the Sunda Strait in Indonesia experienced a tsunami event triggered by

Table 9.1.8.2 List of major tsunamis triggered by Nankai Trough

Year	Known as	Magnitude (M_w)	Casualties	Notes
684	Hakuho Tsunami	8.0	—	The first tsunami recorded in Japan
887	Ninna Tsunami	8.0	—	
1096	Eicho Tsunami	8.0	—	
1361	Shohei Tsunami	8.0	—	
1498	Meio Tsunami	8.2	26,000	
1605	Keicho Tsunami	7.9	5,000	
1707	Hoei Tsunami	8.6	20,000	Mt. Fuji eruption

Table 9.1.8.3 List of tsunamis generated by the collapse of mountain sides

Year	Known as	Casualties	Notes
1596	Keicho Bungo Tsunami	708	Several reports and literature is available, describing that a tsunami took place
1640	—	700	Eruption and collapse of Mt. Hokkaido Komagatake (Hokkaido)
1741	Kanpo Tsunami	2000	Eruption and collapse of Oshima (Hokkaido)
1792	Higo Meiwaku	15,000	Eruption and collapse of Mt. Unzen (Nagasaki)

the eruption of the Krakatau Volcano (for a detailed field survey of this disaster, see Section 2.8).

Another type of tsunami can be generated by atmospheric pressure fluctuations, and one such event was generated in 2022 by the eruption of a submarine volcano that struck Tonga (the *2022 Hunga Eruption Tsunami*). The volcanic explosion caused pressure waves in the atmosphere, which generated sea surface waves in the vicinity of Japan and reached the Japanese archipelago first. Then, a second tsunami was also generated by the volcanic eruptions, impacting particularly countries in the vicinity of Tonga, then propagating across the Pacific Ocean and striking Japan 2 hours later. This was the first such event in 139 years since the Krakatau Volcano eruption in 1983 (see also Section 2.8). Although no official post-tsunami surveys have been conducted at the time this chapter was written (given the lack of access to the country due to the ongoing COVID-19 pandemic), the local maximum tsunami height estimated was 15 m, with 3 recorded casualties in Tonga. Further casualties were reported in Peru, where the tsunami was 2 m high in places and killed 2 people.

Another distinctive tsunami was the one in 1983, caused by an earthquake with a momentum magnitude M_w of 7.7 on a fault in the Sea of Japan (the sea between Japan and South Korea, see the cross-mark in Figure 9.1.8.1). Since most tsunamis in Japan were caused by earthquakes along the trench in the Pacific Ocean side, little attention was paid to the Sea of Japan, though this event generated a wave with a height of 3.5 m that struck Akita Prefecture and killed 100 people (Usami et al., 2013).

Japan has also experienced tsunamis caused by tectonic movements in remote places. The *1960 Valdivia earthquake* in Chile (with a magnitude of 9.5 Mw) triggered a tsunami that travelled across the Pacific Ocean (see also Section 9.5). Although it could be observed along wide areas of the Japanese coast, this event was particularly devastating to the Sanriku region, due to its intricate coastline (known as the "ria coast"), which concentrates wave energy at the end of its many bays. The tsunami was as high as 6.1 m and, due to the lack of warning (given that it was a far-source event no earthquake trembling preceded it), 142 people lost their lives. After this event the Japanese meteorological agency has paid increasing attention to tectonic movements outside of Japan. The *2022 Hunga Eruption Tsunami* in Tonga affected a wide area of the Japanese coast (the highest observed tsunami height was 1.2 m in Amami-Oshima). As mentioned earlier, this eruption caused two different types of tsunami waves, which arrived at different times. Although the Japanese Meteorological Agency did not indicate the exact mechanism of the earlier tsunami, a tsunami warning was issued several hours before the arrival of the first wave and the residents were able to evacuate (there were no human casualties).

9.1.8.2.2 Tsunami countermeasures

Prompt evacuation is essential to save lives, though it can be difficult to appropriately judge whether a place is safe and maintain a high level of awareness given that tsunamis have return periods ranging from decades to hundreds of years. The *2011 Tohoku Tsunami* (see Section 2.6), for example, was the first major event to affect Iwate Prefecture in around 80 years (since the *1933 Showa Sanriku Tsunami*), though for Miyagi and Fukushima Prefectures, the southern of Sanriku region, it was the first time in 1,200 years (since the *869 Jogan Tsunami*). Even within the Sanriku region, there were differences in tsunami awareness amongst residents, which were reflected in the actual evacuation behaviour. Thus, the level of knowledge and overall awareness of individuals and society as a whole can greatly influence the scale of the damage caused by tsunamis.

Since 2011, Japan's awareness about the tsunami potential of earthquakes has changed dramatically, and local governments have been implementing countermeasures against multiple scenarios. However, it should not be forgotten that the 2011 tsunami was the largest in the history of the region, overcoming various types of countermeasures and reaching areas that were considered safe. The Ministry of Land, Infrastructure, Transport and Tourism anticipates the next major tsunami that Japan could experience could be one triggered by an earthquake in the Nankai Trough, and the state of the ongoing preparations and information directed to local residents or foreign tourists is available at its website.

9.1.8.3 *Storm surge*

9.1.8.3.1 *History*

Given that storm surges are generated by typhoons, in Japan they are most likely to take place from July to October (the period that is generally considered to be the typhoon season). In Japan, due to the narrow land shelf, the worst incidents of storm surges and their resulting floods have generally taken place along some of the bays, which also tend to concentrate much of the population and industry of the country. In particular, Tokyo Bay, Osaka Bay, and Ise Bay are shallow bays that have small mouths, which can enhance the resulting storm surge if a typhoon approaches from a particular direction (see Figure 9.1.8.2).

Table 9.1.8.4 provides a list of some of the major storm surges that have affected Japan since the beginning of the 20th century. Among them, the *1959 Isewan Typhoon* highlighted the necessity of improving countermeasures against storm surges in order to protect ports and coastal residents, which began in earnest following this event. Although the *1961 Daini-Muroto Typhoon* was as strong as the *Isewan Typhoon*, the damage was reduced due to the lessons that had been learnt. After that, no significant storm surge occurred for nearly 40 years, until a storm surge in 1999 caused large-scale inundation to Kumamoto Prefecture in Kyushu. The most recent significant flooding due to a storm surge occurred in Osaka Bay, when the *2018 Typhoon Jebi* flooded some airports and other coastal settlements (see Section 3.7).

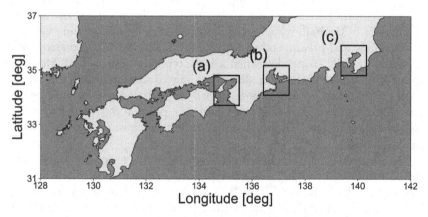

Figure 9.1.8.2 Map of the Japanese bays where major storm surges have taken place. These bays are open to a particular direction: (a) Osaka Bay (open to the south), (b) Ise Bay (open to the south-east), (c) Tokyo Bay (open to the south-west).

Table 9.1.8.4 List of major storm surges in Japan since the beginning of the
20th century

Year/Date	Region	Max surge height (T.P. [m])	Casualties	Notes
1917/10/1	Tokyo	3.1	1,324	
1934/9/21	Osaka	3.1	3,026	Muroto Typhoon
1945/9/17	Kyushu	2.6	3.122	
1959/9/26	Ise	3.9	5,098	Isewan Typhoon
1961/9/16	Osaka	3.4	200	Daini-Muroto Typhoon
1999/9/24	Kyushu	4.5	13	
2004/8/30	Seto	2.5	2	
2018/9/24	Osaka	3.3	14	

Source: The ground elevation in Japan is measured with reference to T.P. (Tokyo Peil), which was
determined by the tidal records at Tokyo Bay during the period between 1873 and 1879, and
corresponds to the mean water level in the bay.

9.1.8.3.2 Storm surge countermeasures

After the *1959 Isewan Typhoon*, coastal dikes and storm surge barriers were
constructed along the Japanese coastline using this typhoon as the "design
typhoon". Recently, the Japanese society as a whole has been paying more
attention to the possibility of a major typhoon causing catastrophic damage, given also an increased awareness about the heavy rains associated with
these weather systems (since 2015, Japan has experienced devastating river
floods almost every year). With improvements in meteorological forecasting,
warnings about potential storm surges are typically issued one or two days
in advance of landfall, which results in preventive measures being taken by
many public and private companies (e.g., trains in the Tokyo metropolitan
area are typically suspended when a large typhoon approaches). However,
the risk of a potential storm surge flooding coastal areas still exists, and
local governments are developing hazard maps that show the location and
spatial extent of the flooding that could take place in the "worst scenario"
(see Sections 8.3 and 7.3), taking into account the most critical phenomena
in a given area.

The Japanese central government's manual for making hazard maps
introduces the concept of "scenario typhoon", which assumes the worst scenario for each given area. These "scenario typhoons" assume a simple vortex and a straight path, and justify the combination of parameters and path
that maximizes the damage (for the vortex model, the Rankin's vortex is
implemented, defined using three independent parameters: translate speed,
central pressure, and maximum wind radius). In the case of Tokyo, the

largest values ever observed are utilized for each parameter; that is, a translate speed of 73 km/s and maximum wind radius of 75 km, similar to the 1959 *Isewan Typhoon*, and the central pressure of 910 hPa, similar to the *Muroto Typhoon*. According to the hazard map of Tokyo, low-lying parts of the city could be inundated to a height of almost 10 m above ground level, particularly along Koto and Arakawa wards (which due to land subsidence in the 20th century are currently below mean sea level).

However, such parameters might not actually represent those that might lead to the worst actual damage. In the case of Tokyo, the fastest typhoon (with a translate speed of 73 km/h) is supposed to cause the worst damage, though this might overlook different patterns of storm surge damage that have been observed elsewhere, such as the slow-moving hurricane Harvey in 2017, which caused prolonged flood damage over Texas, USA (Blake and Zelinsky 2018). Thus, it is clear that to ensure a given settlement is adequately protected it is necessary to consider more detailed scenarios, taking into account the recent improvement in wave simulations models and changes in typhoon characteristics due to climate change.

9.1.8.4 Conclusions

This chapter introduced two of the major coastal hazards in Japan, namely tsunamis and storm surges, and possible countermeasures against them. More than 1,000 years of tsunami records in Japan show that most of these events are caused by periodic earthquakes in the Sanriku region and the Nankai Trough. Although coastal structures have been improved in recent times, individual evacuation still plays an important role in preserving the lives of coastal residents. The cornerstone of tsunami risk prevention should thus always remain the maintenance of a high level of awareness about such hazards, and to take appropriate evacuation actions when an event takes place.

Storm surges, particularly along bays, can take place when a typhoon approaches. Tokyo Bay and Osaka Bay, which have a large metropolitan area situated in the vicinity of the coast, are particularly at risk of suffering urban flooding, even though the number of casualties due to storm surges has been greatly reduced due to the improvement of coastal structures since the *1959 Isewan typhoon*. In terms of storm surge countermeasures, updating the currently used "worst scenario" to account for the possible intensification of tropical cyclones is important, in order to improve the resilience of communities against such events and ensure their long-term sustainability.

9.2 SOUTH ASIA

9.2.1 Bangladesh

Khandker Masuma Tasnim

Weathernews, Chiba, Japan

9.2.1.1 Introduction

Bangladesh is one of the most vulnerable countries in the world to climatic hazards. The coastal area of the country is characterized by its low elevation (typically below 10 m above mean sea water level, Bernard et al. 2021, Becker et al. 2019) and high population density (1,240 people per square km, World Bank, 2021), and is often exposed to coastal disasters such as tropical cyclones and storm surges. Nicholls et al. (1995) reported that about 42% of the casualties associated with tropical cyclones have occurred in Bangladesh over the course of the last two centuries.

In most cases the reason for such large-scale casualties and destruction are storm surges (Ikeda 1995). Since 1970, four severe cyclones with maximum wind speeds greater than 61 m/s and with storm surges over 4 m in height have hit Bangladesh (namely the cyclones on November 1970, April 1991, May 1997 and November 2007, see Karim and Mimura, 2008). Table 9.2.1.1 shows a list of severe cyclone events that have hit Bangladesh between 1970 and 2020. Due to the relatively high tidal range experienced along the coastline of the country, the effect of these storm surges can be particularly severe when it coincides with a high tide (IWM 2005). The *1970 Bhola cyclone* caused the highest water level of 10 m above MSL, which coincided with a high spring tide that resulted in large-scale inundation to low-lying coastal areas (Karim and Mimura 2008). However, during the last 20 years the human loss due to coastal disasters has been reduced significantly, mostly due to the implementation of early warning and large-scale evaluation systems along coastal regions. Despite this, economic losses have been increasing due to increased development of economic activity in the coastal areas.

9.2.1.2 Case study of a recent coastal disaster: cyclone Sidr (November 2007)

The most recent cyclone event that caused large-scale devastation in Bangladesh was the *2007 Cyclone Sidr*, one of the 10 strongest to hit the country after 1876 (Hasegawa 2008). It made landfall as a very severe cyclone at approximately 15:00 to 17:00 UTC on 15 November across the Barisal coast, with a maximum wind speed of 59 m/s and sea-level pressure of 944 hPa (Akter and Tsuboki 2012, Hasegawa 2008). Across 30 districts of Bangladesh, 8.9 million people were affected by this cyclone, 3363 people were reported dead and 871 people missing (Hasegawa et al. 2008). The estimated economic loss was more than US$ 3.1 billion (Akter and Tsuboki 2012).

Table 9.2.1.1 Severe cyclonic storms that have hit Bangladesh in the last 50 years

Date of occurrence (DD.MM.YY)	Nature/name	Landfall area	Maximum wind speed (km/h)	Tidal surge (m)	Central pressure (mb)
23.10.70	Severe Cyclonic Storm of hurricane intensity	Khulna-Barisal	163	NA	NA
12.11.70	Severe Cyclonic Storm with a core of hurricane wind	Chittagong	224	3–10	NA
28.11.74	Severe Cyclonic Storm	Cox's Bazar	163	2.7–5.2	NA
09.11.83	Severe Cyclonic Storm	Cox's Bazar	136	1.5	986
24.05.85	Severe Cyclonic Storm	Chittagong	154	4.6	982
29.11.88	Severe Cyclonic Storm with a core of hurricane wind	Khulna	160	0.61–4.4	983
29.04.91	Severe Cyclonic Storm with a core of hurricane wind	Chittagong	225	3.7–6.7	940
02.05.94	Severe Cyclonic Storm with a core of hurricane wind	Cox's Bazar-Teknaf Coast	220	1.5–1.8	948
25.11.95	Severe Cyclonic Storm	Cox's Bazar	140	3	998
19.05.97	Severe Cyclonic Storm with a core of hurricane wind	Sitakundu	232	4.6	965
27.09.97	Severe Cyclonic Storm with a core of hurricane wind	Sitakundu	150	3–4.6	NA
20.05.98	Severe Cyclonic Storm with a core of hurricane wind	Chittagong Coast near Sitakunda	173	0.91	NA
15.11.07	Severe Cyclonic Storm with a core of hurricane wind (Sidr)	Khulna-Barisal Coast near Baleshwar river	223	4.6–6	942
30.05.17	Severe Cyclonic Storm (Mora)	Chittagong-Cox's Bazar Coast near Kutubdia	146	NA	NA

Source: Bangladesh Meteorological Department.

Figure 9.2.1.1 Left) Sand bags at the coastal side of a dyke at West Kuakata after Sidr; Centre) An old-style primary school that doubles as a cyclone centre in Sarankhola (Shibayama et al. 2009) Right) A newly built multipurpose disaster shelter.(Photo Source: World Bank 2021.)

Soon after Cyclone Sidr, The Japan Society of Civil Engineers (JSCE) conducted two field surveys of the affected area, with the coastal engineering team being headed by the lead editor of this book, Tomoya Shibayama. According to their report, inundation height in the riversides near the landfall location was higher than along some of the coastal areas. Run-up flow along the river branches and waterways caused large-scale inundation (5.5 to 7 m) near Burishar river (Somboniya, Naltona and surrounding areas) and around 6 to 6.5 m near Baleshwar river (Southkhali and Sarankhola). Among the coastal areas, Kuakata suffered the most, being located on the right-hand side of the path of Cyclone Sidr. The maximum inundation height observed in West Kuakata was 5.6 m. Erosion of the river bank and coastal embankments also being quite significant, with Figure 9.2.1.1(a) showing the damage to a coastal embankment in West Kuakata. Even though the height of the storm surge level resulted in the overtopping of the coastal embankments at many locations in Kuakata, the existence of these structures arguably played an important role in minimizing damage to life and property (JSCE 2008; Shibayama et al. 2009; Tasnim et al. 2015).

9.2.1.3 Current state of disaster management preparedness

Over the past two decades, Bangladesh has been making efforts to establish a comprehensive disaster management system from the national to the community level. An early warning system was first implemented in the country after the 1970 cyclone, which caused the death of 300,000–500,000 people (Landsea et al. 2006). A subsequent cyclone in 1991 caused the death of 138,000 people, due to failure to correctly warn residents of coastal areas in time and the lack of cyclone shelters (Paul et al. 2010). The reduction in casualties in subsequent events (Sidr: 3363 death, Aila: 330 death) has been credited to improvements in disaster risk management, including improved forecasting, warning and evacuation systems (GOB 2008, Ahmed et al. 2014, Rowsell et al. 2013). Early warning dissemination of information regarding cyclones in Bangladesh has hugely improved over the last decade through CPP (Cyclone Preparedness Program) volunteers helping local communities

(Rowsell et al. 2013). The number of CPP volunteers has increased significantly from 20,000 (1991) to 43,000 (2007), together with the number of cyclone shelters (2,400 emergency shelters of various kinds had been constructed by 2010, Dasgupta et al. 2010). Figure 9.2.1.1(center) and 9.2.1.1(right) show an old-style and new-style cyclone shelter in Bangladesh, respectively.

The construction of resilient structures and houses is also steadily proceeding, especially after cyclone Sidr, with the United Nations Development Program (UNDP) building over 9,000 disaster-resilient houses in the hardest-hit areas (UNDP 2010). Construction of new cyclone shelters and rehabilitation of some existing shelters is also being contemplated by the World Bank's Multipurpose Disaster Shelter Project (World Bank 2021).

Nevertheless, even though cyclone shelters are being constructed and rehabilitated, there are still a number of factors which hinder the willingness of people to evacuate, including:

- the location and distribution of shelters do not always correlate well with demographic density, and in some places there is a shortage in the capacity of shelters,
- the design of the shelters does not take into account the difference in the potential surge height distribution. In some areas the storm surge could reach the 2nd floor of a shelter (where people are taking refuge).
- some shelter structures are old, and rehabilitation work is necessary.
- shelters are only for people. Residents hesitate to go to shelters and abandon their livestock, given the necessity to maintain their livelihoods after the storm.

9.2.1.4 Challenges posed by climate change to coastal resilience in Bangladesh

Climate change is now one of the most pressing issues for Bangladesh, and in order to cope and adapt to the unexpected conditions created by climate change local communities in Bangladesh are investing in multiple alternative livelihood strategies based on their indigenous knowledge and coping mechanisms. However, the ongoing nature of climate change will continue to create new risks against which these communities might show lower levels of resilience. Moreover, it is expected that the intensity of typhoons will increase in the future (Knutson et al. 2021 and Section 7.2.), which will generate higher storm surges that, compounded with ongoing sea-level rise, will result in much higher water levels during the passage of these events.

Current disaster management strategies might not be enough to overcome such challenges, requiring additional adaptation strategies to be implemented. In this sense, rural infrastructure development and arrangements for pre-planning evacuation by deploying buses, trucks and motorized boats to transport people are very important. It is worth noting that the development of transport infrastructure can be seen as a no-regrets strategy, as it can also serve to stimulate the local economy (a co-benefit to the improvement of disaster resilience). Additionally, it is important to understand that cyclone shelters reduce death, but do not protect property. Hence, for a cyclone-prone country, considering the large-scale economic losses (which can be significant, for cyclone Sidr the total loss was $1.2 billion, or 2.8 % of GDP (Rowsell et al. 2013)) due to frequent storms the construction of coastal embankments of sufficient height might be necessary (which could effectively turn the entire delta into polders, similar to the Netherlands). These structures may face overtopping during very severe cyclones with a return period of 100 years or more, but for moderate to severe frequent events they are expected to significantly reduce economic losses. Eventually, a full multi-layer safety system might be put in place, similar to that present in other countries (Esteban et al. 2013). If coastal areas can be protected from the damage due to natural disasters, the economic activity will flourish, increasing employment and allowing the community to improve its resilience through local empowerment programs.

9.2.1.5 Conclusions

Even though Bangladesh seems to have learnt lessons from previous experiences with storm surges, its existing cyclone warning and evacuation plans might not be sufficient if large-scale inundation occurs (together with high wind speeds and rain), such as in the case of cyclone Yolanda in the Philippines (Takagi et al. 2014, see Section 3.5). Extreme cyclone events in the future will likely be much stronger than at present, and may remain powerful longer after landfall (see Section 7.2). In such cases, existing cyclone centres may not be able to withstand very high wind speeds for a longer period of time. Furthermore, in Bangladesh riverbank erosion and flooding during storms is also a common problem. The construction of more resilient shelters to sustain stronger winds in coastal and riverside areas, and the renovation of the existing shelters should clearly be addressed to improve the resilience of coastal communities against such events.

9.2.2 Natural hazards in Bhutan Himalaya: climate change impacts, efforts, and challenges

Cheki Dorji, Karma Tempa, and Leki Dorji

Royal University of Bhutan, Phuentsholing, Bhutan

9.2.2.1 Introduction

Bhutan lies in the Himalayan region, with its territory being mostly composed of rugged and mountainous terrain with altitudes ranging from less than 100 m in the south to more than 7,500 m in the north, within a north-south distance of less than 175 km. Due to its topographical and fragile geological settings (Tempa & Chettri 2021; Tempa et al. 2021a), combined with extreme meteorological events (Dikshit et al. 2019), landslides in the Himalayan region can have profound impacts on the people and economy of the countries in the region (Dikshit et al. 2020). In recent times there has been an increase in rainfall in parts of the Bhutan Himalaya, and particularly the central and southern belt has seen an exponential increase in precipitation, which has led to an increasing number of occurrences of landslides (Chettri et al. 2022). The year 2021 witnessed an alarming number of landslide events that caused substantial damage to road infrastructure and claimed the lives of several people. The road damage and subsequent roadblocks were caused by slope failures, rockfalls, rock mass slides, and debris flow, and in some locations road sinking and cracking.

The present study correlates the landslide events of 2021 with rainfall patterns to formulate a Landslide Hazard Scenario Map (LHSM). Such a map could assist the Department of Roads (DoR) to prioritize resource allocation at the Regional Office (RO) and the Sub-Divisions (SD) level to facilitate and provide safe transport services and reduce the time during which roads are blocked. The flood hazards due to Glacial Lake Outburst Floods (GLOF) are also highlighted in this chapter, in order to provide an overview of the risks present in the Bhutan Himalaya due to the impact of changing climatic conditions.

9.2.2.2 Methodology

A detailed inventory of landslides between April and October 2021 was identified and the hazard scenarios associated with changing rainfall patterns were estimated to understand the possible future impact of climate change. The landslide data was extracted from the DoR daily online data on roadblock status and some landslide events recorded in the print media.

The monthly rainfall data from 10 regional stations (which covers all the areas in the vicinity of landslides) were collected from the National Centre for Hydrology and Meteorology (NCHM). A statistical relationship was developed between the monthly accumulated rainfall data and the corresponding landslide frequency. As a result, the spatial cumulative annual rainfall map of Bhutan was prepared and the landslide inventory was superimposed to produce the Landslide Hazard Susceptibility Map (LHSM) of 2021.

9.2.2.3 Landslides

The majority of the 158 major landslides at 80 locations recorded in the Bhutan Himalaya in the year 2021 were along road corridors between the mountains (see Figure 9.2.2.1), with most events taking place between April and October.

The highest number of landslides (17 events) were recorded at Boxcut, 15 km away from Gelephu (Figure 9.2.2.2(i)) along the Gelephu–Trongsa

Figure 9.2.2.1 Landslides events of 2021 along the road corridor in Bhutan. Top left: Kurizampa–Nganglam PNH (30 km), 27/05/2021. Top right: Kanglung–Trashigang PNH (3 km), 26/08/2021. Bottom left: Nganglam–Kurizampa PHN (33 Km), 22/10/2021. Bottom right: Gelephu–Trongsa PNH (Boxcut, 15 km), Multiple.

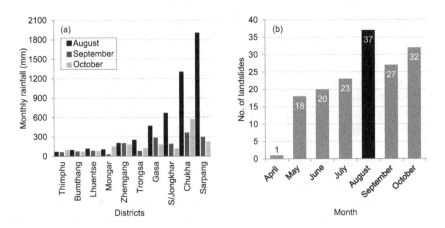

Figure 9.2.2.2 (a) Monthly rainfall in 10 regional districts for August, September, and October. (b) Number of landslides that occurred from April to October 2021.

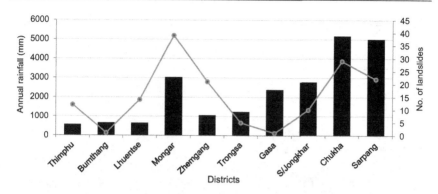

Figure 9.2.2.3 Landslides under 10 regional districts corresponding to the accumulated annual rainfall.

Primary National Highway (PNH), which falls under Mongar district (Dzongkhag), followed by 12 landslide events at the Namling area (Figure 9.2.2.2(f & h)), along the Simtokha–Trashigang PNH. The Kurizampa–Nganglam, and Tingtibi–Panbang PNH also suffered frequent slides throughout the year. All the landslides caused substantial or complete damage to road infrastructure including pavement, retaining walls, bridges, and crash barriers, which caused socio-economic interruptions and economic losses. They can also result in human losses, and the event on 12th June in Laya in the Thimphu district claimed the lives of 10 people. Similarly, the 29th June landslide at Bangay (Pasakha, Chukha) and the 26th August landslide at Khoraypam (Pemagatshel) claimed the lives of two and three people, respectively.

The monthly accumulated rainfall was higher in some of the districts (Figure 9.2.2.2(a)) in August compared to September and October. Sarpang district received the highest rainfall in August (1,904 mm), followed by Chukha (1,303 mm) and Samdrupjongkhar (667 mm), which falls within the southern belt of the country. In the north, Gasa received much lesser precipitation (289 mm) than other districts.

Most landslides took place due to the combined effect of continuous heavy rainfall along newly constructed roads in fragile topographical and geological conditions (e.g., Gyalposhing–Nganglam and Tingtibi–Panpang PNH). Figure 9.2.2.3 indicates that Chukha, Sarpang and Mongar districts recorded a higher number of landslide events in 2021, which correlates well with the rainfall recorded in each of them (see also Figure 9.2.2.2(b)). In 2021, a total of 37 landslides were observed in August, followed by October and September, with 32 and 27 landslide events, respectively.

9.2.2.3.1 The 2021 Landslide Hazard Scenario Map (LHSM)

The LHSM 2021 (Figure 9.2.2.4) was developed to illustrate the landslide hazards according to rainfall patterns in Bhutan Himalaya. The accumulated

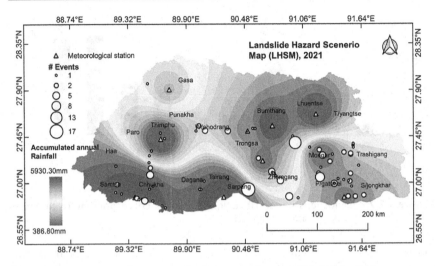

Figure 9.2.2.4 The Landslide Hazard Scenario Map (LHSM) 2021 of Bhutan.

annual rainfall of 10 regional meteorological stations was interpolated in QGIS (QGIS Development Team 2021) using the inverse distance weighted (IDW) method to obtain the spatial distribution, and the 2021 landslide inventory was superimposed on it. As a result, the map shows the spatial correlation between the amount of cumulative rainfall and the frequency of landslides in a particular location.

As shown in Figure 9.2.2.4, the southern part of the country, such as Samtse, Chukha, Sarpang, Pemagatshel, and Samdrupjongkhar, experience higher precipitation and thus more landslide events (61 landslides). Places like Sarpang. Trongsa, Mongar, Zhemgang, and Trashigang have cumulative rainfall as high as 5,930 mm, and accordingly the highest number of land-slides (65 events). Gasa region, located at the extreme north of the greater Himalaya, also recorded 2,390 mm of accumulated annual rainfall, with the highest taking place in July (472 mm). Currently, Namling area in Mongar and Boxcut area (15 km from Gelephu) in Sarpang are at high risk of suffer-ing from landslides. The newly constructed Kurizampa–Nganglam PNH (98.00 km) and Tingtibi–Panbang PNH (77.53 km) are also at risk due to the rough terrain and unstable slopes they traverse.

9.2.2.3.2 Response efforts and challenges

At present, the road network development program, operation, and mainte-nance (O&M) in Bhutan is undertaken by the Department of Roads (DoR), under the Ministry of Works and Human Settlement (MoWHS), consist-ing of nine Regional Offices (RO) and about 19 Sub-Division (SD) across the country, with their Head Quarters (HQ) at Thimphu (Figure 9.2.2.5). The RO and the SD facilitate the road clearing works under their respective

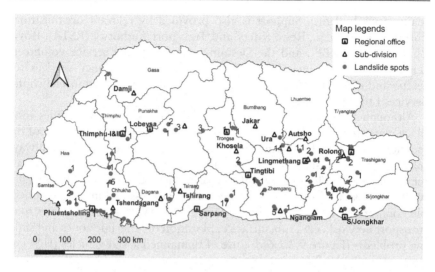

Figure 9.2.2.5 The location of the regional offices, sub-divisions, and the land-slide occurrences in 2021.

Figure 9.2.2.6 Efforts to mitigate landslide hazards through government and public participation in post-disaster activities.

administrative jurisdictions. In 2021, tremendous response efforts were undertaken by DoR for clearing roadblocks, realignment, and improvising temporary measures by deploying heavy equipment and human resources (Figure 9.2.2.6(a & b)). According to the records of the information disseminated to the general public, the road clearance work ranged between 3 and 6 hours in most cases, up to 12 hours in some locations, and in some cases over 48 hours or a week. The DoR informs the public through an official app known as the Bhutan Road Safety App, a Government to Citizen (G2C) initiative, their Facebook portal, and national print media (daily news bulletin). Support is also provided by relevant organizations during disasters, e.g., Road Safety and Transport Authority (RSTA), Royal Bhutan Police (RBP), and the Desuup-trained national service volunteers (Figure 9.2.2.6(b)), including private companies. Emergency services for rescue and evacuation are also supported by the Royal Bhutan Helicopter Services Ltd (Figure 9.2.2.6(c)).

Although the nearest SD could reach the places where landslides took place, persistent heavy rainfall hampered the clearing works in some of the locations. The rough terrain and the topographical settings have always made it difficult to successfully implement emergency works during a disaster, given also the limited availability of modern technology. Post-disaster works are also undertaken simultaneously to address and implement possible countermeasures against future landslide hazards along with the road transport network, e.g., sub-surface exploration of landslide zones and sinking problems (Figure 9.2.2.6(d)), use of unmanned aerial vehicles (UAV) or drones (Figure 9.2.2.6(d)) and seismic refraction tomography (SRT) surveys to characterize the landslide-prone area at Boxcut in Sarpang.

9.2.2.4 Flood hazards

Flooding is a recurrent phenomenon during the monsoon season, and the six major flood events recorded in modern times are the 1994 Lugye GLOF, the 2000 monsoon floods in Phuentsholing, the 2004 Eastern Bhutan monsoon floods, the 2009 floods induced by Cyclone Aila, the 2015 Lemthang lake outburst flood, and the 2016 southern Bhutan monsoon floods. In most settlements fertile agricultural fields are located along the main drainage basins, meaning that low-laying valleys are at high risk of suffering flooding, as shown in Figure 9.2.2.7.

Flash floods are highly localized, and establishing a close network of monitoring stations would be useful to forecast future events. Currently, monitoring stations are few and far apart, which makes assessment and evaluation of flood hazards difficult. Further, most rivers run through deep gorges and ravines, and past events indicated that flash floods in Bhutan occur mostly in tributaries, sweeping boulders and debris downstream.

Figure 9.2.2.7 Flooding in (a) Omchu, (b) Amochu (c) Lhuentse.

The government has recommended developing flood early warning systems along flood-prone rivers and establishing a national network of weather stations. Most weather stations are now automated to provide near real-time information, and some prediction models have been tested in selected river basins. The National Centre for Hydrology and Meteorology has a 24/7 monitoring system in place, with staff trained in satellite rainfall estimation methods.

9.2.2.5 Glacial Lake Outburst Floods (GLOF) hazard

Glacial Lake Outburst Floods (GLOF) occur when a body of water that is contained by a glacier or terminal moraine is released, which is a growing problem in Bhutan due to the retreat of glaciers due to climate change, particularly in the western part of the country (Mool et al. 2001). It is reported that Bhutan experienced GLOF events in 1957, 1960, 1968, and 1994. The northern region of Bhutan is characterized by numerous snow-clad mountains and glacial lakes (Alam and Murray 2005), and 20% of the country is above 4,200 m, and thus permanently covered with snow and ice, which form glaciers and glacial lakes. Of the 2,674 glacial lakes in the country, 24 lakes have been identified as potentially dangerous, posing a threat of GLOF at any time. Data from the Kathmandu-based International Centre for Integrated Mountain Development (Mool et al. 2001) shows that between 1980 and 2010 glacial lakes in Bhutan increased by 8.7%, while the glaciers shrank by 22% (see also Gardelle et al. 2011; Gurung et al. 2017).

The most devastating GLOF occurred in October 1994 due to the partial burst of Luge Tsho in eastern Lunana, which caused enormous damage to property and loss of life in Punakha and Wangdue Valley, as shown in Figure 9.2.2.8. The increasing risks from GLOF in Bhutan have thus become

Figure 9.2.2.8 Damages to arable land by GLOF.

Figure 9.2.2.9 Physical mitigation measures to reduce the water level of a glacial lake.

an urgent environmental and economic issue due to climate change, and as a result the physical countermeasures to lowering the water level in the glacial lake have been initiated, as shown in Figure 9.2.2.9.

9.2.2.6 Conclusions

Natural hazards in Bhutan include landslides, floods due to rainfall, and Glacial Lake Outbursts (GLOF). The cause of landslides along the road corridors between the mountains are related to difficult topographical and geological conditions, and triggered by rainfall and human activities. While the roadblocks due to landslides are usually cleared within a day or two, efforts to mitigate them through research that aims to formulate appropriate long-term countermeasures are still lacking.

Annual floods also disrupt the life of settlements, and affect the fertile agricultural fields in low-laying valleys, which has become a perennial problem. Without proper monitoring stations and a flood early warning system, the risk posed by these events is still significant. Glacial Lake Outburst Floods (GLOF) are a major concern, as numerous rivers are glacial-fed and thus have the potential to result in a major disaster. Due to enhanced glacier melting due to temperature rises as a result of climate change, glaciers shrank by 22% and lake enlargements have been observed. The manual lowering of water levels in some of these glacial lakes has commenced, though this should be continued in other places in order to adapt to future climate change and avoid potential future disasters.

9.2.3 Coastal disasters and mitigation measures in Sri Lanka

Nimal Wijayaratna

University of Moratuwa, Moratuwa, Sri Lanka

9.2.3.1 Introduction

The coastline of Sri Lanka is about 1,620 km long, including bays and islands but excluding lagoons (see Figure 9.2.3.1). It has a very narrow continental shelf to the east, south and west, but widens when it merges with that of India to the north. Sand spits and dunes, barrier beaches, large bays, shallow beaches, developing beach sections, and low population density are the salient features of the coastlines to the north and east. The south and west coastlines include many river outlets, steep and eroding beaches, and are characterized by higher population density and the presence of extensive beach protection measures and other infrastructure facilities such as harbours. The coastal environment is influenced by the proximity to the Indian subcontinent, the relative position of Sri Lanka on the northern Indian Ocean and the monsoon wind patterns over the island.

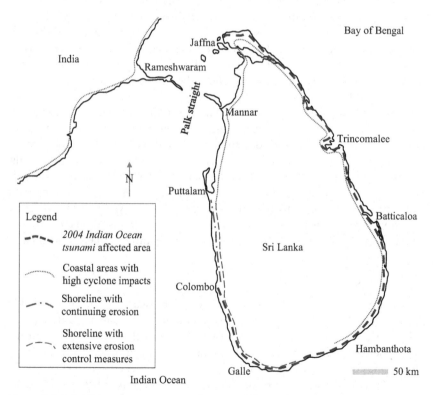

Figure 9.2.3.1 Disaster-prone coastal areas of Sri Lanka.

The coastal zone, comprising its coastal communities, built environment and eco-systems, is exposed to a wide range of hazards arising from natural phenomena and anthropogenic activities. Coastal disasters can be generally divided into two categories: catastrophic events with a rapid onset, like tsunamis and storm surges, and slow-onset processes such as the impacts due to erosion and accretion of sediments, and anthropogenic threats to the coastal ecosystem.

9.2.3.1.1 Tsunamis in Sri Lanka

Historical records of tsunamis in Sri Lanka are very vague given the long return periods associated with them (in contrast to other countries such as Japan and Indonesia, see Sections 2.1, 2.6 and 9.1.), with many of them being but folklore stories about a quick and powerful flooding due to the rising sea. However, based on sediment cores retrieved from Karagan Lagoon at Hambanthota in the southeast shoreline of Sri Lanka (see Figure 9.2.3.1), Jackson et al. (2014) showed that there have been several tsunamis in the past few thousand years. Sediment deposits indicate that these tsunamis have return periods of once in 181 to 517 years to 1045 ± 334 years, with a mean recurrence interval of 434 ± 40 years during the Holocene period. The first historical recorded tsunami to have affected Sri Lanka was on 27th August 1883, due to the volcanic eruption of Krakatoa (see Section 2.8). Though an unusual sea water level movement was observed, the water level fluctuations were not severe and there was no significant inundation or records of damage (Hettiarachchi et al. 2013). The second and most devastating tsunami experience was the 26th of December 2004 Indian Ocean *Tsunami* where 38,000 people lost their lives in Sri Lanka alone, and the economic loss was estimated to about US$ 1 billion (SLCZ&CRMP, 2018, see also Section 2.1). This was the worst natural disaster ever in Sri Lanka, in terms of lives lost and the extent of destruction, with about 5% of the total population (1 million people) directly affected, and coastal ecosystems (corals, wetlands, riverine and lagoon ecosystems, and coastal landscape) and infrastructure being devastated. The long-term consequences were also significant and sometimes profoundly affecting coastal areas for years, with saltwater intrusion into the drinking water sources and farmlands, damages to wetland habitats, and sand deposits on coral reefs leading to mass mortality.

9.2.3.1.2 Cyclones and storm surges

Only a limited number of the atmospheric low-pressure systems generated in the northern Indian Ocean have an impact on Sri Lanka, as its relative proximity to the equator places it on the outer limits of the cyclone zones of the Bay of Bengal and the Arabian sea. As a result, during the 20th century only 16 cyclones have made landfall in the country (Srisangeerthanan et al. 2015).

In 1964, over 1000 lives were lost due to a cyclone (storm ID 1964352N05093, not named) which had a maximum wind speed of about 160 km/h as it approached Sri Lanka. The surge height was as high as 4 m at some places on the northwest coastline, according to anecdotal evidence and model studies (Murthy et al. 2004). In 1978, another cyclone (storm ID 1978323N08092, not named) with a maximum wind speed of 145 km/h made landfall near Batticoloa on the east coast, generating a storm surge of 1.5 to 2 m, claiming 915 lives and devastating houses and infrastructure throughout the island. In 1992, a cyclonic storm (storm ID 1992311N06107, not named) claimed 4 lives and damaged about 29,000 houses along the southeast coastline. A cyclone in the year 2000 (storm ID 2000358N08086, not named) was the strongest tropical cyclone to strike Sri Lanka since 1978. It made landfall along the east coast close to Trincomalee with a maximum wind speed of 165 km/h, killing at least nine and making 500,000 homeless. The cyclone in 2003 (storm ID 2003129N05091, not named) did not make landfall, but rather followed a south-to-north path parallel to the eastern coastline and made its landfall in south India. However, due to heavy rainfall, floods, and landslides over 250 lost their lives in Sri Lanka (Srisangeerthanan et al. 2015).

9.2.3.1.3 Coastal pollution due to marine accidents

As Sri Lanka is located on one of the main east/west shipping routes, around 200 to 300 ships pass the waters around the island daily, and in the course of the last decades there have been several accidents leading to the spill of oil and other hazardous materials. MV X-press Pearl, a container ship carrying 1,486 containers filled with chemicals (including 25 tons of nitric acid) caught fire and sank in July 2021 close to Colombo port, devastating the west and south coastlines of the country. In September 2020 MT New Diamond, carrying over two million barrels of crude oil, caught fire while approaching the eastern coastline of the country. Though the cargo tanks were not damaged, an oil slick could be observed as the vessel was towed away from the coastline. In April 2009 chemical tanker MT Gramba, carrying 6,250 tons of sulphuric acid, was towed off to about 60 nautical miles off the east coast of Sri Lanka before it sank (Kelley 2009). In 2006, MV Amanat Shah, a ship carrying thousands of tons of timber, sank off the southwest coast, loaded with 176 tons of fuel, which were released and polluted a 15 km stretch of shoreline. In 1999, MV Meliksha sank off the southeast coastline, together with over 16,000 tons of chemical fertilizer and 200 tons of fuel (Kulathilaka 2018).

9.2.3.1.4 Coastal erosion

Coastal erosion, as a result of both natural causes and anthropogenic interventions along the shoreline, is an ongoing problem in Sri Lanka, leading to widespread socio-economic and environmental problems. This erosion has been particularly severe along the south and west coastlines, though its impacts have been controlled to a certain degree through nine major coastal erosion management programs from 1987 to 2016, at a total cost of US$ 40 million (SLCZ&CRMP, 2018). Despite this, erosion is still taking place along the northwest coastline, particularly during the monsoon season.

9.2.3.2 Coastal disaster mitigation measures

Before 2005, disaster risk management in Sri Lanka was limited to reactive response and relief distribution (Figure 9.2.3.2). Preparation for imminent disasters and proactive approaches that would reduce the impacts of disaster were limited, and coastal defences were limited to a "hold the line" approach, through the construction of hard structures such as revetments, seawalls and groynes. Disaster impact reduction was not typically on the list of priority projects, and coastal communities were not well prepared for the rapid onset of disasters. The unpreceded devastation experienced during the *2004 Indian Ocean tsunami* (see Section 2.1.) exposed how unprepared the country was and, as a result, the disaster management procedures were changed.

In 2005, to enhance the disaster resilience of the country and to manage all phases of a disaster response more efficiently, the Disaster Management Act was enacted, which would later become a national policy. This policy is based on the Hyogo framework for action for achieving disaster resilience in vulnerable communities, with its most notable result being the establishment of the disaster management centre, which became the national focal point for disaster risk management. Immediately after the *2004 Indian Ocean Tsunami*, the effectiveness of early warning in mitigating the number of casualties resulting from a given event has been clearly recognized (see also Section 8.1). As a result, measures have been taken to transform the technical warning received from the regional tsunami service providers of the Indian Ocean tsunami warning system into a public warning that can be effectively acted upon by the coastal population. The installation of 74

Figure 9.2.3.2 Reactive response is not adequate for a speedy recovery: A bridge in Galle damaged due to the *2004 Indian Ocean Tsunami* is being replaced with a temporary bridge to re-establish connectivity.

tsunami early warning repeater towers, collaboration with television, radio, telephone, and internet providers for issuing warnings, conducting annual evacuation drills, preparation of hazard and evacuation maps, and the installation of evacuation signs, are some important steps that had been taken on this regard. Several tsunami early warnings had been issued through this system, and coastal communities successfully evacuated during the earthquakes that took place in the Sunda Trench, Indonesia, (on 28 March 2005, 12 September 2007 and 11 April 2021) though those tsunamis have had no impact on Sri Lanka.

A more comprehensive disaster management program was put in place during the period 2014 to 2018, which set up legal and institutional systems to minimize disaster risk through capacity building of the vulnerable communities. A coastal hazard profile assessment (cyclone, tsunami, sea surges, sea-level rise and coastal erosion) was carried out between 2015 and 2017. With the adoption of the Sendai framework in 2018, Sri Lanka is now starting to implement the mitigation and preparedness concepts of the disaster risk reduction framework. With this objective, the National Disaster Risk Management Plan (2018–2030) has been drafted, taking into consideration the four priorities outlined in the Sendai framework for disaster risk reduction: understanding disaster risk, strengthening disaster risk governance, investing in disaster risk reduction, and enhancing disaster preparedness for an effective response.

As a result, early warnings are now issued for storm surges, high winds, waves, and cyclones, mainly through radio and television broadcasts. No hard coastal defences have been built to protect against tsunamis or storm surges, but some strict regulatory measures are in place to manage coastal activities. The application of setback standards, establishing conservation areas and enforcing disaster risk reduction by design for new constructions are some of the measures that are currently being enforced. Beach nourishments and sand engine concepts are also being considered as environmentally friendly erosion mitigation measures. The overall aim of these measures is to protect lives and properties from rising sea levels, storm surges and coastal erosion and to minimize public investment in coast protection measures.

9.2.3.3 Conclusions

The shoreline of Sri Lanka is vulnerable to a number of coastal hazards, including tsunamis, cyclones, storm surges, pollution due to maritime accidents and beach erosion. Due to the adverse impacts of anthropogenic activities disasters are becoming more frequent, threatening communities that are dependent on the ocean for their livelihoods. Disaster management mechanisms in the last two decades were limited to reactive response and relief distribution, though slowly more proactive approaches are emerging focusing on capacity building of the vulnerable community, early warning, application of setback lines and delimitation of conservation areas. Such "soft measures" are seen as cost effective, applicable over a wide range of disaster conditions and adjustable over time according to lessons learned after each event.

9.3 WEST ASIA

9.3.1 A study on the tropical cyclones in the North Arabian Sea and the Gulf of Oman

Mohsen Soltanpour and Zahra Ranji

K.N. Toosi University of Technology, Tehran, Iran

9.3.1.1 Introduction

Tropical Cyclones (TCs) are intense, rotating low-pressure systems that generate very large waves in warm tropical areas (see also Section 7.2). The strong winds generated by TCs rotate inwards in a counter-clockwise direction in the northern hemisphere. The Arabian Sea and the Bay of Bengal are two major basins which are affected by TCs in the North Indian Ocean, although these storms are generally more destructive around the Bay of Bengal than the Arabian Sea. This chapter offers an overview of TCs that have affected this area in recent years, providing guidance on how to simulate such events using state-of-the-art atmosphere-wave modelling. As there is a risk that future TCs will increase in intensity as a consequence of global warming (see Section 7.2), recent trends and future potential risks will also be discussed.

9.3.1.2 Recent and historical TCs in the North Arabian Sea and the Gulf of Oman

The Arabian Sea is located in the northwestern part of the Indian Ocean, bordering the Gulf of Oman, which runs to the Persian Gulf through the Strait of Hormuz (Figure 9.3.1.1). Three different data sources regarding TC information over the Arabian Sea are available. The Indian Meteorological Department (IMD 2020) offers an extensive data summary of TCs, including their monthly and annual frequency, direction, and speed. This dataset also provides the historical tracks of cyclones starting from 1891. The International Best Track Archive for Climate Stewardship (IBTrACS) (Knapp et al. 2010) of the World Meteorological Organization (WMO) provides more details, including the TC's minimum pressures and maximum wind speeds from 1990, as well as the tracks from 1842. The Joint Typhoon Warning Center (JTWC) (Chu et al. 2002) is another resource, which compiles the intensities of TCs from 2001 and track data from 1945.

   Considering the differences in TCs between 2001–2019 and 1982–2000, Deshpande et al. (2021) reported a rise of 150%, 262%, and 18% in the frequency, duration, and accumulated energy of severe TCs in the Arabian

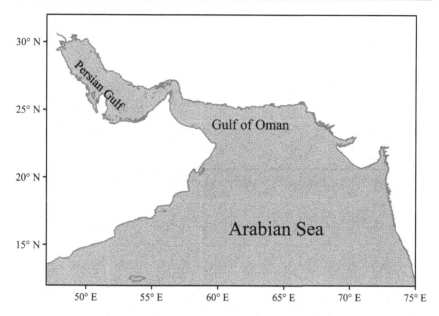

Figure 9.3.1.1 Location of the Arabian Sea and the Gulf of Oman.

Sea, respectively. They noted that these changes are related to enhanced mid-level relative humidity and moist static energy, which depend on the sea surface temperature and heat potential. Previously, several studies attributed this increase in intensified TCs to an upward trend in anthropogenic black carbon and sulfate emissions (Evan and Camargo 2011). Figure 9.3.1.2 presents the 5-year average of TC frequencies in the Arabian Sea from 1890 to 2020, indicating an increase in the frequency of more intense TCs in recent decades.

TCs in the Arabian Sea tend to either travel west towards Oman or re-curve north to strike Pakistan or India (Dibajnia et al. 2010). June is the most frequent month of cyclogenesis, with each of the periods of May–June and October–November contributing about 40% of the total number of storms, i.e., depressions, cyclonic storms, and super cyclonic storms (Soltanpour et al. 2021). The datasets also reveal that the penetration of TCs further north close to Iran is rather infrequent. Figure 9.3.1.3 shows the historical and recent TC tracks that entered the Gulf of Oman, with recent information on these events being presented in Table 9.3.1.1. Although TCs normally do not attack the Iranian coastline, existing data shows that their danger cannot be ignored. The historical tracks show two TCs that crossed the Arabian Peninsula towards the west of the Gulf of Oman and the Persian Gulf. However, TCs have not been considered in past hindcast studies in the Persian Gulf, e.g., Dibajnia (2010), due to their very rare occurrence and also concerns regarding the accuracy of early historical data. If TCs intensify in the future due to climate change (see Section 7.2.) and enter the Persian

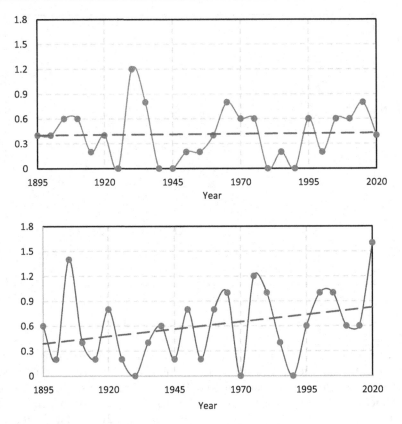

Figure 9.3.1.2 5-year average of TC frequencies in the Arabian Sea. Left: cyclonic storms. Right: severe cyclonic storms.

Gulf, they would have the potential to generate very large waves, much higher than those of any local storm.

Cyclone Gonu in 2007 was the most intense TC on record in the Arabian Sea (Dibajnia et al. 2010). It had an unusual path, travelling much farther west and north than typical TCs in the Arabian Sea. Three years later, in June 2010, the powerful cyclonic storm Phet (category 4) appeared in the same area. Phet had a very rare track, distinguishing it from other TCs over the Arabian Sea, with two landfalls over Oman and Pakistan, and the longest track in recent years (Figure 9.3.1.3).

Cyclone Yemyin, which happened shortly after Gonu in June 2007, was the deadliest cyclone so far in the 21st century. The cyclone formed over the Bay of Bengal, crossed the Indian subcontinent and moved into the Arabian Sea, killing 140 people in India, while it was still in the Bay of Bengal, 730 in Pakistan, and 113 in Afghanistan. After its first landfall near Kakinada, the system began to weaken over India due to land interaction and wind shear, but intensified again when it started moving over the Arabian Sea.

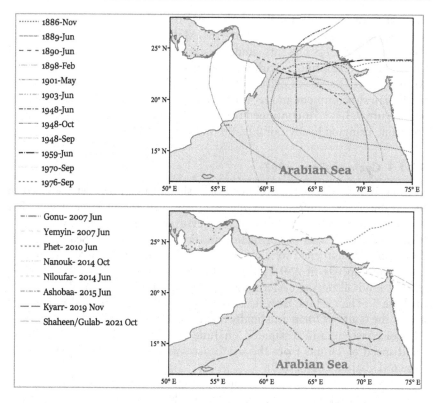

Figure 9.3.1.3 Historical (top) and recent (bottom) tracks of TCs over the north Arabian Sea that entered the Gulf of Oman.(Data from IMD (2020) and IBTrACS (Knapp et al. 2010).)

Table 9.3.1.1 Characteristics and damage of recent TCs to have affected the north Arabian Sea and the Gulf of Oman

Year	Name	Wind speed (km/h)	Min pressure (hPa)	Fatalities	Damage (USD)
2007	Gonu	270	920	78	$4.4 billion
2007	Yemyin	95	986	983	$2.1 billion
2010	Phet	230	970	44	$780 million
2014	Nanouk	85	986	—	—
2014	Niloufar	215	950	4	—
2015	Ashobaa	85	990	—	—
2019	Kyarr	250	922	5	—
2021	Shaheen/ Gulab	130/85	986/992	34	$269 million

Traveling northwest off the Pakistani coast, Yemyin made its second landfall at about 0300 UTC June 26 along the Makran coast, near Ormara and Pasni. Figure 9.3.1.3 presents similar historical cyclones that have traversed the Indian subcontinent without complete dissipation and entered the Arabian Sea. The recent cyclone Shaheen/Gulab in 2021 similarly crossed the Indian subcontinent and re-intensified as it propagated over the Arabian Sea (Figure 9.3.1.3). It continued travelling and made landfall on the Omani coastline, bringing heavy rainfall and duststorms to Iran.

9.3.1.3 Cyclone Gonu 2007

In early June 2007, super cyclone Gonu, the strongest TC yet recorded in the Arabian Sea, entered the Gulf of Oman, generating large waves that hit the coasts of Iran and Oman. Gonu evolved into a low-pressure system from a persistent area of convection in the eastern Arabian Sea on June 1, 2007, and strengthened to a super cyclonic storm (category 5) with maximum wind speeds of 250 kPh (140 knots) by June 4. The cyclone made its first landfall on the easternmost tip of the Arabian Peninsula in Oman on June 5. The cyclone then decreased in intensity as it moved northward across the Gulf of Oman and finally made its second landfall along the south Iranian coastline and dissipated on June 7 (Figure 9.3.1.4). A maximum wind speed of 16 m/s from the southeast was measured on the evening of June 6 (Dibajnia et al. 2010).

An Oceanor Buoy and an Acoustic Doppler Current Profiler (ADCP), AWAC by Nortek Company, had been installed at a depth of 30 m in front of Chabahar Bay, prior to the arrival of cyclone Gonu (Figure 9.3.1.4). The

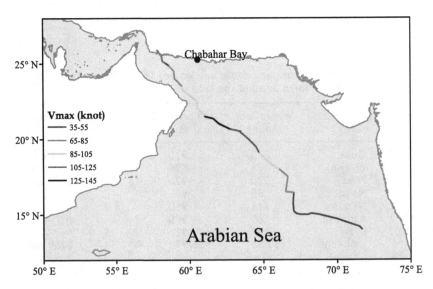

Figure 9.3.1.4 Track and maximum velocity (Vmax) of cyclone Gonu.

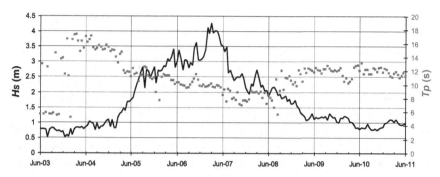

Figure 9.3.1.5 Time series of significant wave height (*Hs*) and peak period (*Tp*) of waves in front of Chabahar Bay at 30 m water depth during the passage of cyclone Gonu (Dibajnia et al. 2010).

mooring system of the buoy could not resist the large waves, but the AWAC continuously recorded the directional waves. It was equipped with a vertically oriented transducer to directly measure the free surface using the Acoustic Surface Tracking (AST) method. Figure 9.3.1.5 presents a time-series plot of the measurements of the significant wave height and peak wave period. It can be observed that the maximum significant wave height recorded at this station was about 4.2 m, with an associated period of 10 s.

9.3.1.4 Numerical simulations

The development and application of numerical models for the hindcasting and forecasting of TCs, storm surges, and storm-induced waves have significantly progressed in recent decades (see also Sections 5.2, 7.2 and 7.3). Alongside field measurements, such tools can help to clarify the associated mechanisms regarding the development of TC and their impacts.

Field measurements (see Sections 3.1–3.7) can be utilized to improve the accuracy of numerical simulations through data assimilation techniques. Soft computing is another technique that utilizes the advantages of long-term measurements to train models and provide accurate wave forecasts at a lower cost and time, compared to numerical simulations. Artificial Neural Networks (ANNs), Fuzzy Inference Systems (FIMs), and regression trees are also popular for predicting waves (Mafi and Amirinia 2017), and such techniques have been implemented in the Bay of Bengal to predict storm surge and inundation (Sahoo and Bhaskaran 2019).

Existing literature shows that attempts have been made to simulate TC-induced waves over the Arabian Sea. Using the Simulating WAves Nearshore (SWAN) model (Booij et al. 1999), Mashhadi et al. (2015) successfully simulated the wave heights and mean wave directions of Gonu, though their results underestimated the peak periods. Dibajnia et al. (2010) also simulated the wave heights of different TCs, including Gonu, using the WAVEWATCHIII (WWIII) wave model, though they did not provide a comparison of the wave periods. Bakhtiari et al. (2018) employed the spectral wave module of MIKE21 software (MIKEbyDHI 2017) to estimate the significant wave heights, peak wave periods, and wave directions of several TCs, including Niloofar, Gonu, and Phet. Vieira et al. (2020) simulated the waves generated by cyclones Gonu, Phet, and Kyarr over the Gulf of Oman, and suggested including background swell waves in future simulations.

Although the parametric models were mostly utilized in previous studies to generate TC wind fields, it has been demonstrated that blended wind fields, generated by reanalysis datasets outside the effective radius of TC, and a parametric model inside the radius, lead to a better simulation of waves and surges (Murty et al. 2020). This approach is also applied for forecasting purposes in the Hurricane WRF (HWRF) model, which modifies the vortex structure using a bogus procedure and blends the new vortex with GFS data to provide initial and boundary conditions of the forecast model (Biswas et al. 2018).

In a more comprehensive approach, the accuracy of the atmospheric field can be increased by dynamical downscaling using the Weather Research and Forecasting (WRF) model (see also Sections 7.2 and 7.3). The wind field with higher spatial and temporal resolution can then be employed to simulate TC-induced waves (Soltanpour et al. 2021). The WRF model was first validated for three cyclones (Gonu, Phet, and Ashobaa) using the TC tracks, intensities, and measured local winds. Figure 9.3.1.6 shows the simulation

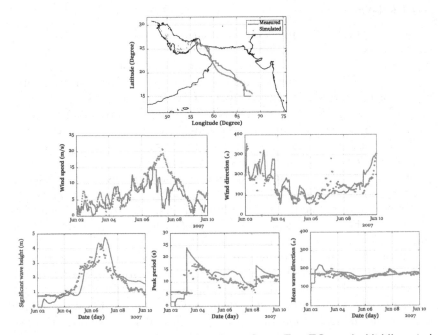

Figure 9.3.1.6 Simulation results of cyclone Gonu. Top: TC track. Middle: wind field. Bottom: wave parameters (Soltanpour et al. 2021).

results of cyclone Gonu. Soltanpour et al. (2021) concluded that dynamical downscaling did not contribute to a significant improvement in wave simulation. However, using WWIII data for adopting boundary conditions in deep waters, and selecting the best physics schemes of SWAN, the simulation results were improved (in terms of simulating the peak period). Soltanpour et al. (2021) noted that for climates with bimodal behaviour of the waves, the Westuysen scheme could better reproduce wave parameters and wave spectra. Bimodal spectra, including swells generated by TCs and local seas, as well as uni-modal spectra comprising seas generated by TCs, were also adequately reproduced.

9.3.1.5 Conclusions

The higher frequency of storms penetrating north towards the Gulf of Oman and the generation of more intense cyclones in recent years indicate that there is the possibility that the risks due to such events will increase in the future, probably due to the effects of global warming and climate change. Past tracks show that TCs might also be able to cross the Arabian Peninsula and enter the Persian Gulf, although the level of this threat is still unknown and further research is necessary.

Efforts should be made to increase the awareness of local communities in neighbouring countries about the risks posed by such events. Overall preparedness needs to be improved to increase the resiliency of the areas that could be affected and ensure public safety. The accurate real-time forecasting of TCs in the Arabian Sea is also essential to warn and prevent communities in advance, minimizing the risks to lives, property and infrastructure.

9.4 NORTH AMERICA

9.4.1 Tsunami hazards in Canada

Ioan Nistor

University of Ottawa, Ottawa, Canada

Jacob Stolle

Institut National de la Resercher Scientifique, Québec City, Canada

9.4.1.1 Introduction

The coastal environment is a significant part of the Canadian landscape, as the country is not only bordered by the world's three main oceans (Atlantic, Arctic, and Pacific) but also encompasses several important internal marine bodies, such as the Gulf of St. Lawrence and Hudson Bay, as well as the Great Lakes. The economic, environmental, and recreational value of these environments is compounded by the fact that 25% of the Canadian population lives within the coastal zone. However, these areas face inherent

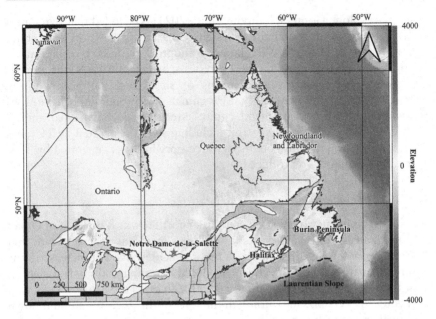

Figure 9.4.1.1 Tsunami events (red star) in East Canada. Bathymetry and topography from the GEBCO database (IOC et al. 2003).

hazards due to their proximity to the shoreline, which can manifest in the form of extreme flooding events caused by storm surges, hurricanes, and tsunamis, potentially affecting millions of Canadians.

The vulnerability of Canadian coastal communities is further accentuated by ongoing and future climatic changes (Manson 2005). The National Research Council of Canada (NRC), as well as several levels of government, have recently expressed the intention to include the effects of climate change in upcoming versions of the National Building Code of Canada (NBCC). However, these changes have not yet been implemented (Barrett and Hannoush 2016). Tsunamis, which occur less frequently than storm surges and hurricane events, can be significantly more devastating (see Section 5.1) and are also not currently captured within the NBCC. The following sections look at several case studies of tsunamis (shown in Figure 9.4.1.1) in Canada and how they can inform hazard assessments in future design standards.

9.4.1.2 Case studies

9.4.1.2.1 The 1908 Notre-Dame-de-la-Salette Tsunami

The south-eastern Canadian landscape has been significantly influenced by the last Ice Age, and a variety of natural hazard risks arise from the deposition of unconsolidated sediments by retreating glaciers and the draining of the Champlain Sea. These regions are prone to erosion and slumping failure caused by natural factors (waves, river erosion, rainfall, etc.), human intervention (in-filling), as well as strong vibrations (blasting, earthquakes, etc.) (Thakur et al. 2017). The *1908 Notre-Dame-de-la-Salette landslide tsunami* is a stark example of the challenges in identifying and designing for tsunami hazards in this region.

Notre-Dame-de-la-Salette is located on the Lièvre River in a valley formed by clay deposited by retreating glaciers (Ells 1908). Lièvre River has a maximum water depth of about 7 m near the village and the river banks rise up to 18 m above the low water level, with an average slope of 30° (Locat et al. 2017). The tsunami was generated by a landslide that occurred on April 6th, 1908 and was thought to have taken place as a result of the formation of a fissure and infiltration of rain water into the underlying clay layer (Ells 1908). The landslide occurred in two stages, with an estimated slide volume of 1.20×10^6 m^3 (Locat et al. 2018). The debris from the slump travelled up to 250 m upstream and 200 m downstream, partially blocking and redirecting the river (Ells 1908), as well as displacing a large volume of water that generated the tsunami.

The disaster resulted in the death of 27 people and the destruction of 15 buildings, with an associated run-up of approximately 15 m (Ells 1908). However, it is the co-hazards associated with the tsunami that make this event rather unique. In their post-event report, Ells (1908) noted that the tsunami occurred before the ice melted and an approximately 0.50 m thick ice sheet lay over the river on the day of the event. The damage to the buildings was primarily associated with the impact of the ice floes on the structures, similar to debris hazards observed in other events (Stolle et al. 2020). Locat et al. (2017) modelled the event using OpenFOAM and obtained a reasonable representation of the wave run-up, but noted that the influence of the ice cover on the generated waves is still largely unknown.

9.4.1.2.2 The 1917 Halifax port explosion

The Halifax Port Explosion represents a different kind of tsunamigenic event: a human-induced explosion. It occurred on December 6th, 1917 when two ships (namely, *Mont Blanc* and *Imo*) collided in Halifax Harbor. The Mont Blanc was carrying ammunitions and the resulting explosion released 1.21×10^{10} kJ into Halifax Harbor (Greenberg et al. 1993). Surface explosions initially formed a cavity which continued to expand (allegedly the harbour bottom was briefly visible during the Halifax explosion) followed by a rapid collapse of the cavity and the formation of a vertical jet, which eventually fell due to gravity, forming the wave (Van Dorn et al. 1968). The explosion, coupled with the resulting tsunami, resulted in the death of over 2000 people – 200 directly attributed to the tsunami – as well as the complete destruction of a Mi'kmaq indigenous settlement, which was directly across the bay from the explosion (Scanlon 1998). The horizontal inundation extent of the tsunami was over 200 m from the shoreline and its maximum run-up was estimated to be 16 m (Greenberg et al. 1993).

The Halifax Port Explosion is a strong reminder of the necessity to consider a wide variety of potential sources when assessing tsunami risks, as some of them are not necessarily directly related to earthquake events (see also Section 5.1 or 5.3). Water waves generated by explosions have been studied in other contexts including hydrovolcanic eruptions, projectiles, and asteroid impacts (Slaughter et al. 1978), though they are rarely included in tsunami hazard assessments.

9.4.1.2.3 The 1929 Burin Peninsula

The Atlantic coast of Canada is generally viewed as being less susceptible to tsunami hazards, due to it presenting limited seismic and landslide activity in comparison to the West coast (Clague et al. 2003). However, the most catastrophic earthquake-generated tsunami in Canada occurred on November 18, 1929 impacting the Burin Peninsula in Newfoundland, and resulting in the death of 28 people (Ruffman 1996).

The tsunami was generated by a complex M_w 7.2 strike-slip earthquake south of Newfoundland that resulted in a submarine landslide, displacing on the order of 100 km³ of sediment along the Laurentian slope seismic zone (Figure 9.4.1.1). Residents on the Burin Peninsula were the hardest hit and eye-witness reports noted three wave pulses on top of a longer wave. The mechanism of the landslide is still being debated, as there is evidence of localized translational landslides and larger rotational slumping, contributing to the wave profile as a whole (Løvholt et al. 2019). The Burin tsunami is the only known landslide tsunami that was discernable at transoceanic distances, as it was recognized in Europe as well as the Caribbean (Fine et al. 2005). The Burin Peninsula again shows the importance of considering compounding hazards (earthquake–landslide–tsunami), which can be difficult to capture with current hazard assessment methodologies.

9.4.1.3 Tsunami risk for the Canadian West Coast

Canada is also at risk of being impacted by extreme seismogenic tsunamis (Leonard and Bednarski 2014). The West Coast (British Columbia) of the country has historically experienced major tsunamis, though damage to man-made structures has not been reported mainly due to the low population density at their time of occurrence (Clague et al. 2003). There is an existing threat of a major earthquake and accompanying near-field tsunami taking place at the Cascadia Subduction Zone (CSZ), where the Juan de Fuca Plate is dipping beneath the North American Plate (Figure 9.4.1.1). Several communities along the western seaboard of Vancouver Island, such as Tofino, are thus at risk of suffering from a tsunami, be it from the nearshore Cascadia Fault or from distant sources such as southern Alaska.

In fact, the last major earthquake event, which occurred at the CSZ on January 26, 1700, is known to have generated a massive tsunami wave measuring approximately 10 m along the west coast of North America (e.g., Atwater and Yamaguchi 1995). According to Satake et al. (2003), the event is classified as a full rupture event with a fault length of 1,000 km, and the estimated range of the moment magnitude (M_w) was between 8.8 and 9.3. It has been predicted that the next M_w 9.0 event has a 17% possibility of occurrence in the next 50 years and 25% in the next 100 years (Jonston and Lence 2012).

Local earthquake tsunamis could also be generated from the Explorer-North America plate boundary, but these sources are currently poorly understood (Leonard et al. 2013). Wave heights between 7.5 cm and 10.8 cm were recorded in Tofino in 2004 from a tsunami generated by an M_w 6.6 earthquake originating at the Explorer-North America boundary (Rabinovich et al. 2003). Additionally, three tsunamis have been recorded originating from the Queen Charlotte fault. The first was in 1949, which was the largest earthquake actually recorded in Canada, having an Mw of 8.1. In 2001, a smaller earthquake, M_w 6.1, caused wave heights between 11.3 cm and 22.7 cm on the outer coast of Vancouver Island (Rabinovich et al. 2008). Finally, in 2012, an Mw 7.7 subduction earthquake caused run-ups between 3 m and 13 m to the (unpopulated) region to the west of Haida Gwaii (Leonard and Bednarski 2014). This was the first time a thrusting movement was recorded in this fault, which is primarily strike-slip.

Local crustal earthquakes could also produce a tsunami within inland water-ways or fiords (Clague et al. 2003, Leonard et al. 2013). Such an earthquake with an M_w 7.3 occurred in 1946 in central Vancouver Island, which triggered landslides and submarine failures that generated a tsunami with a maximum wave height of between 1 m and 2 m in the Strait of Georgia – between Vancouver Island and mainland British Columbia (Mosher et al. 2004).

Far-field tsunamis are more frequent – 11 of 16 recorded tsunamis between 1964 and 2007 were far-field – but generally less damaging (Leonard et al. 2013). Two earthquake tsunamis have originated from the Aleutian Trench in recent history. The first, in 1946, was produced by an M_w 8.6 earthquake, with the tsunami completely destroying the Scotch Cap Lighthouse on Unimak Island in Alaska but not causing any damage in Canada (Bodle 1946). However, the tsunami originating from the *Great Alaska Earthquake* of 1964 (Mw 9.2), which killed 139 individuals in Alaska alone, was the most devastating in Canadian history. The wave reached the Canadian coast 4 hours later (Clague et al. 2003, Stephenson et al. 2010), causing the greatest damage to inlets rather than the main coast. Port Alberni incurred the most damage with 260 damaged houses, including 60 extensively damaged and 2 were that were swept into the ocean. The tsunami waves were augmented in Port Alberni due to the high tides and the combined effects of the topographic amplification in Barkeley Sound, and the resonance magnification in the Alberni inlet; resulting in a magnification factor of 3 to 4 between the entrance of Barkley Sound and the head of the Alberni inlet (Figure 9.4.1.2). The maximum run-up in Port Alberni was

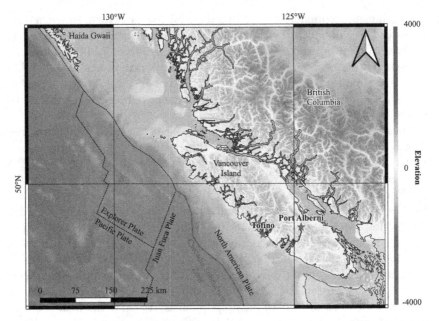

Figure 9.4.1.2 Cascadia Subduction Zone and location of the District of Tofino on Vancouver Island. Bathymetry and topography taken from the GEBCO database (IOC et al. 2003). Tectonic plate data taken from Bird (2003).

10.3 m, approximately 6 m less than the estimated run-up from the 1700 event. The village of Hot Spring Cove was also destroyed and houses were swept out of their foundation in Zeballos. The total damage cost in Canada was $10 million in 1964 dollars, with the highest wave being recorded in North Haida Gwaii (8 m), and 2.4 m in Tofino.

The *1960 Chile tsunami*, originating from an M_w 9.5 earthquake (see Section 9.5) caused some damage to Vancouver Island and Haida Gwaii, with 1.3 m wave heights being recorded in Tofino (Leonard et al. 2013). It should be noted that the recorded wave heights were 1.1 m shorter than those of the 1964 tsunami due to the longer distance travelled and the angle of arrival.

9.4.1.4 Conclusions

There are a number of potential tsunami sources that could affect the Canadian (and North American in general) coastlines. The American Society of Civil Engineers (ASCE) recently published a new standard for the design of tsunami-resilient infrastructure (ASCE 2016) with a hazard assessment component based around Probabilistic Tsunami Hazard Analysis (PTHA). PTHA quantifies probabilities regarding tsunami hazards (predominantly inundation limits) based on potential source scenarios. Both the east and west coast of the United States are exposed to similar tsunami hazards as

Canada – though as of now the PTHA does not include Canada. Research has been initiated to address this shortcoming and apply it to the west coast of Canada in particular.

The ASCE exclusively considers tsunamis generated from earthquake events though, as the historical record shows, that may not adequately capture the vulnerability of Canadian communities. While designing tsunami-resilient infrastructure across the coastline might not be feasible, improving the understanding of vulnerability and providing tools to communities so that they can try to increase their resilience to such events should be a priority (regardless of hazard). Examples of this can be seen across the country, including mapping of landslide risks by the Ministry of Public Security (Quebec) and the creation of web platforms identifying climate adaptation strategies of other communities (Resilience – C University of British Columbia and Natural Resource Canada Adaption Platform).

9.4.2 Analysis of beach nourishment and coastal structures: shoreline stabilization alternatives for a highly erosive beach in the USA

Michael B. Kabiling

Taylor Engineering, Inc., Jacksonville, FL, USA

9.4.2.1 Introduction

Hurricanes, one of the most damaging types of natural disasters in the United States, currently cause about $28 billion in annual damage (0.16% of US gross domestic product), which is projected to increase to $39 billion (0.22% of US GDP) by 2075 (CBO 2021). NOAA (2020) reports that coastal shoreline communities account for 40% of the total jobs in the United States and contribute 46% of the GDP of the country. Coastal erosion, often the most visible sign of hurricane damage, results in approximately $500 million in property and land losses per year. As a result, the US government spends about $150 million annually on beach nourishment and other shoreline erosion control measures to mitigate the effects of erosion (NOAA 2013) and protect coastal communities. One such coastal protection project is the Fort Pierce Shore Protection Project (FPSPP). Located in St. Lucie County, Florida, USA, the FPSPP nourishes a 2.09 km stretch of the Atlantic Ocean shoreline south of Fort Pierce Inlet. Constructed in 1971, the project initially involved the placement of about 550,000 m³ of sediment dredged from an offshore borrow site. Originally authorized for 10 years, the shore protection project received extensions to later years and provided placement of an additional 264,500 m³ in 1980, 634,600 m³ in 1999, 471,000 m³ in 2005, and 385,200 m³ in 2007 on the beach south of the inlet (known as "south beach"). Partial south beach nourishments also occurred in 1987, 1989, 1990, 1995, 1998, 2003, 2004, 2009, 2011, 2012,

2013, 2014, and 2015, with the volume of the placements varying from approximately 15,300 m³ to 382,300 m³.

Historical measurements after each nourishment indicate the beach fill erodes non-uniformly, with a hotspot along the northernmost 0.7 km, requiring nourishment after about two years of normal wave regimes. However, storms can quickly erode the beach fill, which then requires re-nourishment earlier than the normal two-year interval. Longshore transport carries most of the eroded fill to the south, creating a strong feeder-beach effect. This chapter describes a study that evaluated a "coastal structures" alternative that combines beach re-nourishment and coastal structures to produce more uniform erosion throughout the beach nourished project area, thus increasing the nourishment interval. Such work is of interest not only to the USA but also to other countries severely affected by coastal erosion, such as Vietnam or Thailand (see Section 9.1). The study also included a "no structures" alternative to serve as a baseline condition for the evaluation of the performance of the coastal structures shoreline stabilization alternative. To ascertain this, the study used several state-of-the-art numerical models to evaluate the two-dimensional effect of "no structures" and "with structures" alternatives on hydrodynamics, waves, sediment transport, and beach morphology.

9.4.2.2 Shoreline stabilization alternatives

The study compared the baseline alternative (only beach re-nourishment with no structures) to the coastal structures alternative to evaluate the structures alternative's effectiveness in reducing beach erosion and increasing the beach nourishment interval. The coastal structures alternative (see Figure 9.4.2.1), included the same baseline beach nourishment and construction of five T-head groins (T1 – T5), a T-head weir (W6), and a break- . water (B7) approximately 480 m (T5), 550 m (W6), and 630 m south of the jetty, respectively.

9.4.2.3 Numerical modelling

Taylor Engineering (2011) describes the model calibration/verification, and the parameters adopted for the calibrated HD, SW, and ST models for the Fort Pierce Inlet and Fort Pierce Shore Protection Project area. This study extended the Taylor Engineering (2011) existing MIKE21 two-dimensional hydrodynamic (HD), wave (SW), and sediment transport (ST) models and added a hybrid shoreline morphology (SM) model to simulate water surface elevation, flow velocity, sand transport, erosion, deposition, and shoreline movement. Danish Hydraulic Institute (2016) provides documentation for the SM model numerical schemes.

The MIKE21 FM fully integrated models represent portions of the Atlantic Ocean, Fort Pierce Inlet, and the Intracoastal Waterway (IWW)

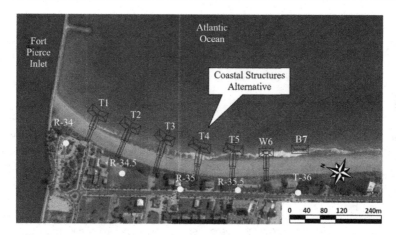

Figure 9.4.2.1 Layout of coastal structures (T1 to B7) and beach monuments (R-34 to T-36).

as two-dimensional waterways. The HD, SW, ST, and SM models link dynamically:

(1) the SW model computes and feeds back wave radiation stresses to the HD model to affect the HD model flow computation;
(2) the HD model computes the tide- and wave-generated flow velocities that drive sediment transport;
(3) the ST model computes sediment movement that results in either erosion or deposition; and
(4) the SM model uses the ST model computed sediment transport to compute the bed and shoreline changes that affect subsequent SW and HD model computations (i.e., the SM model feeds back bed level changes to the SW and HD models).

The fully dynamic integration of the wave calculation, flow calculation, sediment transport, and bed changes makes the MIKE21 FM a state-of-the-art modelling system that can simulate nearshore, inlet, and estuarine hydraulics and morphological processes.

9.4.2.3.1 Application of effective waves

To save computational resources, effective (or representative) wave and tide conditions represented offshore wave conditions and corresponding tide levels for the multi-year sediment transport and morphology simulations. A MIKE Littoral Processes (LP) FM model, calibrated using the Taylor Engineering (2004) 1967–1999 sediment budget average annual longshore sediment transport rates at locations 2.7 km north, 0.8 km south, and 2.5 km south of the inlet provided the means to estimate long-term southward-directed net longshore transport (Figure 9.4.2.2) at the north and south of Fort Pierce Inlet.

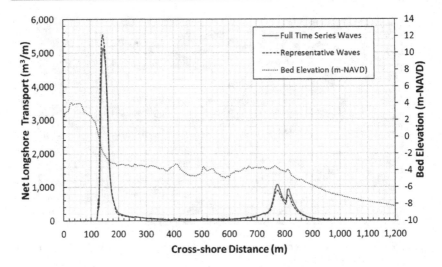

Figure 9.4.2.2 Comparison of cross-shore distribution of longshore transport one year after 2007 beach nourishment. The horizontal axis represents the cross-shore distance from a location just behind the dune. Waves propagate from right to left in this figure.

Using an empirical proportionality coefficient K value of 0.08 in the Coastal Engineering Research Center (CERC) formula, Taylor Engineering (2004) also provides a CERC formula calculated southward net longshore transport rate of 166,673 m³/yr near the project site for the period 1976–1995. Taylor Engineering (2004) also provides a Kamphuis formula calculated southward net longshore transport rate of 212,928 m³/yr. Thus, the calibrated LP model produces a net longshore transport rate that is consistent in magnitude and direction with those calculated from the CERC and Kamphuis net longshore transport equations.

9.4.2.3.2 Sediment transport model

Multiple sources provided spatially varying sediment median grain sizes that could be used in the ST model:

a) May 1991 beach sediment samples along profiles at monuments R-31 (median grain size of 0.22 mm) and T-36 (median grain size of 0.29 mm) south of the inlet;

b) beach sediment samples along transects at monuments R-35, T-41, and R-46 in 2002 provided composite median grain sizes of 0.20 mm, 0.14 mm, and 0.13 mm, respectively;

c) US Army Corps of Engineers (USACE) vibracores data provided a median sediment grain size of approximately 0.15 mm at the nearshore areas below mean low water.

Note that "beach monuments"[1] are locations with fixed latitude and longitude, numbered consecutively and spaced approximately 305 m along the beach.

The ST model calculated the resulting transport of non-cohesive materials based on HD model flow conditions and SW model wave radiation conditions. Combined currents from tides and waves were applied as the main driving mechanism for sediment transport.

9.4.2.3.3 Shoreline morphology model

The June 2007 post-fill survey and the July 2008 (taken one-year after the post-fill survey) cross-shore transects surveys at monuments R-34 to R-46 provided the SM model calibration data of mean high water (MHW) locations. The SW model applied at the offshore boundary of the LP model provided effective (or representative) waves based on the hourly USACE Wave Information Study Station 63449 hindcasted significant wave height, peak period, and mean wave direction, and the HD model applied at the offshore boundary of the representative water levels based on National Oceanic and Atmospheric Administration (NOAA) Station 8722105 (Vero Beach Ocean) predicted tides.

Table 9.4.2.1 compares the modelled and measured onshore (negative) and offshore (positive) net movement of the MHW shoreline one year after the 2007 beach nourishment. The SM model provided good estimates of the erosion (shoreline recession) at monuments R-35, T-36, T-37, R-38, and T-40, where model results are within 6.7 m of the measurements. The model

Table 9.4.2.1 Measured and modelled MHW onshore (Negative) and offshore (Positive) net movement for 2007 beach nourishment

Monument	Measured (m)	Modeled (m)	Difference (m)
R-34	−78.9	−120.1	−41.1
R-35	−33.2	−39.9	−6.7
T-36	−26.5	−26.8	−0.3
T-37	−10.1	−13.1	−3.0
R-38	−8.5	−11.6	−3.0
R-39	−11.3	16.5	27.7
T-40	−8.5	−4.3	4.3
T-41	−8.8	4.6	13.4

[1] In Florida beach monuments are designated locations (with permanent latitude, longitude) spaced approximately 1,000 ft apart and numbered consecutively (in clockwise direction) within a county, and used by the Florida Department of Environment to reference beach locations.

overestimated erosion at monument R-34 and provided small accretion instead of erosion at monuments R-39 and T-41. The SM model results are generally consistent with the observed erosion pattern from monuments R-34 to R-38. The discrepancies between the modelled and measured shoreline movement are acceptable given the limitations in the available model input data as (a) no complete information exists on the sediment median size distribution throughout the project and nearshore areas and (b) the offshore waves and tides applied in the model are not from direct measurements but come from wave hindcasts and tide predictions. As the model will be mostly used to evaluate the effect of coastal structures on shoreline movement south of monument R-34 and it showed good agreement with observations, its use in this area is warranted.

9.4.2.4 Results of the analysis

Figure 9.4.2.3 shows a comparison of the shoreline position and bed levels for the baseline and the coastal structures alternatives three years after beach nourishment. The long-term multi-year model simulations that included

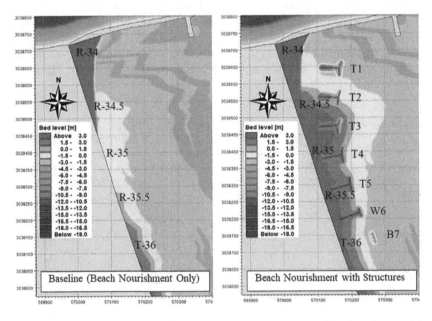

Figure 9.4.2.3 Modeled shoreline morphology three years after beach nourishment without coastal structures (left), and with coastal structures (right).

normal tides, waves and storm conditions show a baseline shoreline movement pattern that is very similar to the general historical pattern observed in the project area:

a. general temporal progressive shoreline retreat (erosion) north of monument R-35, as waves breaking nearshore erode the shoreline and consequently allow succeeding waves to break closer to the shoreline
b. good comparison with the high levels of erosion observed in the field between monuments R-34 and R-35
c. lesser shoreline retreat at monument R-35 and very small shoreline movement south of monument T-36 (relative to those north of monument T-36).

With the coastal structures, the model results show the shoreline almost fully eroded behind structures T1 and T2 while structures T3–T5, W6, and B7 retain the beach nourishment longer and thus provide shoreline stabilization from monuments R-34.5 to T-36. Notably, as the shoreline is stabilized north of monument T-36, the results also show a decrease in shoreline erosion from monuments T-36 to R-38 (note that R-38 is outside of the figure) when compared to the baseline shoreline erosion. Essentially, the results show the coastal structures will retain beach fill longer than in the baseline simulations, resulting in lower erosion rates from 0.8 to 6.4 km south of the jetty. The coastal structures alternative could thus feasibly extend the beach nourishment interval to four years. Nevertheless, finding an optimal design for the shoreline stabilizing measures would require evaluation of different combinations of coastal structures and refining their locations and geometry.

9.4.2.5 Conclusions

The Fort Pierce Shore Protection Project enables regular beach nourishment at the highly erosive beach south of Fort Pierce Inlet to maintain protection from coastal erosion that can take place due to hurricanes and other high-energy wave events. The ultimate aim of the project is to mitigate potential future damage to coastal communities in the area and the local economy. The construction of coastal structures, combined with beach nourishment, appears to be a good solution to stabilize the shoreline, and could be applied to other coastlines suffering from erosion elsewhere (such as Vietnam or Thailand, see Section 9.1).

9.5 SOUTH AMERICA

Rafael Aránguiz

Universidad Catolica de la Santisima Concepcion, Concepcion, and Research Center for Integrated Disaster Risk Management (CIGIDEN), Santiago, Chile

9.5.1 Overview of the South American continent

South America is the fourth largest continent, extending from the Gulf of Darien in the north, on the Panama–Colombian border, to Tierra del Fuego in the south. The continent is composed of 12 independent countries, namely, Argentina, Bolivia, Brazil, Chile, Colombia, Ecuador, Guyana, Paraguay, Peru, Uruguay, Suriname and Venezuela, and an overseas French Department: French Guiana (See Figure 9.5.1). It has a total area of 17,840,000 km² and a population of approximately 423 million. Brazil is the largest country, which covers 47.3% of the land area or the continent and has around half of its total population. The continent's location on the South American Plate and along the subduction zone of the Nazca Plate creates a complex geography, which includes mountains and highlands, river basins and coastal planes (NationalGeographic, 2012). The primary mountain system is The Andes, which runs for 8,850 km along the western edge of the continent and has hundreds of volcanic peaks. In addition, there are

Figure 9.5.1 Map of South America.

Source: modified from GoogleEarth.

three important river basins, namely the Amazon, Orinoco and Paraná (see Figure 9.5.1). The Amazon is considered to be the largest river in the world, with its basing covering 39% of the total continental area. Finally, coastal planes are located on the northeastern coast of Brazil and the western Pacific coasts of Perú and Chile. In the western coastal plain, it is possible to find the driest region of the planet, the Atacama Desert, which is a rich source of minerals for the Chilean economy (NationalGeographic, 2012).

9.5.2 Natural hazards

Due to its geographical characteristics, the people of South America are at risk of natural phenomena such as storms, earthquakes, volcanic eruptions, floods, landslides, sea-level rise and tsunamis, among others. This, combined with the extreme vulnerability of many settlements, results in large economic and human costs when a disaster strikes. In fact, fundamental development problems such as fast and unregulated urbanization, persistence of poverty, environmental degradation, inefficient public policies and infrastructure problems, contribute to the vulnerability of the population of South America (Nunes, 2011). It is believed that the human toll may increase in the coming years due to the persistence of widespread poverty, continuing demographic growth and migration towards coasts and megacities (Charveriat, 2000). In fact, most South-American cities lie in regions prone to events of geological or hydrometeorological nature, and the physical growth of urban environments has taken place in areas at risk of suffering from floods or mass land movements (Nunes, 2011).

9.5.2.1 Hurricanes

The northern coast of South America, facing the Caribbean Sea, is exposed to cold fronts, tropical storms, and hurricanes that can cause great damage to physical, economic, biological and social systems (Devis-Morales et al., 2017). Cold fronts come from higher latitudes and typically occur between December and March. These events can cause flooding not related to local meteorological conditions, but due to the arrival of heavy swell waves (Silva et al., 2014). In fact, Cartagena de Indias, Colombia, has been affected by several events in the past, and the extreme vulnerability of coastal areas, in combination with sea-level rise and coastal erosion, is of great concern in the future (Andrade et al., 2013). Hurricanes take place between June and November and maintain their position and trajectory in northern parts of the Caribbean basin due to the existence of warm ocean temperatures. Those events may affect coastal areas due to the strong winds, heavy rains and storm waves they generate (Rangel-Buitrago & Anfuso, 2015). Severe weather conditions related to tropical storms occur along the northern coast of South America, mainly the Colombian and Venezuelan offshore basins

(Devis-Morales et al., 2017). For example, Tropical Storm Bret (August 1993) was one of the deadliest disasters to have taken place in Venezuela, with the storm generating winds gusts of up to 85 km/h that destroyed the roofs of many houses. However, it was the heavy rain that proved to be the most destructive, causing mudslides and flooding. As a result, 173 deaths were reported in Venezuela, 500 were injured and 11,000 were left home-less (Tropical Storm Bret (1993), 2021). Nevertheless, it is also worth noting that vulnerability to hurricanes has decreased significantly in South America and the Caribbean in the last 40 years due to the improvement of early warning systems, reforestation of coastal areas and construction of shelters (Silva et al., 2014)

9.5.2.2 Extratropical cyclones

Extratropical cyclones take place in the southern part of South America, with the more extreme events having huge socioeconomic repercussions on some countries, due to their associated strong winds, heavy rain and storm surges. Three regions of extratropical cyclones formation have been identi-fied. One region is located on the Atlantic Ocean over the coastal regions of Argentina, Uruguay and South Brazil, a second region is in the Pacific Ocean west of the Andes, and a third region is located north of the Antarctic Peninsula (Mendes et al. 2010). The cyclogenesis activity is concentrated in the Southern Hemisphere winter season (July to September), though in the Antarctic Peninsula region it is fairly active throughout the year. It is known that extratropical cyclones make important contributions to the annual pre-cipitation over the South Atlantic Ocean (Reboita et al. 2018). In addition, the cyclones over the coast of Argentina, Uruguay and Brazil generate sig-nificant sea waves that propagate into the Plata River basin that can cause flooding in low laying areas of Buenos Aires, Argentina (da Rocha et al. 2004). Similarly, high sea waves frequently affect the coast of Brazil, and can lead to significant flooding in Rio de Janeiro. As an example, a storm that took place between 9–11 August 1988 (though it was not felt over the continent) generated sea waves of 3 m and flooding in some coastal areas along the Brazilian coast from 22° to 32°S, causing significant dam-age to infrastructure and one death (Innocenti and Caetano Neto, 1996). Significant cyclones and destructive storms have also been observed over the South Pacific Ocean, affecting the coast of Chile. One of the most destruc-tive events occurred on 8 August 2015, when a low-pressure system asso-ciated with strong winds of up to 110 km/h generated offshore waves of above 7 m along the coast of Central Chile, causing significant damage to public and private infrastructure as the waves hit during high tide (Winckler et al., 2017). Six people were reported dead and around 60,000 people were affected. Figure 9.5.2 shows images of the wave impact of the August 2015 storm to the Valparaiso Region.

Figure 9.5.2 Pictures of the impact of the August 2015 Storm to the Valparaiso Region, Chile. a) Overtopping and damage of the breakwater at the Yacht Club at Higuerillas, Concon. b) Overtopping at Avenida Altamirano, Valparaiso. c) Wave impact at a coastal dike of Avenida Perú, Viña del Mar. d) Overtopping and inundation at Avenida Perú, Viña del Mar. (Photos by Patricio Winckler).

9.5.2.3 Coastal erosion

Coastal erosion has been observed in some areas due to human activities such as the offshore mining of sand and the destruction of dunes as a result of tourist activity. For example, in Northeast Brazil some constructions have caused interference in the circulation of coastal sediment and affected the sediment balance. In addition, when human activity is combined with destructive meteorological events, the natural recovery may be affected and coastal erosion can progressively worsen, as observed in Santa Catarina, Brazil (Silva et al., 2014). All these effects may be enhanced by the influence of climate change, since wave climate, winds and rainfall may shift in the near future. Recently, Winckler et al. (2020) found that climate-driven changes may negatively impact several processes along the Chilean coastal zone, such as coastal erosion, impact on wetlands and operational downtime in ports. Winckler et al. (2020) observed that rainfall shows a negative trend, which can cause a reduction in river discharge and subsequently reduce the availability of coastal sediments along central Chile. On that ground, the change in wave climate can play an important role in coastal erosion.

Observations show that the significant wave height in Central Chile has increased by up to 0.4 m in the period between 1980 and 2015 (Winckler et al., 2020). In addition, there has been a southward rotation of the mean wave direction of 3.5° in the same period of time (Winckler et al., 2020), which can significantly affect the longshore sediment transport and equilibrium of beaches. Moreover, a similar pattern of change has been found in extreme wave heights along the western coast of South America (Camus et al., 2017), while the number of events per year shows an increase, which could have significant consequences on port operability and coastal flooding.

9.5.2.4 Earthquakes and tsunamis

Since the Nazca plate subducts under the South American plate, the Pacific coast of South America is exposed to the effect of both earthquakes and the tsunamis that can be generated by them. A database of earthquakes in Latin America and the Caribbean is presented in Cardona et al. (2018), who compiled and standardized such information so that it can be used in vulnerability models. In a similar manner, Soloviev & Go (1975) compiled a complete catalogue of tsunamis on the eastern shore of the Pacific Ocean. Several events, as early as 1515, were reported in Colombia, Ecuador, Perú and Chile in the course of the centuries.

The largest event ever instrumentally recorded was the Mw 9.5 1960 earthquake in southern Chile (Barrientos and Ward, 1990). The rupture area extended from Arauco Peninsula in the north (37°S) to Taitao Peninsula in the south (46°S), and significant changes in land surface were observed along 1,000 km. This event generated a destructive tsunami that affected the entire Pacific Ocean, including Hawaii, New Zealand and Japan. Inundation heights of 4.2 and 4.9 m were measured in the Japanese cities of Onagawa and Ofunato, respectively (Takahashi, 1961). In Chile, average inundation heights were around 10 m, and the maximum run-up was 20–25 m on the southern shore of Mocha Island.

Paleotsunami records have demonstrated that similar events have taken place on average every 285 years (Cisternas et al., 2005). Historical records show that at least 36 large earthquakes have affected the coast of Chile in the last 500 years, and 25 of them have generated destructive tsunamis. For example, the earthquake of July 1730 had an estimated magnitude of 9.1–9.3 (Carvajal et al., 2017) and the tsunami caused severe damage to coastal settlements in central Chile. A more recent event took place in February 2010 and had a magnitude Mw 8.8, with the rupture area being located just north of the 1960 event and extending for approximately 500 km (see also Section 2.4). The tsunami had a diverse effect along the coast, such that some areas near the epicentre experienced run-up in the order of 5 m only, while other areas exceeded 15 m, and a maximum run-up of 29 m was measured in Constitución (Fritz et al., 2011). The damage due to the earthquake and tsunami was estimated to be 30 billion US dollars, which is equivalent to 18% of the GPD of Chile (Contreras & Winckler, 2013). Moreover, despite the large number of historical events that have taken place, the 2010 tsunami demonstrated that Chile had a high level of vulnerability, due to the fact that the tsunami warning system did not work properly, inundation maps were not in the public knowledge and urban planning did not consider tsunami hazards in urban development.

Even though the coast of Chile has been demonstrated to generate the largest magnitude earthquakes in South America, the coast of Peru and Colombia have also experienced significant earthquakes and tsunamis in the past. In 1996, a 7.5 magnitude earthquake off the coast of northern Peru generated a tsunami that affected the Chimbote area, with waves of up to 5.14 m on the northern shore of Chimbote bay. A total of 12 people were killed by the tsunami and 57 were injured (Kulikov et al., 2005). Similarly, in June 2001 a tsunami was generated by a magnitude Mw 8.4 earthquake off the coast of southern Peru. It was reported that 80 people were killed and 200,000 people were affected by the earthquake alone. In addition, the tsunami claimed at least 23 additional lives (Kulikov et al., 2005), with the average tsunami height in Camaná province being 5 m and the inundation penetrating more than 1 km inland (Okal et al., 2002). Another two earthquakes with estimated magnitudes of 8.5 and 8.3 occurred in 1906 and 1979, respectively, off the coast of Colombia. These two events generated large tsunamis that caused severe damage and the loss of life along the coast of Colombia and Ecuador (Sanchez-Escobar et al. 2020). The 1979 Tumaco tsunami reached an inundation height of up to 8 m and claimed the lives of more than 300 people in the southern coast of Colombia (Adriano et al., 2017).

9.5.3 Conclusions

In conclusion, due to its complex geography there are a variety of types of natural hazards that can affect the various countries in South America. While the north of the continent is predominantly affected by tropical storms/hurricanes, the east coast is mostly at risk of floods and storm surges, and in the west coast tsunami risks are higher than those of any other type of hazard. The intensity of the disaster depends on the vulnerability of the affected area, and South America has shown to have a high level of physical and socio-economical vulnerability. This situation is enhanced by fast and unregulated urbanization, increasing poverty, environmental degradation, and inefficient public policies. The risk of suffering coastal disasters could be increased in the near future, not only due to the progression of climate change in the course of the 21st century, which may increase the frequency of extreme events, decrease of rainfall and increase of coastal erosion, but also due to the ongoing demographic growth and migration towards coasts and mega-cities. This makes it imperative that improved disaster risk management practices are implemented, in order to improve the resilience of coastal settlements and improve their long-term sustainability.

9.6 EUROPE

9.6.1 Estonia

Martin Mäll

Yokohama National University, Yokohama, Japan

9.6.1.1 Introduction

Estonia is a small low-lying coastal country located in Northern Europe. It is one of the nine countries that borders the Baltic Sea – a semi-enclosed, almost tideless brackish sea with a surface area of 377,000 km^2 (415,000 km^2 if Kattegat is included), and an average depth is 55 m. The Baltic Sea and Estonia lie in the so-called North Atlantic storm track, where during the colder half of the year extra-tropical cyclones (ETCs; also known as "winter storms") form over the North Atlantic Ocean in the mid-latitudes and travel over the Scandinavian peninsula towards higher latitudes to the northeast. Such weather systems dominate much of the disaster risk that exists in Estonia (and many other European countries, such as Germany, see Section 8.2), which can result in the inundation of low-lying coastal and urban areas, coastal erosion, damage to infrastructure and the loss of life.

As ETCs mostly form during the winter season, the impact of such storms can vary depending on several conditions. For instance, regarding coastal erosion processes the presence of sea ice can potentially lead to higher erosion when it is not fully developed (transportation of ice floes during stormy conditions and due to the open water leaving the coast exposed), or help protect the coastline if there is a fully developed ice sheet. The semi-enclosed nature of the Baltic Sea can also keep water levels above average when prevailing wind conditions force water into the Baltic Sea from Kattegat through the Danish Strait (Figure 9.6.1.1; left). This was indeed what happened during the most severe ETC event in the Estonian recorded history: The 2005 January storm "Gudrun" (also known as Erwin in the British Isles).

9.6.1.2 General history of coastal disasters

Among the countries in the Baltic Sea, Estonia has the 2nd highest storm surge recordings, at +2.75 m (during the 2005 ETC Gudrun) as, due to its geographic layout and the general direction of approaching storm tracks (see Figure 9.6.1.1(b)), it is one of the most prone areas to such phenomena, only after St. Petersburg (recorded maximum of +4.21 m). The most vulnerable area in Estonia is the small summer resort city of Pärnu (Though winterstorms can cause occasional flooding in other Estonian cities as well, e.g., Haapsalu, Kuressaare and Narva-Jõesuu). While Pärnu only has around 40,000 residents, it is visited by hundreds of thousands of visitors during the warmer half of the year (Parnu, 2017). The city is located at the end of the narrow Pärnu

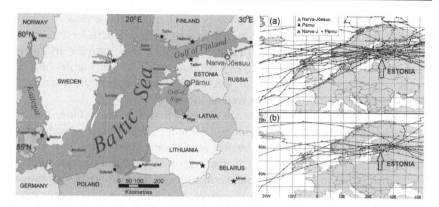

Figure 9.6.1.1 Left: General area map of the central Baltic Sea. Right: a) Tracks of cyclones that caused maximum annual storm surges in Pärnu and Narva-Jõesuu between 1950 and 2016, in Estonia and b) tracks of 12 most notable cyclones in Estonia. (Figures after Suursaar et al. (2018).)

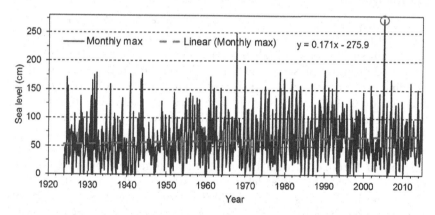

Figure 9.6.1.2 Variations in observed monthly maximum sea levels (1924–2014) at Pärnu tide gauge. The storm surge generated by Gudrun (January 2005) is marked with a circle. The long term trend in sea-level rise is linear at 1.7 mm/year; no land uplift correction is introduced. (After Mäll et al., 2017.)

Bay and, due to its shape, it offers ideal conditions for the build-up of storm surges. Indeed, historically this has been the coastal area most influenced by extreme weather events (other notable locations include Narva-Jõesuu, close to St. Petersburg). Continuous sea-level measurements at Pärnu started in the autumn of 1923 (Figure 9.6.1.2. In the capital Tallinn near-continuous measurements began already in 1842; Suursaar & Sooäär, 2007).

In Pärnu, the critical water level mark is at 160 cm, at which point water starts flowing into streets (thus this level represents the beginning of the risk for inundation). Between 1923–2015 (as recorded from the Pärnu tide gauge, see Figure 9.6.1.2), there have been 21 individual events that have passed this critical line. However, in terms of truly impactful events there have been two notable events that can be considered extreme in Pärnu: +2.53 m in October 1967 and +2.75 m in January 2005 (see Figure 9.6.1.2). In the history of Estonia these two events are the most notable in terms of their impact on communities, as well as to coastline.

However, extreme surge events are not only driven by the cyclone activity alone. As mentioned, an important component to reach high surge levels is also the Baltic Sea's background water level. For instance, during the 2005 storm the high cyclonic activity in the region caused the background sea level to be +0.70 m above the long-term average. The situation was similar for the 1967 October storm. In recent years a notable winter storm event was the 2013 October storm St. Jude (+1.44 m in Pärnu), though no flooding took place as it was the first storm of the season and the background sea level had not increased (Mäll et al., 2020).

9.6.1.3 Damage due to ETC Gudrun in 2005

As explained earlier, storm Gudrun in 2005 was the worst storm to have affected the country, and took people and authorities by surprise as the previous high surge event had taken place 38 years earlier (the 1967 October storm; +253 cm). The sea water intruded up to 1 km from the coastline, flooding dense urban areas and spas that tend to be close to the beach. As a result, one person lost their life and economic losses due to wind and surge damage reached ~0.7% of the GDP of the country (Tõnisson et al. 2008, Suursaar et al. 2006). Aside from flooding many urban areas, Gudrun also caused significant changes to the coastal geomorphology in many sites over all of Estonia, particularly to the areas that were most exposed to high winds, increased water levels and resultant high wave activity (e.g., Saaremaa island; Tõnisson et al. 2008). The storm event started long discussions regarding how to better prepare against disaster and how the coastal zone should be managed. It became clear that extreme events can and will happen in Estonian coastal waters, and better management approaches are necessary to mitigate and reduce disaster risks brought on by these events. Global warming and potential future relative sea-level rise (Estonia is still undergoing postglacial land uplift) are other factors that need to be taken into consideration as, in the worst-case scenarios, storm surge heights could be as high as 3–4 m in Pärnu by the end of the 21st century (Mäll et al. 2017; Mäll et al., 2020).

9.6.1.4 Countermeasures employed along the Estonian coastline

The current countermeasure strategies tend to follow a mixture of approaches, though there is a strong emphasis on soft measures, including raising the awareness of people in coastal communities. The issue for the case of Pärnu (and by extension other coastal locations in Estonia) is the economic feasibility of implementing hard measures, such as the construction of a massive dike (a 3.2 m high, 7.35 km long dike along the shoreline, and 3 km along the riverbank upstream, has been considered). Another concern with the implementation of such a massive project would be its environmental impact. Thus current countermeasures focus on soft adaptation measures such as inundation maps, real-time and 36 h forecast of sea levels, banning construction in low-lying areas with high risk, and establishing a minimum ground floor height of 3 m, among other measures. For the protection and restoration of sandy beaches, natural approaches such as beach nourishment are currently considered to be the best approaches (Tõnisson et al., 2018, see also Section 9.4).

All such strategies represent a big step forward in Estonia's disaster prevention and coastal management practices. However, in the wake up of climate uncertainty (e.g., local future sea-level rise, shifts in climatology, intensity and frequency of storm events) it is possible that Estonia might have to re-evaluate its strategies to adapt to the changing climate of the planet.

9.6.1.5 Conclusions

Estonia is a small Northern European country that shares its Western and Northern borders with the nearly tideless Baltic Sea. The most dangerous natural hazards in the country are related to winter storms (extratropical cyclones) that occur during the colder half of the year. These storm events are known to cause wind damage, though occasionally stronger events can have a notable impact on the coastline and its communities. One such event was the 2005 storm Gudrun, that caused extensive flooding in Pärnu city (+2.75 m) and started discussions on the improvement of coastal management and disaster prevention measures in Estonia. Currently the country mostly relies on soft and smart measures to deal with extreme events, though in the wake of climate uncertainty it might become increasingly necessary to develop other strategies to protect coastal communities.

9.6.2 Germany

Kristina Knüpfer
Waseda University, Tokyo, Japan

9.6.2.1 Introduction

North-Western Europe, including Germany, is strongly influenced by extratropical cyclones (ETCs), which can cause high waves and storm surges during the winter season (Feser et al., 2015). This, together with sea-level rise and coastal erosion (which in the Baltic is estimated at an average of 40 cm per year), compound the risk of flooding-related hazards in the region (Sterr, 2008). Located between the North and Baltic Seas and regularly exposed to ETCs, the German coastline as a flat landmass with tidal regimes has a long history of adaptation to and living with flooding from storm surges.

The two coasts of Germany differ in their geomorphological evolution (see Figure 9.6.2.1), with the Baltic side (2,100 km in length) covered in ice during the recent Weichselian glaciation period, while the North Sea side (1,600 km) remained glacier-free (Sterr, 2008). Overall, the coastal zone is mostly shallow, with wetlands and dunes lining the shore. On the Baltic side, an extensive number of peninsulas and islands, which also result in more steep coastal sections, protect the coast from the open sea. This coastline has a micro-tidal regime, with a water level difference of only 0.1–0.2 m (see also Section 9.6 on Estonia), while the North Sea side has a tidal range of 1.5–4 m (Sterr, 2008), which also affects the daily water level in Hamburg.

The dominant mitigation measure, especially along the flatter North Sea coast, is the maintenance of a dyke line (560 km and 1,340 km along the Baltic and the North Sea coasts, respectively, see Sterr, 2008). A relevant exception concerns the Halligen, a group of islands off the North Sea coast, which traditionally avoid dykes and instead build on warfts (which are basically foundations made from water-tight materials, which raise the ground floor level of a building, so that it remains above the waterline during storm surges), which can be supplemented by temporary flood walls. The Intergovernmental Panel on Climate Change 5th Assessment Report (IPCC 5AR) mid-scenario for 2100 forecasts a sea-level rise of 60 cm for the German coastline, as well as an increase in the mean tidal range of 0.15 cm per year and a mean high water level increase of 0.65 cm per year (Sterr, 2008). At the same time, sea-level rise is expected to cause further marsh

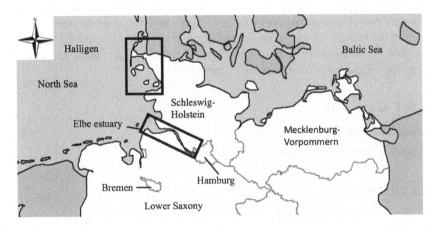

Figure 9.6.2.1 The five federal states comprising Northern Germany and their location between the North and Baltic Seas.(Base map source: Postmann, 2004. GNU Free Documentation License Version 1.2.)

subsidence, which may also cause existing dykes to sink due to their weight (Niemeyer et al., 2016). While the main strategy of the coastal states in northern Germany is to preserve the current dyke line, this will likely require the reinforcement of physical structures and the revision of coastal protection strategies to adapt to both sea-level rise and land subsidence.

The Free and Hanseatic City of Hamburg is a city-state and one of the five federal states in the north of Germany (see Figure 9.6.2.1). The HafenCity, a new quarter in Hamburg, provides a case study of how modern developments in coastal protection can help to build resilience against storm surges in Germany. It is also worth noting that, despite an assumed sea-level rise of 60 cm in the region, Hamburg is preparing its structures to withstand a rise of 80 cm by 2100 (Niemeyer et al., 2016).

9.6.2.2 Storm surges in Hamburg

Despite being situated over 140 km away from the North Sea, the location of Hamburg in the Elbe estuary means that storm surges are pushed up into the city regularly between September and April. In this tidal environment, the difference between normal water levels and flooding in the city is measured at St. Pauli, with the normal water level denoted as NN (the German Ordnance Datum) (see Figure 9.6.2.2).

A storm surge is defined as water levels raising above +3.4 m NN, with + 4.5 m NN being classified as a severe storm surge and +5.5 m NN as a very severe storm surge event (Hamburg, 2021). The city's modern protection countermeasures are comprised of dykes (75%), flood protection walls, sluices, barrages, pumping stations and gates where traffic routes cross the dyke line (Hamburg, 2012). Figure 9.6.2.3 shows an example of a flood gate designed to close the dyke line directly adjacent to the Elbe river, and a sluice gate situated in the city centre.

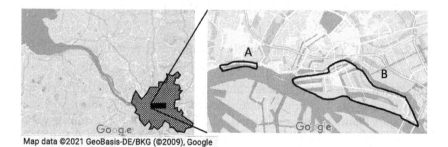

Map data ©2021 GeoBasis-DE/BKG (©2009), Google

Figure 9.6.2.2 Left: Hamburg and the Elbe estuary, connecting the city with the North Sea. Right: The embankment area around St. Pauli (A) and the HafenCity (B).

Source: Google Maps, 2021a.

Figure 9.6.2.3 Left: Modern flood gate integrated into a historic building at St. Pauli-Landungsbrücken (Source: Onnen, 2014). Right: The Rathausschleuse sluice gate next to the town hall

Source: Hamburg, 2012; Reich, 2009. All images reproduced with permission of the respective copyright owners.

The first recorded storm surge in the North Sea coast dates to 17th February 1164 ("The First Julian Flood") and is estimated to have killed 20,000 people in the wider region (Egidius, 2003). Adapting to regular storm surge events over the centuries, Hamburg began methodically documenting events from 1750. Since then and until around 1850 frequent storm surge events and flood damages were recorded, which were alleviated by a calm period until the 17th February 1962 storm surge event ("Second Julian Flood"), which marked the beginning of a recent period of increased wave heights, at least partially due to an intensification of the North Atlantic Oscillation and dredging of the Elbe river channel for shipping. Damages have been mitigated by improved protection measures, which mainly focus on dyke reinforcement (von Storch and Woth, 2006). Table 9.6.2.1 summarises five key events in recent history that illustrate this development and the corresponding protection measures, raising the dyke line from a height of +5.4 m NN in 1825 to up to +9.2 m NN by 1976.

9.6.2.3 The HafenCity: construction and protection on the dyke's waterside

The HafenCity masterplan was published in the year 2000 and revised in 2006 (HafenCity Hamburg, 2006), describing the development of a new quarter in Hamburg over a period of 25–30 years on previously vacant islands in the Elbe river. The plan would increase the area of the inner by 40%, aiming to host 14,000 residents, 45,000 workers, 5,000 students and ca. 80,000 tourists per day (HafenCity Hamburg, 2017). The islands where the HafenCity would be built on sat unprotected in front of the dyke line, with a historic elevation of between +4.5 and –6.5 m NN. To be able to

Table 9.6.2.1 Five recent storm surge events in Hamburg

Date	Storm	Max. height (+ m NN in St. Pauli)	Description
3rd–5th 02.1825	Unnamed	5.24	• Called the "Great Hallig Flood" as it temporarily submerged most of the Hallig islands and warfts. • Recorded 800 deaths. • Highest recorded event in St. Pauli until then. • Dykes were raised from +5.4 m NN to +5.7 m NN.
16th–17th. 02.1962	Vincinette	5.7	• Called the "Second Julian Flood". • Several dykes broke • Recorded 340 deaths. • Dykes were raised to +7.2 m NN. • Centralisation of flood protection at city level.
6th .11. to 17th .12. 1973	28 storm surge events in total	5.33	• Longest chain of storm surge events recorded to date. • Six events were categorised as very severe. • Damage mostly limited to the harbour area.
1st–5th 01.1976	Capella	6.45	• The highest recorded flood level in Hamburg to date. • Dykes were raised to +8 m NN and +9.2 m NN.
4th–10th 12.2013	Xaver	6.09	• Second highest event recorded in Hamburg to date. • Dykes under +9.2 m NN continue to be raised.

Sources: Hamburg, 2021

quickly begin development and due to the cost associated with re-routing the dyke line, the city decided to revise its flood risk management plan to introduce the warft concept as a physical flood protection countermeasure, as well as non-physical "Flutschutzgemeinschaften" (flood protection communities), which are centrally legislated by the Hamburg government, but the responsibility of each property owner to manage. Among the responsibilities are to ensure that occupants are flood prepared (for example through drills), alerted to take appropriate actions in a timely manner before and during storm surge events, and that the flood protection gates are closed when necessary. In addition to these measures, property owners in the HafenCity are also required to permanently employ a flood protection expert, who confirms both that the necessary physical protection measures have been taken and that the occupants of the property are prepared to adequately respond to storm surge mitigation measures (Hamburg, 2012).

Developing the area according to the warft concept essentially means protection through elevation, and requires buildings to be constructed so that the road and pedestrian pathway level is raised to +8.3 m NN (+7.5 m NN pre-2012) through a warft floor, which itself stands on 20 m long poles that bypass the clay sediment onto the sand foundations (HafenCity Hamburg, 2017). The ground floor level of buildings cannot be used for living or sleeping and, if the warft has openings below +8.3 m NN (to accommodate garages or shops), these must have built-in flood gates (Hamburg, 2012; HafenCity Hamburg, 2021). Due to the elevation provided by the warft, both firefighters and citizens can easily access all plots in any weather condition (Walraven and Aerts, 2008). An exception to this is the Sandtorkai plot, one of the earliest developed, which directly neighbours the Speicherstadt, a UNESCO cultural heritage warehouse district that remains at the historic quay level (UNESCO, 2015). Figure 9.6.2.4 provides an aerial overview of the area, including the Kibbelsteg bridge, which serves as evacuation and access routes to the Sandtorkai plot on the mainland, as its access road remains at the historic quay level.

Figure 9.6.2.5 shows how the warft concept is realized in the HafenCity at the Sandtorkai plot, while keeping the waterfront accessible to the public.

As one of the earliest completed plots, Sandtorkai has been exposed to several severe and very severe storm surge events, such as those following the ETCs Xaver in 2013, which generated a storm surge of up to +6.09 m NN, and Herwart in 2017 (+5.49 m NN) (Innenbehörde, 2017). Figure 9.6.2.6 shows the flooded street level of the Speicherstadt (at the

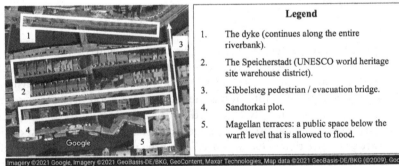

Figure 9.6.2.4 The area around Speicherstadt and Sandtorkai within the HafenCity.

Source: Google Maps, 2021b.

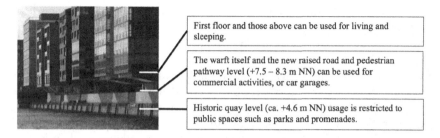

First floor and those above can be used for living and sleeping.

The warft itself and the new raised road and pedestrian pathway level (+7.5 – 8.3 m NN) can be used for commercial activities, or car garages.

Historic quay level (ca. +4.6 m NN) usage is restricted to public spaces such as parks and promenades.

Figure 9.6.2.5 HafenCity warft structured build at the Sandtorkai.

Source: Janß, 2016. Reproduced with permission of the copyright owner.

Figure 9.6.2.6 Flooding of Sandtorkai after ETC Herwart. a) The incline from the historic to the modern HafenCity ground floor level. b) The warft-level flood gates protecting the car garages from flooding. c) Pedestrian perspective from warft level, looking toward Kibbelsteg bridge.

Source: Beyer, 2017. Reproduced with permission of the copyright owner.

historic quay level), in contrast with the Sandtorkai warft (with closed flood protection gates for the garages), the Kibbelsteg bridge and the generally raised street level in the HafenCity proper.

9.6.2.4 Lessons and outlook in the context of future sea-level rise

The mitigation measures in the HafenCity at least partially rely on the cooperation of private residents in preparing their buildings for storm surge events, as well as on mobile facilities such as gates. Therefore, issues with damage have to date mainly consisted of construction site flooding (equipment being submerged), and delayed or omitted actions, such as leaving cars in areas that are expected to flood, as well as facility malfunctions, such as gates not closing properly (Welt, 2007, 2013). However, even during one of the highest recorded storm surges in Hamburg in recent times (+6.09 m NN in 2013), where the mobile flood protection gates on dyke-crossing bridges were closed (i.e., cars could not cross them to go to or from the Hafencity), commuting routes via the pedestrian bridges (such as Kibbelsteg), as well as the underground trainline (U4) remained uninterrupted (Welt, 2013). Overall, by relying on a raised ground level and active and timely participation by citizens in storm surge protective measures, the HafenCity shows how valuable proximity to the water (culturally and in terms of landscaping) can be maintained even as sea levels rise. At the same time, formulating appropriate disaster risk management strategies is crucial to ensure the continuous improvement of forecasting and hazard communication between agencies and citizens.

One advantage of dykes (compared to warfts) is that they can be reinforced without necessarily impacting other built structures and, while plots developed in the HafenCity pre-2012 only needed to set the ground floor at a height of +7.5 m NN, due to updated sea-level rise predictions this requirement was raised to +8.3 m after 2012 (Mees et al., 2013). It is already planned that this should consistently be raised to +9.2 m (Hamburg, 2012), raising the question of how the warfts can adapt to higher storm surges. First concepts regarding how to achieve long-term resilience were put forward in 2011, ranging from influencing the flow velocity of the river upstream through sandbanks and re-routing of tributaries, to "flying warfts" (HafenCity Hamburg and IBA, 2011). The flying warft concept proposes to connect the HafenCity buildings through bridges at rooftop level. The rooftops themselves would ideally be regularly used, public spaces that would also become evacuation routes during storm surges if the ground floor level were to be flooded. Further, the concept proposes that first floors should be designed to be wet rooms, so that they are not damaged by occasional flooding and can easily be cleaned afterwards.

9.6.2.5 Conclusions

Overall, cities such as Hamburg aim for daily life to not be disrupted by storm surges and to maintain access to the waterfront, while also staying adaptable to changing natural environments and hazards. As the approaches of different regions are influenced by various factors such as the desire to protect cultural heritage and lifestyles, as well as based on the specificities of the respective coastal environment and weather patterns, hazard protection and adaptation measures will differ between regions. The case of Hamburg in northern Germany shows how communities can remain close to the water and allow some flooding to take place while maintaining a safe environment that minimizes disruption to human activities and economic losses.

9.6.3 The UK

Ravindra Jayaratne

University of East London, London, UK

9.6.3.1 Introduction

Natural hazards in the UK range from localized issues such as coastal erosion, to regional problems that include coastal and riverine flooding to high-impact, low-frequency events such as space weather (Met Office, 2022). The impacts of these natural hazards can vary from disruption to infrastructure and transport links, to more significant effects on human welfare, health and even the loss of lives. It has been recently estimated that about 3,000 km of UK beaches are eroding due to the rising sea levels and increased levels of storminess (Williams, 2020). England and Wales experience faster erosion rates (e.g., 0.1 m/year) than Scotland, due to the isostatic rebound effect (crustal movement) and hard geology in the latter.

The UK has suffered floods throughout its history and there have been 914 coastal flood events documented between 1014 and 2020 (SurgeWatch, 2022) The 1607, 1953, 2013–2014 and 2020 floods have been categorized as the most catastrophic that the country has suffered, and will be discussed in detail later in this chapter. The risk due to tsunamis does not seem particularly high, and the only event that has been recorded to affect the British Isles was the 8.5–9.0 Mw 1755 *Lisbon Earthquake* tsunami (which affected the Cornish, Walsh and Irish coastlines). In order to protect the UK coastlines from natural hazards driven by hydro-meteorological factors, hard sea defences such as seawalls, rock armoured revetments, timber-groynes and breakwaters have been constructed in the most vulnerable locations. In terms of soft measures, a managed retreat that employs an appropriate defence system (e.g., riprap) and sand engines are now popular not only to maintain stable beaches and reduce flood levels but also to improve biodiversity.

9.6.3.2 Episodic natural disasters

9.6.3.2.1 The 1607 floods

Horsburgh and Horritt (2007) confirmed through numerical modelling that an extremely high tide superimposed with a storm surge was likely responsible for the 1607 flooding event. Eye witnesses reported that gale-force winds were the main cause of the disaster that killed more than 2,000 people around the Bristol Channel. The seawall at Burnham on Sea was overtopped by the storm waves. There are no records of the actual heights of the waves but historic indicators such as marks in nearby buildings (e.g., churches) and flood plaques show that the flood height must have been around 7.8 m and extended inland in Devon and Wales up to 6.4 km, and in Somerset up to 22.5 km. The total area of flooded land was estimated to have been 520 km², consisting predominantly of farmlands at the time.

9.6.3.2.2 The 1755 Lisbon earthquake and tsunami

Tsunamis are not common in the UK coastal waters, although the country was nevertheless affected by the *1755 Lisbon Earthquake and Tsunami*. This tsunami, triggered by an earthquake of 8.5–9.0 magnitude on the Richter scale, caused enormous damage and the loss of many lives. It was generated about 200 km off the Atlantic coast of Portugal, devastating Lisbon, Cape St. Vincent and some parts of Spain, Morocco, and finally propagated towards the English and Irish coastlines. Since there were no comprehensive post-disaster field measurements taken after this event, tsunami wave heights have been documented by eye witness accounts or estimated through numerical simulations. The tsunami wave height in Lisbon was reported to have been 6 m, whereas run-up heights at Cape St. Vincent were estimated to have been over 15 m according to historical data. The wave heights in Spain and Morocco were over 10 m. The number of casualties in Lisbon itself was about 900, and the wave penetrated about 250 m into the city (Baptista et al., 1998).

Since there is a possibility that other tsunamis could be generated from the Azores–Gibraltar Fracture Zone, North Sea, Canary Islands or Western Celtic Sea, Defra conducted a comprehensive study in 2005 regarding the potential impacts that such events could have on the population and infrastructure of the UK. Based on their numerical simulations, a tsunami generated from the Azores–Gibraltar Fracture Zone could have a wave height of 5 m and reach Cornwall, North Devon, Bristol Channel and South Wales in 5–8 hours (Defra, 2005). HR Wallingford (2006) reported that the largest wave height (4 m) from such sources could be expected in the southwestern coast of the UK, whereas along most of the Cornish and Irish coastlines the wave heights could be 1–2 m. This indicates that given the time it would take for the tsunami to travel to the UK, it is likely that the public could be warned to evacuate by employing a suitable early warning system.

9.6.3.2.3 The storm of 1953

An Extra-Tropical Cyclone (ETC) caused one of the worst storm surges in recent British history, generating a storm surge about 5 m high in the North Sea and badly affecting the east coast of England, especially the south of Yorkshire. This event resulted in 440 deaths, the evacuation of over 32,000 people, the inundation of over 1,000 km² of land (160,000 acres) and damage to 24,000 properties in the UK alone (Hall, 2015). The impact was extremely severe due to the lack of flood warning systems, the breaching of about 1,200 sections of coastal defences and severe erosion. This event led the British Government to start planning the construction of a new flood barrier in London against future flood events.

At the time of this flood event, four UK National tidal gauges were operational. As stated by SurgeWatch (2022), the skew surge at Lowestoft would have been greater than 2.3 m, generating water depths with a recurrence interval of about 1: 400 years. Moreover, the significant wave height was estimated to have been greater than 7 m off the Norfolk coast, which has been classified as a 1: 50-year wave height.

9.6.3.2.4 The 2013–2014 winter storms

This storm was the worst coastal flood event in the UK since the storm of 1953, generating a skew surge of over 1 m in the Irish Sea and up to 2 m in the North Sea. Lowestoft, in the northeast of England, experienced the highest skew surge at 1.93 m, estimated to be a 1: 200 years return period, and many parts of the UK recorded significant wave heights as large as 8–10 m in the open sea (Sibley et al., 2015).

Flood warnings were issued by the Environment Agency (EA), UK, before the incident, with the water breaching a number of coastal defences in the east and southwest of the country. As a consequence of this the EA (2016) reported that 8,342 residential properties, 4,897 businesses and 45,000 hectares of farmland were flooded. The damage to transport links, both rail (160 km) and roads (190 km), was significant. Noticeable coastal erosion took place along the east coast of England. Due to the existence of modern state-of-the-art flood warning systems in the UK people were well prepared for such events (in comparison to the 1953 event) and the number of casualties was significantly reduced to 17 (Met Office, 2014).

9.6.3.2.5 The 2020 floods

Six years after the catastrophic winter storms in 2013–2014, strong winds coupled with torrential rain (a month's rainfall in 18 hours) were brought about by storm "Ciara" to the English and Welsh coastlines in February 2020, causing disruptions to transport links and electricity outrages (Met Office, 2020).

Some parts of the Lake District and Snowdonia showed a record high rainfall of 80–150 mm, with the highest record at the Lake District being 180 mm. This event caused disruptions to land and air transport links, power cuts to 675,000 homes and flooded over 500 residences in riverine areas of Cumbria, Lancashire, Manchester and Yorkshire (Met Office, 2020; BBC, 2020).

9.6.3.3 Present countermeasures, and climate change effects

The coastal defence systems in the UK use either soft or hard engineering defences, or combinations of both. These range from beach stabilisation by offshore sand dredging, managed retreat by allowing the shoreline to move inland, constructing reinforced sand dunes with vegetation, to typical mass reinforced concrete seawalls, rock-armoured revetments, earth embankments, rock-armoured breakwaters, timber groynes, flood gates and barriers.

The UK Committee on Climate Change (2018) evaluated the coastal risk of flooding and erosion due to changing climate and reported that the country is not prepared to face the impact of increasing sea-level rise. It is further reported that current coastal management practices are unsustainable although, with a change in social attitudes, sustainable coastal adaptation could be possible. In order to achieve these goals, the UK Environment Agency, Defra and the local governments need to be enabled by the national government, and long-term funding/investment should be made available.

9.6.3.4 Conclusions

The major hydro-meteorological natural hazards in the UK consist of storm surges generated by low-pressure systems, high winds, extreme tidal conditions and torrential rains. Floods and erosion are the two major impacts that can be brought about by storm surges, which can claim lives and affect the local economy. There is a low probability of tsunamis affecting the island, though the risk from some far-field sources located in the Azores–Gibraltar Fracture Zone cannot be entirely dismissed.

UK government organisations such as Environment Agency (EA) and the Department for Environment, Food and Rural Affairs (Defra) have formulated a number of soft and hard engineering countermeasures in order to tackle these problems. Despite challenges regarding the complexity and unpredictability of natural hazards, EA and Defra continue to carry out regular maintenance as well as improvements to existing flood defences and erosion protection structures.

9.7 AFRICA

9.7.1 Tanzania

Joel Nobert and Philip Mzava
University of Dar es Salaam, Dar es Salaam, Tanzania

9.7.1.1 Introduction

Increased urban development and the subsequent increase in surface water runoff have increased the risks posed by urban flooding to both sustainable economic development and human safety all around the globe, particularly in the developing world (Miller and Hutchins, 2017). In the period

2018–2020 severe cases of flooding were reported in different parts of the world, including Zimbabwe, Namibia, Malawi, Tanzania, Somalia, and Kenya in Africa; Brazil, Colombia, Bolivia, Mexico, and the United States in South and North America; and Cambodia, China, India, Nepal, Pakistan, Bangladesh and Vietnam in Asia. These recent flooding events have highlighted the significant negative impacts that floods can have on human welfare (Ding et al., 2013; Marsh et al., 2016; NASEM, 2019).

It is estimated that around half of the world's population currently lives in urban areas, with this ratio being expected to increase up to around 70% by 2050 (Li *et al.*, 2015). Urban areas are centres of economic and political activities, though recently there have been growing concerns over urban-scale problems in relation to climate change and the need to take mitigation actions to reduce greenhouse gas emissions. However, adapting to global climate change at the urban scale is more relevant to local administrative authorities assigned with the task of designing and implementing possible responses. This will likely facilitate decisions that are more appropriate to tackle the associated risks and formulate appropriate adaptation at the local urban level (Hunt and Watkiss, 2011). Climate change impacts are potentially significant and unique, and/or aggravated in urban areas. For instance, rainwater infiltration into the soil is inhibited by built-up land cover in urban areas, which leads to increased stress on urban drainage systems that can increase the vulnerability of other infrastructure. Also, the high population densities of urban areas make flooding events potentially more severe and catastrophic (Lindley *et al.*, 2006). Badana and Nobert (2020) assessed the flood risk in Luvuvhu River, Limpopo province, South Africa, indicating that settlements in urban areas are the most vulnerable to flood events, followed by cultivated lands.

Urban areas of Dar es Salaam, the main commercial city in Tanzania, have experienced increased cases of flooding in recent years, including the 2015, 2018, 2019 and 2020 events. Ngailo *et al.* (2016) ranked flooding second after epidemics among the top ten natural disasters that threaten the economy of the country. In the past, cases of extreme rainfall and flooding have resulted in casualties within the urban areas of Dar es Salaam, given that it has the highest population density in Tanzania ($\approx$ 3,133 people/km^2). Following the floods of April 2018 in Dar es Salaam, household losses were estimated to be over 100 million USD, equivalent to 2–4% of the region's Gross Domestic Product (GDP) (Erman *et al.*, 2019). Although flooding in urban Dar es Salaam can be attributed to different factors, including the lack of or poor storm water drainage systems, river valley encroachment, or the poor dumping of solid waste (Sakijege *et al.*, 2012), rainfall remains the major driving force for surface-water runoff generation.

Similar cases of flooding have also been reported in the Morogoro region of Tanzania. According to the Morogoro local government, in 2014 alone almost 9,000 residents were affected by flooding. Kilosa district alone had an estimated 7,600 residents that were left homeless by a recorded 628.3

m³/s flood, damaging economic and social infrastructure and resulting in food insecurity. The socio-economic activities practised by households which reside by the Mkondoa River can be explained as one of the reasons behind the flooding (Paavola, 2008), as intensified agriculture has led to a depletion of natural resources such as forest and river banks, and encroachment onto the water system. Despite the risks to these urban areas of Tanzania, flood maps that detail the expected inundation extents according to the severity of flooding are lacking in most cases. Such information can be useful for engineering design and the construction of infrastructure that can be used for developing detailed emergency plans.

9.7.1.2 Flooding in Msimbazi river catchment, Dar es Salaam

9.7.1.2.1 Description of the study area

Dar es Salaam is a coastal city located in the eastern part of the Tanzanian mainland between longitudes 38°59′18.37″–39°33′29.55″ E and latitudes 6°35′17.48″–7°11′18.92″ S. With an area of 1,630 km², it occupies 0.19% of the Tanzanian mainland, stretching about 100 km between the Mpiji River to the north and beyond the Mzinga River to the south, with the Indian Ocean to the east. Its coastal area is comprised of sand dunes and tidal swamps. The central urban portion of Dar es Salaam is located within the Msimbazi catchment and is frequently affected by flooding. This catchment has an area of approximately 266 km², starting from the highlands of the Pwani region, running through the central urban part of Dar es Salaam and draining its water into the Indian Ocean (see Figure 9.7.1.1).

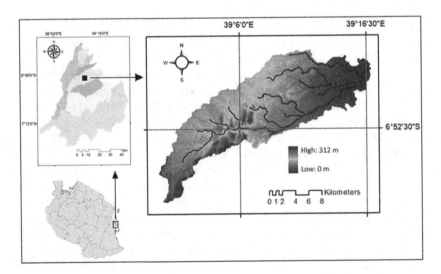

Figure 9.7.1.1 River catchments across Dar es Salaam region, Tanzania, highlighting the Msimbazi catchment.

9.7.1.2.2 Trends of extreme rainfall in the Msimbazi River Catchment

Since extreme rainfall events are directly linked to flooding, understanding these is crucial when trying to explain the increased cases of flooding in the urban area of Dar es Salaam in recent years. A significant increasing trend of annual and seasonal maximum rainfall has been observed in the Dar es Salaam urban area during the period 1967–2050 (Mzava et al., 2020). Also, it has been observed that rainfall events with a 2 to 10-year return period in the Dar es Salaam urban area cause pluvial and fluvial flooding, which is indicative of the issues that exist with the drainage systems in the area. Future projections of extreme rainfall in the Dar es Salaam urban area show a 20–25% intensification of extreme rainfall magnitudes in the period 2018–2050 relative to those of the past (Mzava et al., 2020). According to these findings, immediate measures are required to better manage flooding impacts in the Dar es Salaam urban area, including improved engineering designs for drainage systems and other countermeasures.

9.7.1.2.3 Land use/cover changes in the Msimbazi River catchment

Research over the past 36 years regarding land use/cover change in the urban Dar es Salaam area clearly shows how human activities have profoundly changed the land cover, with a clear shift from thick vegetation to an urban built-up environment (Mzava et al., 2019). Future projected land cover shows that the growth in the urban settlement will be continuous, and about 81.04% of the total study area will be urbanized by 2046, which represents an increment of 43.97% and 14.95% from 37.07% and 66.09% coverage of this land cover class in 2014 and 2030, respectively (Mzava et al., 2019). This land cover class is characterized by very low to zero water permeability. Changes in land use/cover in urban Dar es Salaam are most likely the result of human activities related to the increase in total population (due to population growth and the influx of migrants), and the growth of the economy. The combined influence of climatic and anthropogenic factors will affect the hydrologic response of the Msimbazi catchment and more likely lead to intensification worsening of the urban flooding problem in the Dar es Salaam urban area. Strategies and policies for better planning and management of Dar es Salaam urban growth are therefore necessary to diminish land use/cover-related negative impacts to the society and environment.

9.7.1.2.4 Changes in the peak flow magnitudes and flooding behaviour

As stated earlier, the magnitudes of peak flows in the Dar es Salaam urban area are influenced differently by climate and land use/cover, though future climate change is likely to have a greater impact on peak flow magnitudes than land-cover change (Mzava et al., 2021). The coupled effects of both climate and land-cover changes will likely have a much bigger impact on the change in peak flows than any separate scenario of either land use or climate change. From the combined effects of climate and land-cover changes, the magnitudes of mean peak flows were estimated to increase between 34.4–58.6% in the future relative to the past (Mzava et al., 2021).

9.7.1.3 Flood analysis in Kilosa, Morogoro – Mkondoa catchment

9.7.1.3.1 Kilosa district

A total of six districts make up the Morogoro region in Tanzania, which include Morogoro urban, Morogoro rural, Kilosa, Ulanga and Kilombero. Kilosa district has an area of 14,918 km² and is the second largest in the Morogoro region. Figure 9.7.1.2 shows the Mkondoa catchment, which covers the Kilosa district.

9.7.1.3.2 Flood mapping in Kilosa district

Kilosa district has for a long time been known to be at risk of suffering flooding during the rainy seasons, the impact of which has been felt mostly by farmers settling along the Mkondoa River and residents of Kilosa town. In order to assess the risks in the district, mapping of flood extent for different return periods was done through the integration of hydrologic modelling, remote sensing, GIS and hydraulic modelling (Figure 9.7.1.3). There is limited discharge data available for the area, and hence hydrological modelling was needed to extend the flow records and fill the gaps in them.

The hydraulic analysis showed the depths and extent of floods overtopping the downstream end of the embankment on the left side of Mkondoa River along Kilosa town, but also affecting settlements and buildings around the town (such as the district council's office and the railway line that passes along Mkondoa River towards Kilosa Town, see Figure 9.7.1.4).

9.7.1.4 Conclusion

Increases in the cases of urban flooding in Dar es Salaam city and the Morogoro region are partly explained by the increase in annual and seasonal maximum rainfall. Most of the pluvial and fluvial flooding in the study area are happening due to rainfall events with return periods between 2 and 10 years. Non-climatic factors such as land use/land cover changes are increasing impermeable surface extent and sedimentation, which ultimately

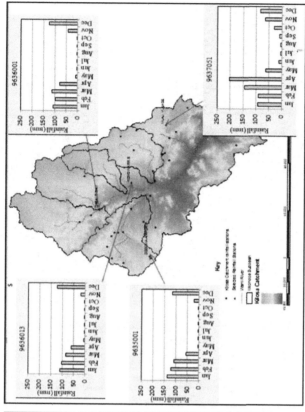

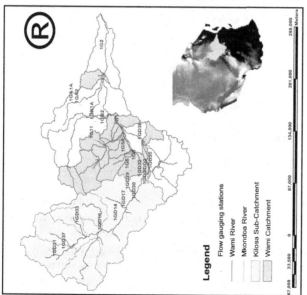

Figure 9.7.1.2 Mkondoa catchment.

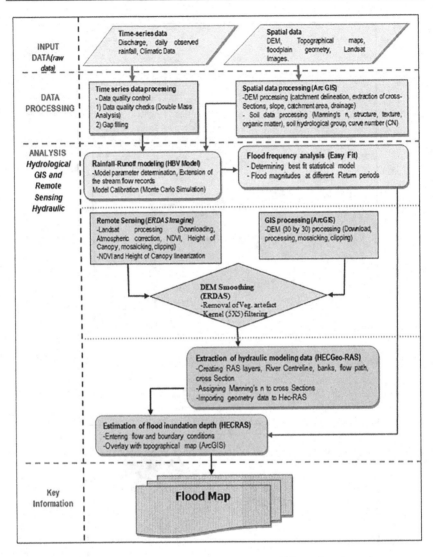

Figure 9.7.1.3 Flowchart for inundation mapping in Kilosa district.

affect the river morphology and hydraulics. Floodplain encroachment was observed to be another important factor behind the reduction in the water carrying capacity of the natural drainage systems in the study area. The coupled effects of climate and land-cover changes will likely have a much larger impact on the change in peak flows than any separate scenario that only considers either land use or climate change, and threaten the sustainability of settlements unless countermeasures are implemented.

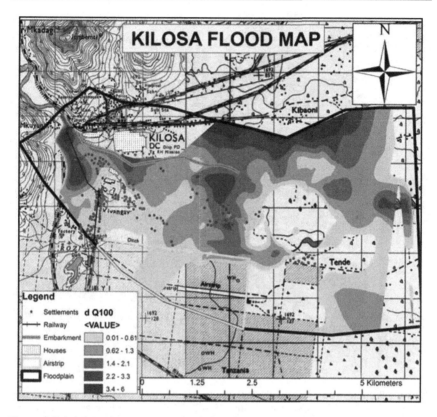

Figure 9.7.1.4 Inundation extents generated by a 100-year flood in the Kilosa Region.

9.8 THE ARCTIC

Martin Mäll

Yokohama National University, Yokohama, Japan

9.8.1 Introduction

The Arctic Ocean is centred approximately on the North Pole and forms the smallest of the oceans on the planet (14,090,000 square km²). It has only slightly more than one-sixth the area of the next largest, the Indian Ocean, and is five times larger than the largest sea, the Mediterranean. The Arctic Ocean and its marginal seas – the Chukchi, East Siberian, Laptev, Kara, Barents, White, Greenland, and Beaufort seas (Figure 9.8.1), are the least-known basins and bodies of water in the world as a result of their remoteness, hostile weather, and perennial or seasonal sea ice cover. The deepest water depth measured in the Arctic is 5,502 m, while the average water

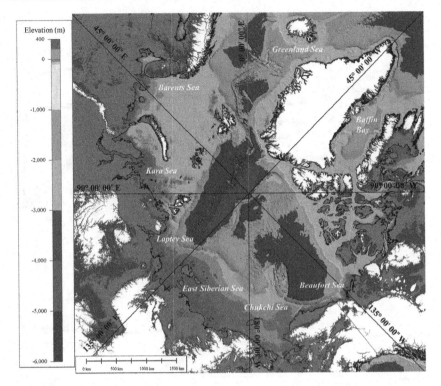

Figure 9.8.1   Elevation map of the Arctic Ocean region and its surrounding topography. Map created from IBCAOv3 data. The white areas show elevations higher than 400 m. Most of the coasts in the Arctic are relatively low-lying and subject to erosion from the thawing of permafrost, longer open water season and increased wave action due to climate change.

depth stands at 987 m (Ostenso, 2018). There are eight countries that surround the Arctic Ocean, with up to 10 million people living there (numbers depend on the definition used). While the global population is expected to continue to grow at a rapid pace until the middle of the 21st century, in the Arctic little to no growth is projected (depending on the regions within it both increases and decreases are expected; Heleniak, 2021).

The relatively untouched and less studied Arctic system is undergoing rapid changes, as it exhibits a strong response to global climate change, resulting in a wide range of positive feedback loops (known as Arctic amplification) which have a wide range of local and global climatic implications. The main mechanism for this is the gradual loss of annual sea-ice cover (including the retreat of perennial ice) that exposes more of the dark ocean surface, leading to a reduction in albedo and thus increasing ocean heat absorption, in turn further reducing ice cover (there are also other factors that play a role in the sea ice reduction and warming of Arctic, e.g., increased flow of heat from the tropics to higher latitudes; Voiland, 2013). This ongoing change has led to an increased interest in the Arctic from the governments (both from countries surrounding the Arctic circle and others outside), industry (natural resources, tourism, etc.) and the scientific community. As the annual sea ice extent continues to decrease the access to the Arctic becomes easier, thus opening doors to new navigation and transportation corridors, as well as the possibility to exploit the natural resources there (oil, gas, and fisheries among others).

9.8.2 Met-ocean drivers

The climatology of the region and its variability is heavily influenced by the Arctic Oscillation, which affects the larger wind circulation patterns around the Arctic zone. For instance, during the positive phase the counterclockwise circulating winds are more strongly confined to the polar area, keeping the colder air masses constrained there. However, during the negative phase the wind belt becomes weaker and distorted, allowing the colder air masses to penetrate southward, leading to increased storminess in the mid-latitudes (NOAA, 2019). The Arctic Ocean Oscillation Index indicates the sense and near-surface circulation of the Arctic Ocean by looking at the sea surface height gradients driven by surface winds. The index shows two circulation patterns which alternate at 5- to 7-year intervals: a cyclonic and anticyclonic phase. During the cyclonic phase a counterclockwise circulation of upper ocean and sea ice is dominant due to low sea-level atmospheric pressure (SLP). During the anticyclonic phase, the reverse takes place (clockwise circulation) due to dominant high SLP (Proshutniky et al., 2015).

The Arctic storm climatology is rather varied and less understood than tropical and other storm systems (see Sections 5.2 and 9.6), and the classification of systems is difficult. There is a broad category of polar mesocyclones, and one of the main types of weather systems is known as polar lows. These are usually short-lived (few days) and less than 1,000 km in the horizontal length scale (usually in the range of 400–600 km) and, to be classified as one, they need to have surface winds of gale force above 17 m/s (as high as 33 m/s have been observed; Turner & Bracegirdle, 2006). There are various types of polar lows and even hybrid systems. In the Arctic, 7 categories have been identified by Rasmussen and Turner (2003), which tend to be stronger during the winter season and can bring the worst weather conditions to coastal and island communities.

9.8.3 Emerging Arctic: wave and surge climate

Over the period of 1979–2012 the mean annual Arctic sea ice extent decreased by around 0.45–0.51 million km^2 per decade. For the summer sea ice minimum, the range has likely been 0.73–1.07 million km^2 per decade, with satellite data showing particularly significant losses at the end of the annual melt season in September, when the sea ice extent is at its minimum (IPCC, 2013). Since the whole system is undergoing rapid changes, the resultant complex immediate and future effects have been gaining more attention from a multitude of scientific fields.

However, in terms of coastal engineering, management and disaster risk studies there are many unknowns regarding how the Arctic system responds to these climatic changes and what will be the implications to social and economic activities in the coming years. Most research regarding the output from general circulation models has investigated the changes in the wind and wave climate (e.g., Dobrynin et al., 2012; Khon et al., 2014), by forcing ocean models such as the WAM and WAVEWATCH III to respond to future climate conditions. These studies suggest that there will be a general increase in wind speed, in turn leading to a more energetic wave climate. Historical studies looking into past reanalysis data confirm similar trends happening within the past decades (i.e., swells becoming more prevalent as the spatial and temporal extent of the effective fetch increases as the minimum sea ice extent decreases; Stopa et al., 2016; Wang et al., 2015).

To study the historical and future surge and wave conditions, researchers typically look into observational data sets, though this presents a problem for the case of the Arctic given that data is extremely scarce and often plagued with uncertainties and gaps in measurements (Manson & Solomon, 2007) due to difficulties in operating equipment in such harsh environments. Thus, for the case of the Arctic various datasets and methods are often used in conjunction with one another, including satellite data (reanalysis products), in-situ measurements (seasonal or permanent), numerical models, and even driftwood line elevations from field surveys (e.g., Kim et al., 2021).

From the studies carried out so far it is clear that the open water season along the Arctic shores is getting longer, the effective fetch is increasing, the permafrost is thawing, and potential relative sea-level rise are all causing significant impacts on the coastal communities in the area. While it is known that the changes are taking place at a very rapid pace, little is known as to what will happen regarding the frequency and intensity of weather extremes in the Arctic. However, it is likely that the effect that storms have on coastal communities will increase due to the aforementioned effects, though the changes in the individual cyclogenesis mechanism require further research. One approach that could be used is the pseudo climate modelling methods (as discussed in Section 8.2), and attempts have already been made to apply this method to the Arctic region (Figure 9.8.2; Mäll et al., 2020) and study the surge and wave climate changes for some case study storm events. However, Mäll et al. (2020) highlighted the need for observational data sets to calibrate and evaluate the performance of such models, in order to increase the confidence in them. Future climate condition simulations (e.g., RCP8.5 by ~2050, without the sea ice in Figure 9.8.2) showed little response in the storm intensity, though in the absence of sea ice waves were able to propagate further into the Canadian archipelago.

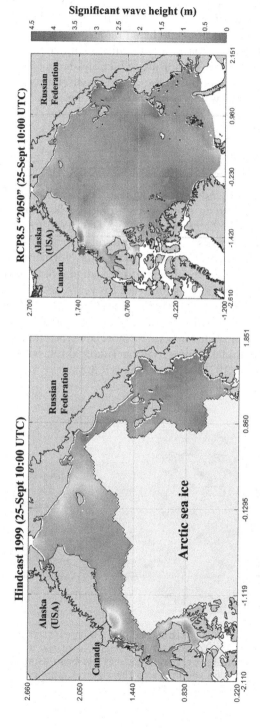

Figure 9.8.2 Left. Hindcast and, Right. Future 2050 RCP8.5 significant wave height simulations from the FVCOM-SWAVE model. The storm event centre and highest wave field can be seen at the border between Canada and USA. The "x" marks the location of the Canadian port hamlet of Tuktoyaktuk.

It is suggested that future individual storm studies should include a number of storm events (depending on data availability), following different tracks to increase the sample pool and better understand how Arctic weather systems develop and impact coastal communities (and change the sea state when considering the potential for future reductions in the extent of sea ice). Additionally, future studies in the Arctic would benefit by considering using specific changes in global warming temperatures (e.g., 1.5 °C, 2 °C, 3 °C, etc.) instead of the often difficult to compare climate change scenarios (Mudryk et al., 2021).

9.8.4 Impact on the coastline and communities

The Arctic coastline is already retreating at an average rate of 0.5 m/year (Lantuit et al., 2012), and in some areas along the Beaufort Sea coastal erosion can be as high as 22 m per year (e.g., Dew Point in Alaska; Jones et al., 2018). Two of the largest coastal communities in the North American continent that face the Arctic Ocean are Barrow in Northern Alaska, USA, and Tuktoyaktuk in the Northwest Territories, Canada. Both have struggled with changing conditions and erosion issues for years, and are facing many challenges to preserve their cultural heritage and homes from the advancing sea. In Tuktoyaktuk geological vulnerability became clear since it was first settled in 1936, though the current rate of shoreline erosion near the hamlet is around 0.8 m/year, and 1.7 m/year along Tuktoyaktuk Island (Canadian Permafrost Association, 2020; see also Figure 9.8.3).

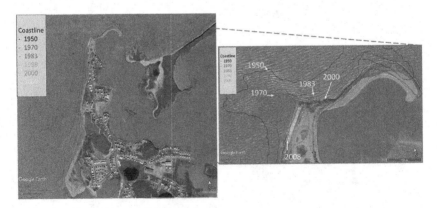

Figure 9.8.3 Coastal erosion in Tuktoyaktuk. Over the years various approaches have been tried to protect the coastline of the area. One of the more successful protective measures has been using concrete slabs (dashed circle on the image to the right). Compiled from data provided by Hynes et al. (2014).

Over the years many buildings and rudimentary coastal defences have been flooded or undermined by the storms. In 1998, the community received 40 concrete slabs that were installed on gravel pads at the northern end of the peninsula (100 m section; Figure 9.8.3). These slabs performed well, though they were not deemed a viable option for the rest of the community due to difficulties in producing them on the site and that placing them requires heavy lift cranes that are not available there (McClearn, 2018). Since then, the community has relied on makeshift approaches by using rip-rap style to protect the coastline. More recently, in 2018 the community – together with Baird Engineering – has been working on erosion mitigation solutions for Tuktoyaktuk, where using rock revetment system appears to be the more feasible approach (Bowling, 2021).

9.8.5 Conclusion

The Arctic is a remote and harsh region in the polar north that is, and has been for some time, undergoing rapid climatic changes. From the coastal engineering and management perspective, these changes are seen through yearly sea ice retreat and extended periods of the open water, which translate to shifts in the regional hydrodynamic regime. A prominent manifestation of that is the increased exposure and impact that wave action and surges are having on the coastline and its communities, through the acceleration of erosion processes.

Coastal settlements along the Arctic coastline have been dealing with these adverse impacts for years. While solutions and opportunities are being tried for remote and small coastal communities such as Tuktoyaktuk, it is increasingly important for civil and coastal engineers to continue improving design methods for such regions. What makes this challenging is the economic feasibility of large-scale hard solutions (in sparse and often spread out communities) and the rapidly changing Arctic system, where even the near future timescales hold many uncertainties.

REFERENCES

Ahmed, M., Rogers, J.D., and Ismail, E.H., 2014. A regional Level Preliminary Landslide Susceptibility Study of the Upper Indus River Basin. *European Journal of Remote Sensing*, 47(1), 343–373. https://doi.org/10.5721/EuJRS20144721

Adriano, B., Arcila, M., Sanchez, R., Mas, E., Koshimura, S., Arreaga, P., & Pulido, N. (2017). Estimation of the tsunami source of the 1979 great tumaco earthquake using tsunami numerical modeling. In *16th World Conference on Earthquake, 16WCEE 2017, S-U1463128263*, pp. 75–82.

Akter, N., and Tsuboki, K., 2012. Numerical Simulation of Cyclone Sidr Using a Cloud-Resolving Model: Characteristics and Formation Process of an Outer Rain band. *Monthly Weather Review*, 140(3), 789–810.

Andrade, C. A., Thomas, Y. F., Lerma, A. N., Durand, P., & Anselme, B. (2013). Coastal flooding hazard related to swell events in Cartagena de Indias, Colombia. *Journal of Coastal Research*, 29(5), 1126–1136. https://doi.org/10.2112/JCOASTRES-D-12-00028.1

Aranguiz, R., Esteban, M., Takagi, H., Mikami, T., Takabatake, T., Gomez, M., Gonzalez, J., Shibayama, T., Okuwaki, R., Yagi, Y., Shimizu, K., Achiari, H., Stolle, J., Robertson, I., Ohira, K., Nakamura, R., Nishida, Y., Krautwald, C., Goseberg, N., and Nistor, I., 2020. The 2018 Sulawesi tsunami in Palu city as a result of several landslides and coseismic tsunamis. *Coastal Engineering Journal (CEJ)*, 62, 445–459.

Aschariyaphotha, N., Wongwises, P., Humphries, U.W., and Wongwises, S., 2011. Study of storm surge due to Typhoon Linda (1997) in the Gulf of Thailand using a three dimensional ocean model. *Applied Mathematics and Computation*, 217(21), 8640–8654.

ASCE, 2016. *ASCE/SEI 7-16: Minimum Design Loads and Associated Criteria for Buildings and Other Structures.* American Society of Civil Engineers, Reston, VA.

ASEP (Association of Structural Engineers of the Philippines), 2016. *National Structural Code of the Philippines 2015 Volume 1 Buildings, Towers, and Other Vertical Structures.* Quezon City, Philippines: ASEP.

Badana, N., & Nobert, J. (2020). *Flood risk assessment in Luvuvhu River, Limpopo province, South Africa. Physics and Chemistry of the Earth.* Elsevier. https://doi.org/10.1016/j.pce.2020.10295

Baker, D.B., Kyaw, T., and Chua, R., 2008. Post-Nargis Joint Assessment. The Tripartite Core Group of the Government of the Union of Myanmar, the Association of Southeast Asian Nations and the United Nations. https://www.recoveryplatform.org/assets/publication/Post-Nargis_Joint_Assessment.pdf. Accessed on 25 September 2021

Bakhtiari, A., Allahyar, M.R., Jedari Attari, M., and Haghshenas, A., 2018. Modeling of Last Recent Tropical Storms in the Arabian Sea. *Journal of Coastal and Marine Engineering*, 1(1), 58–66.

Baptista, M. A., Heitor, S., Miranda, J. M., Miranda, P., & Mendes Victor, L. (1998). The 1755 Lisbon tsunami: Evaluation of the tsunami parameters. *Journal of Geodynamics*, 15(1–2), 143–157.

Barrett, F., and Hannoush, S., 2016. Mitigating the impacts of severe weather, *Reports on the Commisoner of the Environment and Sustainable Development Canada.*

Barrientos, S. E., & Ward, S. N. (1990). The 1960 Chile earthquake: Inversion for slip distribution from surface deformation. *Geophysical Journal International*, 103(3), 589–598. https://doi.org/10.1111/j.1365-246X.1990.tb05673.x

BBC (2020). Storm Ciara: Cumbria begin clean-up after flooding, https://www.bbc.co.uk/news/uk-england-cumbria-51442839 (assessed on 25 March 2022).

Bentley Systems, 2019. *STAAD.Pro CONNECT Edition V22 Update 1 (v.22.01). User Manual.*

Beyer, M. (2017). Sturmflut HafenCity 29.10.2017. https://www.flickr.com/photos/michaelbeyerhamburg/albums/72157688537977534 (accessed on 21 September 2021).

Bird, P., 2003. An Updated Digital Model of Plate Boundaries. *Geochemistry, Geophysics, Geosystems*, 4(3).

Biswas, M.K., Carson, L., Newman, K., Stark, D., Kalina, E., Grell, E., and Frimel, J., 2018. *Community HWRF Users' Guide V4.0.*

Blake, E., and Zelinsky, D., 2018. Hurricane harvey. National Hurricane Center Tropical Cyclone Report. https://www.nhc.noaa.gov/data/tcr/AL092017_Harvey.pdf

Booij, N., Ris, R.C., and Holthuijsen, L.H., 1999. A Third-Generation Wave Model for Coastal Regions: 1. Model Description and Validation. *Journal of Geophysical Research: Oceans* 104(C4), 7649–7666. https://doi.org/10.1029/98JC02622

Bowling, E. (2021). NNSL Media. Cost to save Tuktoyaktuk from climate change: $42 million and rising. https://www.nnsl.com/news/cost-to-save-tuktoyaktuk-from-climate-change-42-million-and-rising/ (accessed on 30 September 2021).

Camus, P., Losada, I. J., Izaguirre, C., Espejo, A., Menendez, M., & Perez, J. (2017). Statistical wave climate projections for coastal impact assessment. *Earth's Future*, 5(9), 918–933. https://doi.org/10.1002/eft2.234

Canadian Permafrost Association (2020). *Virtual annual general meeting*, Arenson, L. U., Rudy, A., & Morse, P. (Eds.). https://canadianpermafrostassociation.ca/userContent/documents/AGM/Abstracts%20AGM%202020.pdf

Cardona, O. D., Ordaz, M., Salgado-Gálvez, M. A., Barbat, A. H., & Carreño, M. L. (2018). Latin American and Caribbean earthquakes in the GEM's Earthquake Consequences Database (GEMECD). *Natural Hazards*, 93, 113–125. https://doi.org/10.1007/s11069-017-3087-9

Carvajal, M., Cisternas, M., & Catalan, P. A. (2017). Source of the 1730 Chilean earthquake from historical records: Implications for the future tsunami hazard on the coast of Metropolitan Chile. *Journal of Geophysical Research: Solid Earth*, 122(5), 3648–3660. https://doi.org/10.1002/2017JB014063

Center for Excellence in Disaster Management and Humanitarian Assistance (CFE-DM), 2017. *Myanmar (Burma) Disaster Management Reference Handbook*. U.S. Department of Defense. https://reliefweb.int/report/myanmar/myanmar-disaster-management-reference-handbook-march-2020. Accessed on 16 September 2021.

Charveriat, C. (2020). Natural disasters in Latin America and the Caribbean: An overview of risk. In *Working Paper 434*. https://doi.org/10.2139/ssrn.1817233.

Chen, C., Liu, H., and Beardsley, R., 2003. An unstructured grid, finite-volume three-dimensional, primitive equations ocean model: Application to coastal ocean and estuaries. *Journal Atmospheric and Oceanic Technology*, 20(1), 159–186. https://doi.org/10.1175/1520-0426(2003)020<0159:AUGFVT>2.0.CO;2.

Chettri, N., Tempa, K., Gurung, L., and Dorji, C., 2022. Association of Climate Change to Landslide Vulnerability and Occurrences in Bhutan, Sarkar, R., Shaw, R., and Pradhan, B. (Eds.), *Impact of Climate Change, Land Use Dynamics Socio-economic and Land Cover, and on Landslides*, pp. 3–38. Springer Nature Singapore Pvt Ltd. https://doi.org/10.1007/978-981-16-7314-6

Choowong, M., Murakoshi, N., Hisada, K., Charusiri, P., Charoentitirat, T., Chutakositkanon, V., Jankaew, K., Kanjanapayont, P., and Phantuwongraj, S., 2008. 2004 Indian Ocean tsunami inflow and outflow at Phuket, Thailand. *Marine Geology*, 248(3–4), 179–192.

Chu, J.H., Sampson, C.R., Levine, A.S., and Fukada, E., 2002. The Joint Typhoon Warning Center Tropical Cyclone Best-Tracks, 1945-2000. NRL/MR/7540-02-16. Washington, DC. https://www.metoc.navy.mil/jtwc/products/best-tracks/tc-bt-report.html

Cisternas, M., Atwater, B. F., Torrejon, F., Sawai, Y., Machuca, G., Lagos, M., Eipert, A., Youlton, C., Salgado, I., Kamataki, T., Shishikura, M., Rajendran, C. P., Malik, J. K., Rizal, Y., & Husni, M. (2005). Predecessors of the giant 1960 Chile earthquake. *Nature*, 437(7057), 404–407. https://doi.org/10.1038/nature03943

Clague, J.J., Munro, A., and Murty, T., 2003. Tsunami Hazard and Risk in Canada. *Natural Hazards*, 28(2), 435–463.

Committee on Climate Change (2018). Managing the coast in a changing climate, UK, https://www.theccc.org.uk/publication/independent-assessment-of-uk-climate-risk/ (assessed on 25 March 2022).

Cong, L.V., Cu, N.V., and Shibayama, T., 2014. Assessment of Vietnam Coastal Erosion and Relevant Laws and Policies, In Thao, N.D., Takagi, H., and Esteban, M. (eds.) *Coastal Disasters and Climate Change in Vietnam: Engineering Planning and Perspectives*. Elsevier, Amsterdam.

Congressional Budget Office, 2021. Potential Increases in Hurricane Damage in the United States: Implications for the Federal Budget. https://www.cbo.gov/publication/51518 Accessed on August 17, 2021.

Contreras, M., & Winckler, P. (2013). Perdidas de vidas, viviendas, infraestructura y embarcaciones por el tsunami del 27 de Febrero de 2010 en la costa central de Chile. *Obras y Proyectos*, 14, 6–19. https://doi.org/10.4067/S0718-28132013000200001

Da Rocha, R. P., Sugahara, S., & Da Silveira, R. B. (2004). Sea waves generated by extratropical cyclones in the South Atlantic Ocean: Hindcast and validation against altimeter data. *Weather and Forecasting*, 19(2), 398–410. https://doi.org/10.1175/1520-0434(2004)019<0398:SWGBEC>2.0.CO;2

Danish Hydraulic Institute, 2016. *MIKE21 Flow Model FM Sand Transport Module, Including Shoreline Morphology User Guide*. Hørsholm, Denmark.

Defra (2005). The threat posed by tsunamis to the UK, Study commissioned by Defra Flood Management and produced by British Geological Survey, Proudman Oceanographic Laboratory, Met Office and HR Wallingford, 167.

Department of Coast Conservation and Coastal Resource Management, Sri Lanka, 2018. Coastal Zone and Coastal Resource Management Plan (SLCZ&CRMP)-2018.

Department of Marine and Coastal Resources: DMCR, 2011. Coastal zone management in Thailand. In *Proceeding of the International Conference on Coastal Erosions*, Bangkok, 27–29 April 2011 (in Thai).

Department of Marine and Coastal Resources: DMCR, 2017. *Guidelines for coastal erosion solutions and protection plan*. DMCR, Bangkok (in Thai).

Deshpande, M., Singh, V.K., Ganadhi, M.K., Roxy, M.K., Emmanuel, R., and Kumar, U., 2021. Changing Status of Tropical Cyclones over the North Indian Ocean. *Climate Dynamics*, 57(11–12), 3545–3567. https://doi.org/10.1007/s00382-021-05880-z

Devis-Morales, A., Montoya-Sanchez, R. A., Bernal, G., & Osorio, A. F. (2017). Assessment of extreme wind and waves in the Colombian Caribbean Sea for offshore applications. *Applied Ocean Research*, 69, 10–26. https://doi.org/10.1016/j.apor.2017.09.012

Dibajnia, M., 2010. An Updated Wave Climate Hindcast for the Persian Gulf. In *9th International Conference on Coasts, Ports & Marine Structures, ICOPMAS*, Tehran.

Dibajnia, M., Soltanpour, M., Nairn, R., and Allahyar, M.R., 2010. Cyclone Gonu: The Most Intense Tropical Cyclone on Record in the Arabian Sea, *Indian Ocean Tropical Cyclones and Climate Change*, pp. 149–157. Springer Netherlands, Dordrecht. https://doi.org/10.1007/978-90-481-3109-9_19

Dikshit, A., Sarkar, R., Pradhan, B., Acharya, S., and Alamri, A.M., 2020. Spatial Landslide Risk Assessment at Phuentsholing, Bhutan. *Geosciences*, 10(131), 1–16. https://doi.org/10.3390/geosciences1004013

Dikshit, A., Sarkar, R., Pradhan, B., Acharya, S., and Dorji, K., 2019. Estimating Rainfall Thresholds for Landslide Occurrence in the Bhutan Himalayas. *Water (Switzerland)*, 11(8), 1–12. https://doi.org/10.3390/w11081616

Ding, G., Zhang, Y., Gao, L. et al. (2013). Quantitative analysis of burden of infectious diarrhea associated with floods in northwest of Anhui Province, China: A mixed method evaluation. *PLoS One*, 8(6), e65112.

Dobrynin, M., Murawsky, J., & Yang, S. (2012). Evolution of the global wind wave climate in CMIP5 experiments. *Geophysical Research Letters*, 39, L18606. https://doi.org/10.1029/2012GL052843

Donghae Regional Office of Oceans and Fisheries, 2018. *The Construction Administration report of Revetment (Zone 2) in the Donghae Port Stage 3.*

Duy, N.T., 2005. The current situation of seawalls/dikes along Vietnam coastline and damages caused by 2005 typhoons. In *Intl. Symposium on Coastal Disasters and Tsunami*, Phuket, Thailand, 20.

Egidius, H. (2003). *Sturmfluten : Tod und Verderben an der Nordseeküste von Flandern bis Jütland*, p. 35. CCV Concept Center Verlag, Varel. ISBN: 9783934606166.

Ells, R.W., 1908. Report on the landslide at Notre-dame de la Salette, Lièvre River, Quebec. SE Dawson.

Environment Agency (2016). Delivering benefits through evidence: The costs and impacts of the winter 2013/14 floods, SC140025/R1, p. 266.

Erman, A., Tariverdi, M., Obolensky, M., Chen, X., Vincent, R. C., Malgioglio, S., Rentschler, J., Hallegatte, S., & Yoshida, N. (2019). *The role of poverty in exposure, vulnerability and resilience to floods in Dar es Salaam.*

Esteban, M., Matsumaru, R., Takagi, H., Mikami, T., Shibayama, T., de Leon, M., Valenzuela, V.P., and Thao, N.D., 2013. *Study on Disaster Information Dissemination and People's Response for Evacuation – The Case of the 2013 Typhoon Yolanda (Haiyan)*, vol. 70.

Esteban, M., Takabatake, T., Achiari, H., Mikami, T., Nakamura, R., Gelfi, M., Panalaran, S., Nishida, Y., Inagaki, N., Chadwick, C., Oizumi, K., and Shibayama, T., 2021. Field survey of flank collapse and run-up heights due to the 2018 Anak Krakatau Tsunami. *Journal of Coastal and Hydraulic Structures*, 1(1).

Esteban, M., Valenzuela, V.P., Yun, N.Y., Mikami, T., Shibayama, T., Matsumaru, R., Takagi, H., Thao, N.D., De Leon, M., Oyama, T., and Nakamura, R., 2015. Typhoon Haiyan 2013 evacuation preparations and awareness. *International Journal of Sustainable Future for Human Security*, 3(1), 37–45.

Evan, A.T., and Camargo, S.J., 2011. A Climatology of Arabian Sea Cyclonic Storms. *Journal of Climate*, 24 (1), 140–158. https://doi.org/10.1175/2010JCLI3611.1

Farquhar, C.R., Deighton, H., Busswell, G., Snaith, H.M., Ash, E., Collard, F., Piolle, J.F., Poulter, D.J.S., and Pinnock, S., 2013. *Globwave: A Global wave data portal.*

Feser, F., Barcikowska, M., Krueger, O., Schenk, F., Weisse, R., & Xia, L. (2015). Storminess over the North Atlantic and northwestern Europe - A review. *Quarterly Journal of the Royal Meteorological Society* 141, 350–382.

Fine, I.V., Rabinovich, A.B., Bornhold, B., Thomson, R., and Kulikov, E.A., 2005. The Grand Banks Landslide-Generated Tsunami of November 18, 1929: Preliminary Analysis and Numerical Modeling. *Marine Geology*, 215(1–2), 45–57.

Fosu, B.O., and Wang, S.Y.S., 2015. Bay of Bengal: Coupling of pre-monsoon tropical cyclones with the monsoon onset in Myanmar. *Climate Dynamics*, 45(3), 697–709.

Fritz, H., Petroff, C., Catalan, P., Cienfuegos, R., Winckler, P., Kalligeris, N., Weiss, R., Barrientos, S., Meneses, G., Valderas-Bermedo, C., Ebeling, C., Papadopoulos, A., Contreras, M., Almar, R., Dominguez, J., & Synolakis, C. (2011). Field Survey of the 27 February 2010 Chile Tsunami. *Pure and Applied Geophysics*, 168, 1989–2010. https://doi.org/10.1007/s00024-011-0283-5

Fu, X., Dong, J., Ma, J. et al., 2009. Analysis and numerical simulation of the storm surge caused by Typhoon Hagupit. *Marine Forecasts*, 26 (4), 68–75 (in Chinese).

Fujiwara, O., Aoshima, A., Irizuki, T., Ono, E., Obrochta, S.P., Sato, Y., and Takahashi, A., 2020. *Tsunami deposits refine great earthquake rupture extent and recurrence over the past 1300 years along the Nankai and Tokai fault segments of the Nankai Trough*, Japan. Quaternary Science Reviews, Volume 227.

Gardelle, J., Arnaud, Y., and Berthier, E., 2011. Contrasted Evolution of Glacial Lakes along the Hindu Kush Himalaya Mountain Range between 1990 and 2009. *Global and Planetary Change*, 75(1–2), 47–55. https://doi.org/10.1016/j.gloplacha.2010.10.003

Global Facility for Disaster Reduction and Recovery (GFDRR), 2018. https://www.giz.de/en/worldwide/352.html

Google Maps (2021a). Figure 2: Left: Hamburg and the Elbe estuary, connecting the city with the North Sea. Middle: The harbour area within Hamburg. Right: The embankment area around St. Pauli (striped) and the HafenCity (grey). https://www.google.co.jp/maps/@53.8868137,9.1410381,9.07z?hl=en&authuser=0 (accessed on 21 September 2021).

Google Maps (2021b). Figure 4: The area around Speicherstadt and Sandtorkai within the HafenCity. https://www.google.co.jp/maps/@53.5427715,9.9887506,16.85z?hl=en&authuser=0 (accessed on 21 September 2021).

Greenberg, D.A., Murty, T., and Ruffman, A., 1993. A Numerical Model for the Halifax Harbor Tsunami due to the 1917 Explosion. *Marine Geodesy*, 16(2), 153–167.

Gurung, D.R., Khanal, N.R., Bajracharya, S.R., Tsering, K., Joshi, S., Tshering, P., Chhetri, L.K., Lotay, Y., and Penjor, T., 2017. Lemthang Tsho Glacial Lake Outburst Flood (GLOF) in Bhutan: Cause and Impact. *Geoenvironmental Disasters*, 4(17), 1–13. https://doi.org/10.1186/s40677-017-0080-2

HafenCity Hamburg and IBA (2011). Stadtküste Hamburg: Herausforderung Stadtentwicklung und Hochwaserschutz. Dokumentation zum HafenCity IBA LABOR vom 4./5./6. Mai 2011. https://www.yumpu.com/de/document/read/9175610/dokumentation-zum-hafencity-iba-labor (accessed on 19 September 2021).

HafenCity Hamburg (2006). Der Masterplan: Neuauflage 2006. https://www.hafencity.com/_Resources/Persistent/c/7/1/d/c71db692487a55836aa17c935cb54d973e403384/z_de_broschueren_24_Masterplan_end.pdf (accessed on 20 September 2021).

HafenCity Hamburg (2017). Themen Quartiere Projekte, Vol. 27, March issue. https://www.hafencity.com/_Resources/Persistent/4/0/b/d/40bdc5b576140f189c5a2ad78964ff8caf79f795/HafenCityProjekte_Maerz_2017_deutsch.pdf (accessed on 19 September 2021).

HafenCity Hamburg (2021). Stadträume. https://www.hafencity.com/stadtentwicklung/stadtraeume#neue-stadttopographie (accessed on 19 September 2021).

Hall, A. (2015). Plugging the gaps: The North Sea flood of 1953 and the creation of a national coastal warning system. *Journal of Public Management and Social Policy*, 22(2), Article 8.

Hamburg (2012). Sturmflutschutz in Hamburg gestern - heute - morgen. Berichte des Landesbetriebes Straßen, Brücken und Gewässer Nr. 10/2012. https://www. hamburg.de/contentblob/3281680/1822cf666737349331ec6e88b8e2ce58/data/ sturmflut-in-hamburg-1.pdf (accessed on 19 September 2021).

Hamburg (2021). Flutkatastrophen: wenn der Sturm kommt. https://www.hamburg. de/hamburger-hafen/4391672/sturmfluten/ (accessed on 19 September 2021).

Hasegawa, K., 2008. Features of Super Cyclone Sidr to Hit Bangladesh in Nov 07 and measures for disaster from results of JSCE investigation. In *Proc. WFEO-JFES-JSCE Joint Int. Symp. on Disaster Risk Management*, Sendai, Japan.

Heleniak, T. (2021). The future of the Arctic populations. *Polar Geography*, 44 (2), 136–152, https://doi.org/10.1080/1088937X.2019.1707316

Hettiarachchi, S.S.L., Samarawickrama, S.P., and Wijeratne, N., 2013. Risk Assessment and Management for Tsunami Hazard – Case Study of the Port of Galle. IOC Manuals and Guides No. 52, Document 6.

Horsburgh, K., & Horritt, M. (2007). The Bristol Channel floods of 1607-Reconstruction and analysis, *Weather*, 61 (10).

HR Wallingford (2006). Tsunamis-Assessing the hazard for the UK and Irish coast. In *Proceedings of the 41st Defra Flood and Coastal Management Conference*.

Hsiao, L., Liou, C., Yeh, T., Guo, Y., Chen, D., Huang, K., Terng, C., and Chen, J., 2010. A vortex relocation scheme for tropical cyclone initialization in advanced research WRF. *Monthly Weather Review*, 138, 3298–3315. https://doi. org/10.1175/2010MWR3275.1

Hunt, A., & Watkiss, P. (2011). Climate change impacts and adaptation in cities: A review of the literature. *Climate Change*, 104, 13–49.

Hynes, S, Solomon, S M, & Whalen, D. (2014). Geological survey of Canada. *Open File*, 7685, 7. https://doi.org/10.4095/295579

Ikeda, K., 1995. Gender Differences in Human Loss and Vulnerability in Natural Disasters: A Case Study from Bangladesh. *Indian J Gen Stud*, 2(2), 171–193.

IMD, 2020. Cyclone Eatlas. Indian Meteorological Department. http://www. rmcchennaieatlas.tn.nic.in

Innenbehörde (2017). Bildergalerie: Sturmflut vom 29.10.2017. https://www. hamburg.de/innenbehoerde/sturmflut/14695186/sturmflut-2017-bildergalerie/ (accessed on 19 September 2021).

Innocentini, V., & Caetano Neto, E. D. S. (1996). A case study of the 9 August 1988 South Atlantic storm: Numerical simulations of the wave activity. *Weather and Forecasting*, 11(1), 78–88. https://doi.org/10.1175/1520-0434(1996)011<0078: ACSOTA>2.0.CO;2

IOC, IHO and BODC, 2003. *Centenary Edition of the GEBCO Digital Atlas*. CD-ROM on behalf of the Intergovernmental Oceanographic Commission and the International Hydrographic Organization as part of the General Bathymetric Chart of the Oceans; British Oceanographic Data Centre, Liverpool.

IPCC (2013). Climate change 2013: The Physical science basis. In *Contribution of Working Group I to the Fifth Assessment Report of the Intergovernmental Panel on Climate Change*, Stocker, T. F., Qin, D., Plattner, G.-K., Tignor, M., Allen, S. K., Boschung, J., Nauels, A., Xia, Y., Bex, V., & Midgley, P. M. (Eds.). Cambridge University Press, Cambridge, United Kingdom and New York, NY, USA, 1535.

Ishibashi, K., 2004. Status of historical seismology in Japan. *Annals of Geophysics*. https://www.annalsofgeophysics.eu/index.php/annals/article/view/3305. Accessed on 25 December 2004.

Jackson, K.L., Eberli, G.P., Amelung, F., McFadden, M.A., Moore, A.L., Rankey, E.C., and Jayasena, H.A.H., 2014. Holocene Indian Ocean Tsunami History in Sri Lanka. *Geology*, 42(10), 859–862. https://doi.org/10.1130/G35796.1

Janß, C. (2016). Am Sandtorkai. Hamburgbilder. https://hamburgbilder.de/am-sandtorkai/ (accessed on 21 September 2021).

Japan International Cooperation Agency (JICA), 2013. Preparatory survey report on the project for establishment of disastrous weather monitoring system in the Republic of the Union of Myanmar https://openjicareport.jica.go.jp/551/551/551_104_12112546.html. Accessed on 27 September 2021.

Japan International Cooperation Agency (JICA), 2015. Country report Myanmar: Natural disaster risk assessment and area business continuity plan formulation for industrial agglomerated areas in the ASEAN region. http://open_jicareport.jica.go.jp/pdf/1000023395.pdf. Accessed on 26 September 2021.

Jones, B. M., Farquharson, L. M., Baughman, C. A., Buzard, R. M., Arp, C. D., Grosse, G., et al. (2018). A decade of remotely sensed observations highlight complex processes linked to coastal permafrost bluff erosion in the Arctic. *Environmental Research Letters*, 13 (11), 1274.

JSCE, 2008. Investigation Report on the Storm Surge Disaster by Cyclone Sidr in Nov 2007 in Bangladesh (Transient Translation) (2008): Investigation Team of Japan Society of Civil Engineers.

Kang, Z., Jin, R., and Bao, Y. Characteristic analysis of cold wave in china during the period of 1951-2006. *Plateau Meteorology*, 2010, 29(2), 420–428.

Karim, M.F., and Mimura, N., 2008. Impacts of Climate Change and Sea Level Rise on Cyclonic Storm Surge Floods in Bangladesh. *Global Environment Change* 18(2008), 490–500.

Kelley, L., 2009. Chem Tanker with Sulphuric Acid Cargo Sinks Off Sri Lanka. ICIS News, https://www.icis.com/explore/resources/news/2009/04/09/9207622/chem-tanker-with-sulphuric-acid-cargo-sinks-off-sri-lanka. Accessed on 26/09/2021

Khon, V. C., Mokhov, I. I., Pogarskiy, F. A., Babanin, A., Dethloff, K., Rinke, A., & Matthes, H. (2014). Wave heights in the 21 century Arctic Ocean simulated with a regional climate model. *Geophysical Research Letters*, 41, 2956–2961, https://doi.org/10.1002/2014GL059847

Kim J., Murphy, E., Nistor, I., Ferguson, S., Provan, M. (2021). Numerical analysis of storm surges on Canada's Western Arctic Coastline. *Journal of Marine Science and Engineering*, 9(3), 326. https://doi.org/10.3390/jmse9030326

Kittitanasuan, W., 2020. Measures for solving Thailand coastal erosion problems. *The National Defence Colledge of Thailand Journal*, 62(3), 65–79 (in Thai).

Knapp, K.R., Kruk, M.C., Levinson, D.H., Diamond, H.J., and Neumann, C.J., 2010. The International Best Track Archive for Climate Stewardship (IBTrACS). *Bulletin of the American Meteorological Society*, 91(3), 363–376. https://doi.org/10.1175/2009BAMS2755.1

Knutson, T.R., Chung, M.V., Vecchi, G., Sun, J., Hsieh, T.-L., and Smith, A.J.P., 2021. *Climate Change is Probably Increasing the Intensity of Tropical Cyclones*. ScienceBrief_Review_CYCLONES_Mar2021.

Kulathilaka, I., 2018. Oil Spills and Their Devastating Effects to Our Oceans. https://www.slycantrust.org/post/oil-spills-and-their-devastating-effects-to-our-oceans. Accessed on 26/09/2021

Kulikov, E. A., Rabinovich, A. B., & Thomson, R. E. (2005). Estimation of tsunami risk for the coasts of Peru and Northern Chile. *Natural Hazards*, 35(2), 185–209. https://doi.org/10.1007/s11069-004-4809-3

Kweon, H.M., Lee, J.R., and Yun, G.H., 2014. A strategic policy for the beach reservation utilizing steel type breakwater. *Journal of Korea Water Resources Association (Water for future)*, 47(2), 87–97 (in Korean).

Kyaw, T.O., Esteban, M., Mall, M., and Shibayama, T., 2021. Extreme waves induced by cyclone Nargis at Myanmar coast: Numerical modeling versus satellite observations. *Natural Hazards*, 106(3), 1797–1818.

Lantuit, H., Overduin, P. P., Couture, N., Wetterich, S., Are, F., Atkinson, D., et al. (2012). The arctic coastal dynamics database: A new classification scheme and statistics on arctic Permafrost coastlines. *Estuaries and Coasts*, 35 (2), 383–400. https://doi.org/10.1007/s12237-010-9362-6

Lee, U.H., Han, S.W., and Lee, J.L., 2020. A layout design system of hard defense and soft defense methods for coastal erosion mitigation. *Journal of Coastal Disaster Prevention*, 7(4), 243–249 (in Korean).

Li, E., Endter-Wada, J., & Li, S. (2015). Characterizing and contextualizing the water challenges on megacities. *Journal of the American Water Resources Association*, 51(3), 589–613. https://doi.org/10.1111/1752-1688.12310.

Li, Q., Dai, Y., Li, Z. et al., 2010. Surface layer wind field characteristics during a severe typhoon 'Hagupit' landfalling. *Journal of Building Structures*, 31(4): 54–61 (in Chinese).

Lin, I.I., Chen, C.H., Pun, I.F., Liu, W.T., and Wu, C.C., 2009. Warm ocean anomaly, air sea fluxes, and the rapid intensification of tropical cyclone Nargis (2008). *Geophysical Research Letters*, 36(3).

Lindley, S. J., Handley, J. F., Theuray, N., et al. (2006). Adaptation strategies for climate change in the urban environment: Assessing climate change risks in UK urban areas. *Journal of Risk Research*, 9(5), 543–568.

Locat, J., Turmel, D., Leblanc, J., and Demers, D., 2018. Tsunamigenic landslides in québec, Landslides and engineered slopes, *Experience, theory and practice*. CRC Press, Boca Raton, pp. 1305–1312.

Locat, J., Turmel, D., Locat, P., Therrien, J., and Létourneau, M., 2017. The 1908 disaster of notre-dame-de-la-salette, québec, canada: Analysis of the landslide and tsunami, *Landslides in Sensitive Clays*. Springer, Berlin, pp. 361–371.

Losada, M., Kobayashi, N., and Martin, F.L., 1992. Armor stability on submerged breakwater. *Journal of Waterway, Port, Coastal, and Ocean Engineering, ASCE*, 118(2), 207–212 (in English).

Løvholt, F., Schulten, I., Mosher, D., Harbitz, C., and Krastel, S., 2019. Modelling the 1929 Grand Banks Slump and Landslide Tsunami. *Geological Society, London, Special Publications*, 477(1): 315–331.

Lu, X., Yu, H., Ying, M., Zhao, B., Zhang, S., Lin, L., Bai, L., and Wan, R., 2021. Western North Pacific tropical cyclone database created by the China Meteorological Administration. *Advances in Atmospheric Sciences*, 38(4), 690–699. https://doi.org/10.1007/s00376-020-0211-7

Mafi, S., and Amirinia, G. 2017. Forecasting Hurricane Wave Height in Gulf of Mexico Using Soft Computing Methods. *Ocean Engineering*, 146(December), 352–362. https://doi.org/10.1016/j.oceaneng.2017.10.003

Mäll, M., Suursaar, U., Nakamura, R., & Shibayama, T. (2017). Modelling a storm surge under future climate scenarios: Case study of extratropical cyclone Gudrun (2005). *Natural Hazards*, 89, 1119–1144. https://doi.org/10.1007/s11069-020-03911-2

Mäll, M., Nakamura, R. & Shibayama, T. (2020a). Surge and wave conditions under warmer Ice-free arctic ocean. *Coastal Engineering Proceedings*, 36, 39.

Mäll, M., Nakamura, R., Suursaar, U., & Shibayama, T. (2020b). Pseudo-climate modelling study on projected changes in extreme extratropical cyclones, storm waves and surges under CMIP5 multi-model ensemble: Baltic Sea perspective. *Natural Hazards*, 102(1), 67–99. https://doi.org/10.1007/s11069-020-03911-2

Manson, G.K., 2005. On the Coastal Populations of Canada and the World. In *Canadian Coastal Conference*, Halifax, Nova Scotia, 1–11.

Manson, G. K., & Solomon, S. M. (2007). Past and future forcing of Beaufort Sea coastal change. *Atmosphere-Ocean*, 45(2), 107–122. https://doi.org/10.3137/ao.450204

Marsh, T. J., Kirby, C., Muchan, K., et al. (2016). *The winter floods of 2015/2016 in the UK- a review*. Centre for Ecology & Hydrology, Wallingford, UK. 37.

Mashhadi, L, Hadjizadeh Zaker, N., Soltanpour, M., and Moghimi, S., 2015. Study of the Gonu Tropical Cyclone in the Arabian Sea. *Journal of Coastal Research*, 313(May), 616–623. https://doi.org/10.2112/JCOASTRES-D-13-00017.1

McClearn, M. (2018). The Globe and Mail. In Tuktoyaktuk, residents take a stand on shaky ground against the Beaufort Sea's advance. https://www.theglobeandmail.com/canada/article-in-tuktoyaktuk-residents-take-a-stand-on-shaky-ground-against-the/ (accessed on 20 March 2019).

Mees, H., Driessen, P., Runhaar, H., 2013. Legitimate adaptive flood risk governance beyond the dikes: The cases of Hamburg, Helsinki and Rotterdam. Individual case study reports. Copernicus Institute of Sustainable Development, Utrecht University, the Netherlands. https://edepot.wur.nl/341526 (accessed on 19 September 2021).

Mendes, D., Souza, E. P., Marengo, J. A., & Mendes, M. C. D. (2010). Climatology of extratropical cyclones over the South American-southern oceans sector. *Theoretical and Applied Climatology*, 100(3), 239–250. https://doi.org/10.1007/s00704-009-0161-6

Met Office (2014). Winter storms, December 2013 – January 2014. https://www.metoffice.gov.uk/binaries/content/assets/metofficegovuk/pdf/weather/learn-about/uk-past-events/interesting/2013/winter-storms-december-2013-to-january-2014---met-office.pdf (assessed on 25 March 2022).

Met Office (2020). The Meteorological Office, UK. https://www.metoffice.gov.uk/weather/warnings-and-advice/uk-storm-centre/storm-ciara (assessed on 25 March 2022)

Met Office (2022). The Meteorological Office, UK, https://www.metoffice.gov.uk/weather/specialist-forecasts/space-weather (assessed on 25 March 2022).

Miller, J. D., & Hutchins, M. (2017). The impacts of urbanization and climate change on urban flooding and urban water quality: A review of the evidence concerning the United Kingdom. *Journal of Hydrology: Regional Studies*, 12: 345–362.

Mikami, T., Shibayama, T., Takagi, H., Matsumaru, R., Esteban, M., Thao, N., De Leon, M., Valenzuela, V., Oyama, T., Nakamura, R., Kumagai, K., and Li, S., 2016. Storm surge heights and damage caused by the 2013 typhoon Haiyan along the leyte gulf coast. *Coastal Engineering Journal*, 58(1), 1640005-1–1640005-27. https://doi.org/10.1142/S0578563416400052

MIKEbyDHI. 2017. *MIKE 21 Spectral Waves FM*. MIKE by DHI, Denmark.

Ministry of Land, Transport and Maritime Affairs, 2011. *The 3rd general planning of the whole country ports (2011~2020)* (in Korean).

Mool, P.K., Wangda, D., Bajracharya, S.R., Kunzang, K., Gurung, D.R., and Joshi, S.P., 2001. Inventory of Glaciers, Glacial Lakes and Glacial Lake Outbrust Floods, *Monitoring and Early Warning Systems in the Hindu Kush-Himalayan Region*. International Centre for Integrated Mountain Development, Nepal.

Mudryk, L. R., Dawson, J., Howell, S. E. L., Derksen, C., Zagon, T. A., & Brady, M. (2021). Impact of 1, 2 and 4°C of global warming on ship navigation in the Canadian Arctic. *Nature Climate Change*, 11, 673–679. https://doi.org/10.1038/s41558-021-01087-6

Murty, P.L.N., Srinivas, K.S., Rao, E.P.R., Bhaskaran, P.K., Shenoi, S.S.C., and Padmanabham, J., 2020. Improved Cyclonic Wind Fields over the Bay of Bengal and Their Application in Storm Surge and Wave Computations. *Applied Ocean Research*, 95(February), 102048. https://doi.org/10.1016/j.apor.2019.102048

Myanmar Information Management Unit (MIMU), 2019. Population in low-lying coastal zone. https://themimu.info/sites/themimu.info/files/documents/Population_Map_Population_in_Low-Lying_Coastal_Area_MIMU1646v01_23May2019_A4_0.pdf. Accessed on 16 September 2021.

Mzava, P., Nobert, J., & Valimba, P. (2019). Land cover change detection in the urban catchments of Dar es Salaam, Tanzania using remote sensing and GIS techniques. *Tanzania Journal of Science*, 45(3), 315–329.

Mzava, P., Nobert, J., & Valimba, P. (2020). Characterizing past and future trend and frequency of extreme rainfall in urban catchments: A case study. *H2Open Journal*, 3(1), 288–305. https://doi.org/10.2166/h2oj.2020.009

Mzava, P., Nobert, J., & Valimba, P. (2021). Quantitative analysis of the impacts of climate and land-cover changes on urban flood runoffs: A case of Dar es Salaam, Tanzania, *Journal of Water and Climate Change*, 12(6), 2835–2853. https://doi.org/10.2166/wcc.2021.026

Nakamura, R., Mall, M., and Shibayama, T., 2019. Street-scale storm surge load impact assessment using fine-resolution numerical modelling: A case study from Nemuro, Japan. *Natural Hazards*, 99(8), 391–422. https://doi.org/10.1007/s11069-019-03791-1

Nakamura, R., Shibayama, T., Esteban, M., and Iwamoto, T., 2016. Future typhoon and storm surges under different global warming scenarios: Case study of typhoon Haiyan (2013). *Natural Hazards*, 82(3), 1645–1681. https://doi.org/10.1007/s11069-016-2259-3

National Academies of Sciences, Engineering and Medicine (NASEM) (2019). Framing the challenges of urban flooding in the United States, The National Academies Press, Washington, DC. https://doi.org/10.17226/25381

NationalGeographic (2012). South America: Physical geography. Resource Library | Encyclopedic Entry. https://www.nationalgeographic.org/encyclopedia/south-america-physical-geography/

Ngailo, T. J., Reuder, J., Rutalebwa, E., Nyimvua, S., & Mesquita, M. (2016). Modelling of extreme maximum rainfall using extreme value theory for Tanzania. *International Journal of Scientific and Innovative Mathematical Research*, 4(3), 34–45.

Nicholls, R.J.N., Mimura, N., and Topping, J.C., 1995. Climate Change in South and Southeast Asia: Some Implications for Coastal Areas. *Journal of Global Environmental Engineering*, 1, 137–154.

Niemeyer, H. D., Beaufort, G., Mayerle, R., Monbaliu, J., Townend, I., Toxvig Madsen, H., De Vriend, H., & Wurpts, A. (2016) Socio-economic Impacts – Coastal Protection. In Quante, M., & Colijn, F. (Eds). *North Sea Region climate change assessment NOSCCA*. Regional Climate Studies, Springer Nature, 475–488.

NOAA (2019). *Arctic oscillation (AO)*.

NOAA Office of Ocean and Coastal Resource Management, 2013. Beach Nourishment: A Guide for Local Government Officials. https://coast.noaa.gov/archived/beachnourishment/html/human/law/index.htm. Accessed on August 17, 2021.

NOAA. 2020. Socioeconomic Data Summary. https://coast.noaa.gov/data/digitalcoast/pdf/socioeconomic-data-summary.pdf. Accessed on August 17, 2020.

Nunes, L. H. (2011). An overview of recent natural disasters in South America. *Bulletin Des Seances de l'Academie Royale Des Sciences d'Outre-Mer*, 57(2), 409–425.

Ohira, K., Shibayama, T., Esteban, M., Mikami, T., Takabatake, T., and Kokado, M., 2012. Comprehensive numerical simulation of waves caused by typhoon using a meteorology-wave-storm surge-tide coupled model. In *Proceedings of the 33rd International Conference on Coastal Engineering 2012, ICCE 2012 (Proceedings of the Coastal Engineering Conference)*.

Okal, E. A., Dengler, L., Araya, S., Borrero, J. C., Gomer, B. M., Koshimura, S. I., Laos, G., Olcese, D., Modesto Ortiz, F., Swensson, M., Titov, V. V., & Vegas, F. (2002). Field survey of the caman, Peru tsunami of 23 June 2001. *Seismological Research Letters*, 73(6), 907–920. https://doi.org/10.1785/gssrl.73.6.907

Onnen, C. (2014). Abb. 7: St. Pauli-Landungsbrücken, absenkbares Flutschutztor zwischen Promenade und Brücken zu den Pontons, ca. 2011. In Will, T., & Lieske, H. (Eds.), *2015. Flood Protection for Historic Sites – Integrating Heritage Conservation into Historic Concepts: Hochwasserschutz in der Hamburger Innenstadt*, pp. 106–112. Hendrik Bäßler Verlag, Berlin, ISBN: 978-3-945880-05-0.

Ostenso, N. A. (2018). Encyclopedia Britannica. *Arctic Ocean*.

QGIS Development Team, 2021. *QGIS Geographic Information System*. QGIS Association. https://www.qgis.org

Parnu (2017). Pärnu linnavalitsus. Pärnu linna külalised perioodil 1.mai - 31. september 2014-2017 POSITIUM MOBIILIPOSITSIONEERIMISE ANDMETEL.PDF. https://parnu.ee/failid/uuringud/P2rnu_kylalised_2014-2017_2.pdf (accessed on 28 October 2021) (in Estonian).

Pattnaik, D.R., and Rama Rao, Y.V., 2008. Track Prediction of very sever cyclone "Nargis" using high resolution weather research forecasting (WRF) model. *Journal Earth System Science*, 118, 309–329.

Paul, B.K., Rashid, H., Islam, M.S., and Hunt, L.M., 2010. Cyclone Evacuation in Bangladesh: Tropical Cyclones Gorky (1991) vs. Sidr (2007). *Environmental Hazards*, 9(1), 89–101.

Postmann, M. (2004). Map of Germany with the boundaries of the Bundesländer. 778~1,104 pixels. Version from 2007-01-09. https://commons.wikimedia.org/wiki/File:Blank_Map_Germany_States.png#file (accessed on 17 September 2021).

Proshutinsky, A., Dukhovskoy, D., Timmermans, M.-L., Krishfield, R., & Bamber, J. L. (2015). Arctic circulation regimes. *Philosophical Transactions of the Royal Society A* 373, 20140160. https://doi.org/10.1098/rsta.2014.0160

Rangel-Buitrago, N., & Anfuso, G. (2015). *Storms in coastal (Colombia) and from Cartagena Zones: Case studies Cadiz (Spain).* Springer, Cham.

Rasmee, P., and Rasmeemasmuang, T., 2013. Comparative evaluation of the protective and remedial measures of coastal erosion problems in Samutprakan Province. In *Proceeding of the 5th National Convention of Water Resources Engineering* (in Thai).

Rasmussen, E. A., & Turner, J. (Eds). (2003). *Polar lows: Mesoscale weather systems in the polar regions.* Cambridge University Press, 602.

Reboita, M. S., da Rocha, R. P., de Souza, M. R., & Llopart, M. (2018). Extratropical cyclones over the southwestern South Atlantic Ocean: HadGEM2-ES and RegCM4 projections. *International Journal of Climatology*, 38(6), 2866–2879. https://doi.org/10.1002/joc.5468

Reliefweb. Thomson Reuters Foundation, 30 April, 2018. With warning drums and river clean-ups, Indonesian women head off disasters. https://reliefweb.int/report/indonesia/warning-drums-and-river-clean-upsindonesian-women-head-disasters

Reich, W. K. (2009). Rathausschleuse in Hamburg. Maritime Photographie. https://maritime-photographie.de/img/1557 (accessed on 19 September 2021).

Rowsell, E.C.P., Sultana, P., and Thompson, P.M., 2013. The "Last Resort"? Population Movement in Response to Climate-Related Hazards in Bangladesh. *Environmental Science & Policy*, 27(Supplement 1), S44–S59.

Ruffman, A., 1996. *Tsunami Runup Mapping as an Emergency Preparedness Planning Tool: The 1929 Tsunami.*

Sahoo, B., and Bhaskaran, P.K., 2019. Prediction of Storm Surge and Inundation Using Climatological Datasets for the Indian Coast Using Soft Computing Techniques. *Soft Computing*, 23(23), 12363–12383. https://doi.org/10.1007/s00500-019-03775-0

Sakijege, T., Lupala, J., & Sheuya, S. (2012). Flooding, flood risks and coping strategies in urban informal residential areas: The case of Keko Machungwa, Dar es Salaam, Tanzania, *Journal of Disaster Risk Studies*, 4(1), 46–55.

Sanchez-Escobar, R., Diaz, L. O., Guerrero, A. M., Galindo, M. P., Mas, E., Koshimura, S., Adriano, B., Urra, L., & Quintero, P. (2020). Tsunami hazard assessment for the central and southern pacific coast of Colombia. *Coastal Engineering Journal*, 62(4), 540–552. https://doi.org/10.1080/21664250.2020.1818362

Satake, K., Aung, T.T., Sawai, Y., Okamura, Y., Win, K.S., Swe, W., Swe, C., Swe, T.L., Tun, S.T., Soe, M.M., and Oo, T.Z., 2006. Tsunami heights and damage along the Myanmar coast from the December 2004 Sumatra-Andaman earthquake. *Earth, Planets and Space*, 58(2), 243–252.

Scanlon, J., 1998. Dealing with Mass Death after a Community Catastrophe: Handling Bodies after the 1917 Halifax Explosion. *Disaster Prevention and Management: An International Journal.*

Shi, H., Li, W., Lv, Y. et al., 2015. Comparative analysis of two severe storm surge of Hainan Province in 2014. *Marine Forecasts*, 32(4), 75–82 (in Chinese).

Shibayama, T., 2008. *Coastal Processes: Concepts in Coastal Engineering and Their Applications to Multifarious Environments.* World Scientific, Singapore.

Shibayama, T., 2015. 2004 Indian Ocean Tsunami. Esteban, M., Takagi, H., Shibayama, T. (Eds.) *Handbook of Coastal Disaster Mitigation for Engineers and Planners.* Butterworth-Heinemann, Oxford.

Shibayama, T., Aoki, Y., and Takagi, H., 2010. Field survey and analysis of flood behavior of storm surge due to cyclone Nargis in Myanmar. *Annu J Civ Eng Ocean JSCE*, 26, 429–434.

Shibayama, T., Takagi, H., and Hnu, N., 2009. Disaster survey after the cyclone Nargis in 2008. In *Asian and Pacific Coasts 2009*, pp. 190–193.

Shibayama, T., Tajima, Y., Kakinuma, T., Nobuoka, H., Yasuda, T., Hsan, R.A., Rahman, M., and Islam, M.S., 2009. Field Survey of Storm Surge Disaster Due to Cyclone Sidr in Bangladesh. In *Proc. of Coastal Dynamics Conference*, Tokyo, 7-11 September 2009.

Silva, R., Martinez, M. L., Hesp, P. A., Catalan, P., Osorio, A. F., Martell, R., Fossati, M., Da Silva, G. M., Mario-Tapia, I., Pereira, P., Cienfuegos, R., Klein, A., & Govaere, G. (2014). Present and future challenges of coastal erosion in Latin America. *Journal of Coastal Research*, 1–16. https://doi.org/10.2112/SI71-001.1

Skamarock, W., Klemp, J., Dudhia, J., Gill, D., Barker, D., Duda, M., Huang, X., Wang, W., and Powers, J., 2008. A description of the advanced research WRF version 3. NCAR Technical Note. http://dx.doi.org/10.5065/D68S4MVH

Slaughter, M.D., Olinger, B., Kershner, J.D., Mader, C.L., and Bowman, A.L., 1978. *Interaction of Explosive-Driven Air Shocks with Water and Plexiglas.* Los Alamos National Lab.(LANL), Los Alamos, NM.

Soloviev, S. L., & Go, C. N. (1975). *A catalogue of tsunamis on the eastern shore of the pacific ocean.* Nauka Publishing House.

Soltanpour, M., Ranji, Z., Shibayama, T., and Ghader, S., 2021. Tropical Cyclones in the Arabian Sea: Overview and Simulation of Winds and Storm-Induced Waves. *Natural Hazards.* https://doi.org/10.1007/s11069-021-04702-z

Srisangeerthanan, S., Lewangama, C.S., and Wickramasooriya, S., 2015. Tropical Cyclone Damages in Sri Lanka. *Wind Engineers, Japan Association for Wind Engineering*, 40(3).

Sterr, H., 2008. Assessment of Vulnerability and Adaptation to Sea-Level Rise for the Coastal Zone of Germany. *Journal of Coastal Research*, 24 (242), 380–393.

Stopa, J. E., Ardhuin, F., & Girard-Ardhuin, F. (2016). Wave climate in the Arctic 1992-2014: Seasonality and trends. *The Cryosphere*, 10, 1605–1629, https://doi.org/10.5194/tc-10-1605-2016

Stolle, J. et al., 2020. Engineering Lessons from the 28 September 2018 Indonesian Tsunami: Debris Loading. *Canadian Journal of Civil Engineering*, 47(1), 1–12.

Suppasri, A., Shuto, N., Imamura, F., Koshimura, S., Mas, E., and Yalciner, A.C., 2013. Lessons learned from the 2011 great east Japan tsunami: Performance of tsunami countermeasures, coastal buildings, and tsunami evacuation in Japan. *Pure and Applied Geophysics*, 170, 993–1018.

SurgeWatch (2022). An interactive database of UK coastal flood events, University of Southampton, https://www.surgewatch.org/ (assessed on 17 March 2022).

Suursaar, Ü., Kullas, T., Otsmann, M., Saaremäe, I., Kuik, J., & Merilain, M. (2006). Hurricane Gudrun and modelling its hydrodynamic consequences in the Estonian coastal waters. *Boreal Environment Research*, 11, 143–159.

Suursaar, Ü., & Sooäär, J. (2007). Decadal variations in mean and extreme sea level values along the Estonian coast of the Baltic Sea. *Tellus A: Dynamic Meteorology and Oceanography*, 59(2), 249–260. https://doi.org/10.1111/j.1600-0870.2006.00220.x

Suursaar, Ü., Sepp, M., Post, P., & Mäll, M. 2018. An inventory of historic storms and cyclone tracks that have caused met-ocean and coastal risks in the eastern Baltic Sea. In Shim, J.-S., Chun, I., & Lim, H. S. (eds.), *Proceedings from the International Coastal Symposium (ICS) 2018 (Busan, Republic of Korea)*, pp. 531–535, Coconut Creek (Florida), ISSN 0749-0208.

Tõnisson, H., Kont, A., Orviku, K., Suursaar, Ü., Rivis, R., & Palginõmm, V. (2018). Application of system approach framework for coastal zone management in Parnu, SW Estonia. *Journal of Coastal Conservation*, 23, 931–942. https://doi.org/10.1007/s11852-018-0637-6

Tõnisson, H., Orviku, K., Jaagus, J., Suursaar, Ü., Kont, A., & Rivis, R. (2008). Coastal damages on Saaremaa Island, Estonia, caused by the extreme storm and flooding on January 9, 2005. *Journal of Coastal Research*, 24, 602–614. https://doi.org/10.2112/06-0631.1

Takabatake, T., Shibayama, T., Esteban, M., Achiari, H., Nurisman, N., Gelfi, M., Tarigan, T.A., Kencana, E.R., Fauzi, M.A.R., Panalaran, S., Harnantyari, A.S., and Kyaw, T.O., 2019. Field survey and evacuation behaviour during the 2018 Sunda Strait Tsunami. *Coastal Engineering Journal*, 423–443.

Takagi, H., Esteban, M., Shibayama, T., Mikami, T., Matsumaru, R., de Leon, M., Thao, N.D., and Oyama, T., 2014. Track Analysis, Simulation and Field Survey of the 2013 Typhoon Haiyan Storm Surge. *Journal of Flood Risk Management*.

Takahashi, R. (1961). A Summary Report on the Chilean Tsunami of May 1960. Report on the Chilean Tsunami of May 24, 1960, as observed along the Coast of Japan.

Tasnim, K.M. Shibayama, T., and Esteban, M., 2015. Observation and numerical simulation of storm surge due to cyclone Sidr 2007 in Bangladesh, *Coastal Disasters: Lessons Learnt for Engineers and Planners*. Esteban, M., Takagi, H., and Nguyen D.T. (eds).

Tasnim, K.M., Shibayama, T., Esteban, M., Takagi, H., Ohira, K., and Nakamura, R., 2015. Field observation and numerical simulation of past and future storm surges in the Bay of Bengal: Case study of cyclone Nargis. *Natural Hazards*, 75(2), 1619–1647.

Taylor Engineering, Inc., 2004. *Fort Pierce Inlet Sand Bypassing Feasibility Study*. St. Lucie County, Florida, Jacksonville, Florida.

Taylor Engineering, Inc., 2011. *Fort Pierce Inlet Sand Bypassing Preliminary Design and Permitting, Model Calibration and Validation, St. Lucie County, Florida*. Jacksonville, Florida.

Tempa, K., and Chettri, N., 2021. Comprehension of Conventional Methods for Ultimate Bearing Capacity of Shallow Foundation by PLT and SPT in Southern Bhutan. *Civil Engineering and Architecture*, 9, 375–385. https://doi.org/10.13189/cea.2021.090210

Tempa, K., Chettri, N., Aryal, K.R., and Gautam, D., 2021a. Geohazard Vulnerability and Condition Assessment of the Asian Highway AH-48 in Bhutan. *Geomatics, Natural Hazards and Risk*, 12(1), 2904–2930. https://doi.org/10.1080/19475705.2021.1980440

Tempa, K., Peljor, K., Wangdi, S., Ghalley, R., Jamtsho, K., Ghalley, S., and Pradhan, P., 2021b. UAV Technique to Localize Landslide Susceptibility and Mitigation Proposal: A Case of Rinchending Goenpa Landslide in Bhutan. *Natural Hazards Research*, 1(2021), 171–186. https://doi.org/10.1016/j.nhres.2021.09.001

Thakur, V., L'Heureux, J.-S., and Locat, A., 2017. Landslide in sensitive clays–from research to implementation, *Landslides in Sensitive Clays*. Springer, Berlin, pp. 1–11.

Thao, N.D., Takagi, H., and Esteban, M., 2014. *Coastal Disasters and Climate Change in Vietnam: Engineering Planning and Perspectives*. Elsevier, Amsterdam.

The Province of Gangwon, 2018. *Coastal Erosion Monitoring of East Sea Report* (in Korean).

Tropical Storm Bret 1993 (2021). In Wikipedia. https://en.wikipedia.org/w/index.php?title=Tropical_Storm_Bret_(1993)&oldid=1049413477 (accessed on 11 October 2021).

Turner, J. & Bracegirdle, T. (2006). Polar lows and other high latitude weather systems. ECMWF seminar on polar meteorology, 4–8 September 2006. https://www.ecmwf.int/sites/default/files/elibrary/2007/12857-polar-lows-and-other-high-latitude-weather-systems.pdf (accessed on 10 September 2019).

UNDP, 2010. Bangladesh: New Homes for Cyclone Sidr Victims. http://www.content.undp.org/go/newsroom/2010/august/bangladesh-un

UNESCO (2015). Speicherstadt and Kontorhaus District with Chilehaus. https://whc.unesco.org/en/list/1467/ (accessed on 20 September 2021).

Usami, T., Ishii, H., Imamura, T., Takemura, M., Matsuura, R., 2013. *"Nihon Higai Jishin Souran" (Comprehensive list of earthquake damage in Japan)*, University of Tokyo Press, Tokyo, 694 (in Japanese).

Valdez, J., Shibayama, T., Takabatake, T., and Esteban, M., 2022. Simulated flood forces on a building due to the storm surge by Typhoon Haiyan. *Coastal Engineering Journal Special Issue* DOI: 10.1080/21664250.2022.2099683.

Van Dorn, W., Mehaute, B., and Hwang, L., 1968. *Handbook of Explosion-Generated Water Waves*. Office of Naval Research, USA.

Van, H.L., 2013. The severe erosion of Vietnam coastal areas, Vietnamese Magazine Broadcasting Station, Sydney (16 September 2013).

Vieira, F., Cavalcante, G., and Campos, E., 2020. Simulation of Cyclonic Wave Conditions in the Gulf of Oman. *Natural Hazards*, 105(2), 2203–2217. https://doi.org/10.1007/s11069-020-04396-9

Vietnam Administration of Seas and Islands (Ministry of Natural Resources and Environment) & Coordinating Body on the Seas of East Asia (COBSEA), 2013. Symposium on national report consultation for the evaluation of coastal erosion in Vietnam. Hanoi, Vietnam. 17 July 2013.

Vishnu, S., Francis, P.A., Shenoi, S.S.C., and Ramakrishna, S.S.V.S., 2016. On the decreasing trend of the number of monsoon depressions in the Bay of Bengal. *Environmental Research Letters*, 11(1), 014011.

Voiland, A. (2013). NASA earth observatory. In a warming world, storms may be fewer but stronger. https://earthobservatory.nasa.gov/features/ClimateStorms (accessed on 27 September 2021).

Von Storch, H., & Woth, K. (2006). Storm surges – The case of Hamburg, Germany. In *Paper for the 2006 ESSP OSC Panel Session on "GEC, Natural Disasters, and Their Implications for Human Security in Coastal Urban Areas"*.

Walraven, A., & Aerts, J. (2008). *Connecting delta cities. Sea level rise and major coastal cities.* Report Free University Amsterdam, the Netherlands, 72.

Wang, X. L., Feng, Y., Swail, V. R., & Cox, A. (2015). Historical changes in the Beaufort-Chukchi-Bering seas surface winds and waves, 1971–2013. *Journal of Climate,* 28(19), 7457–7469.

Welt (2007). Hamburger HafenCity erstmals geräumt. Article from 09.11.2007. https://www.welt.de/regionales/hamburg/article1348583/Hamburger-Hafencity-erstmals-geraeumt.html (accessed on 19 September 2021).

Welt (2013). Hamburg in der Elbe: Diesmal auch die HafenCity stark betroffen, Article from 07.12.2013. https://www.welt.de/print/die_welt/hamburg/article122669661/Hamburg-in-der-Elbe-Diesmal-auch-die-Hafencity-stark-betroffen.html (accessed on 19 September 2021).

Williams, A. (2020). *Rising sea-levels and increased storms pose threat to coastal communities.* Plymouth University, https://www.plymouth.ac.uk/news/rising-sea-levels-and-increased-storms-pose-future-threat-to-coastal-communities (assessed on 24 March 2022).

Winckler, P. W., Contreras, M., Campos, R. C., Beya, J., Molina, M. M., Winckler, P., Contreras-López, M., Campos-Caba, R., Beyá, J. F., & Molina, M. (2017). El temporal del 8 de agosto de 2015 en las regiones de Valparaiso y Coquimbo, Chile Central. *Latin American Journal of Aquatic Research,* 45(4), 622–648. https://doi.org/10.3856/vol45-issue4-fulltext-1

Winckler, P., Aguirre, C., Farías, L., Contreras-López, M., & Masotti, Í. (2020). Evidence of climate-driven changes on atmospheric, hydrological, and oceanographic variables along the Chilean coastal zone. *Climatic Change,* 163(4), 633–652. https://doi.org/10.1007/s10584-020-02805-3

World Bank, 2021. Bangladesh-Multipurpose Disaster Shelter Project: Virtual Implementation Support Review-August 1 to 12, 2021.

Ying, M., Zhang, W., Yu, H., Lu, X., Feng, J., Fan, Y., Zhu, Y., and Chen, D., 2014. An overview of the China meteorological administration tropical cyclone database. *Journal of Atmospheric and Oceanic Technology,* 31, 287–301. https://doi.org/10.1175/JTECH-D-12-00119.1.

Yoon, J.H., and Huang, W.R., 2012. Indian monsoon depression: Climatology and variability. *Modern Climatology,* 13, 45–72.

You, Z., 2016. Coastal disasters of and countermeasures against flooding and erosion in China. *Report of Chinese Academy of Science,* 31(10), 1190–1196 (in Chinese).

Zou, K., 2020. Research on the man-made island's influence on erosion under typhoon-based on the case of Haikou Bay under Kalmeagi. Thesis submitted for applying master degree in South China University of Technology (in Chinese).

Index

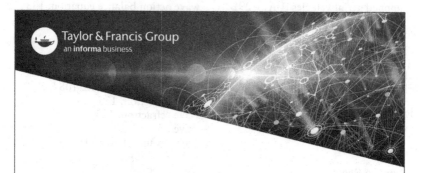